T0122469

Advances in Intelligent Systems and Computing

Volume 1097

The series "Advances in Intelligent Systems and Computing" contains publications on theory, applications, and design methods of Intelligent Systems and Intelligent Computing. Virtually all disciplines such as engineering, natural sciences, computer and information science, ICT, economics, business, e-commerce, environment, healthcare, life science are covered. The list of topics spans all the areas of modern intelligent systems and computing such as: computational intelligence, soft computing including neural networks, fuzzy systems, evolutionary computing and the fusion of these paradigms, social intelligence, ambient intelligence, computational neuroscience, artificial life, virtual worlds and society, cognitive science and systems, Perception and Vision, DNA and immune based systems, self-organizing and adaptive systems, e-Learning and teaching, human-centered and human-centric computing, recommender systems, intelligent control, robotics and mechatronics including human-machine teaming, knowledge-based paradigms, learning paradigms, machine ethics, intelligent data analysis, knowledge management, intelligent agents, intelligent decision making and support, intelligent network security, trust management, interactive entertainment, Web intelligence and multimedia.

The publications within "Advances in Intelligent Systems and Computing" are primarily proceedings of important conferences, symposia and congresses. They cover significant recent developments in the field, both of a foundational and applicable character. An important characteristic feature of the series is the short publication time and world-wide distribution. This permits a rapid and broad dissemination of research results.

** **Indexing: The books of this series are submitted to ISI Proceedings, EI-Compendex, DBLP, SCOPUS, Google Scholar and Springerlink** **

More information about this series at http://www.springer.com/series/11156

Yu-Chen Hu · Shailesh Tiwari ·
Munesh C. Trivedi · K. K. Mishra
Editors

Ambient Communications and Computer Systems

RACCCS 2019

 Springer

Editors
Yu-Chen Hu
Department of Computer Science
and Information Management
Providence University
Taichung, Taiwan

Munesh C. Trivedi
Department of Computer Science
and Engineering
National Institute of Technology Agartala
Tripura, India

Shailesh Tiwari
Computer Science Engineering Department
ABES Engineering College
Ghaziabad, Uttar Pradesh, India

K. K. Mishra
Department of Computer Science
and Engineering
Motilal Nehru National Institute
of Technology
Allahabad, Uttar Pradesh, India

ISSN 2194-5357   ISSN 2194-5365   (electronic)
Advances in Intelligent Systems and Computing
ISBN 978-981-15-1517-0   ISBN 978-981-15-1518-7   (eBook)
https://doi.org/10.1007/978-981-15-1518-7

This Springer imprint is published by the registered company Springer Nature Singapore Pte Ltd.
The registered company address is: 152 Beach Road, #21-01/04 Gateway East, Singapore 189721, Singapore

Preface

The RACCCS 2019 is a major multidisciplinary conference organized to provide a forum for researchers, educators, engineers and government officials involved in the general areas of communication, computational sciences and technology to disseminate their latest research results and exchange views on the future research directions of these fields, exchange computer science and integrate its practices, apply the academic ideas, improve the academic depth of computer science and its application, and provide an international communication platform for educational technology and scientific research for the universities and engineering field experts and professionals.

Nowadays, globalization of academic and applied research is growing with a speedy pace. Computer, communication and computational sciences are the heating areas with lot of thrust. Keeping this ideology in preference, the fourth version of International Conference on Recent Advancement in Computer, Communication and Computational Sciences (RACCCS 2019) has been organized at **Aryabhatta College of Engineering and Research Center, Ajmer, India, during August 16–17, 2019**.

Ajmer is situated in the heart of India, just over 130 km southwest of Jaipur, a burgeoning town on the shore of the Ana Sagar Lake, flanked by barren hills. Ajmer has historical strategic importance and was sacked by Mohammed Gauri on one of his periodic forays from Afghanistan. Later, it became a favorite residence of the mighty Mughals. The city was handed over to the British in 1818, becoming one of the few places in Rajasthan controlled directly by the British rather than being part of a princely state. The British chose Ajmer as the site for Mayo College, a prestigious school opened in 1875 exclusively for the Indian princes, but today open to all those who can afford the fees. Ajmer is a perfect place that can be symbolized for demonstration of Indian culture, ethics and display of perfect blend of wide plethora of diverse religion, community, culture, linguistics, etc., all coexisting and flourishing in peace and harmony. This city is known for the famous Dargah Sharif, Pushkar Lake, Brahma Temple and many more evidences of history.

This is the fourth time Aryabhatta College of Engineering and Research Center, Ajmer, India, is organizing international conference based on the theme of computer, communication and computational sciences, with a foreseen objective of enhancing the research activities at a large scale. Technical program committee and advisory board of RACCCS include eminent academicians, researchers and practitioners from abroad as well as from all over the nation.

RACCCS 2018 received around 160+ submissions from 523 authors of 5 different countries such as Taiwan, Saudi Arabia, China, Bangladesh and many more. Each submission has gone through the plagiarism check. On the basis of plagiarism report, each submission was rigorously reviewed by atleast two reviewers with an average of 2.67 per reviewer. Even some submissions have more than two reviews. On the basis of these reviews, 36 high-quality papers were selected for publication in this proceedings volume, with an acceptance rate of 22%.

We are thankful to the speakers, delegates and the authors for their participation and their interest in RACCCS as a platform to share their ideas and innovation. We are also thankful to Prof. Dr. Janusz Kacprzyk, Series Editor, AISC, Springer, and Mr. Aninda Bose, Senior Editor, Hard Sciences, Springer Nature, for providing continuous guidance and support. Also, we extend our heartfelt gratitude to the reviewers and technical program committee members for showing their concern and efforts in the review process. We are indeed thankful to everyone directly or indirectly associated with the conference organizing team leading it toward the success.

Although utmost care has been taken in compilation and editing, however, a few errors may still occur. We request the participants to bear with such errors and lapses (if any). We wish you all the best.

Ajmer, India   Yu-Chen Hu
                                                                                                    Shailesh Tiwari
                                                                                                      K. K. Mishra
                                                                                                  Munesh C. Trivedi

About This Book

The field of communications and computer sciences always deals with new arising problems and finding effective and efficient techniques, methods and tools as solutions to these problems. This book captures the latest and rapid developments happened in the immediate environmental surroundings of computer and communication sciences.

Nowadays, various cores of computer science and engineering field entered in a new era of technological innovations and development, which we are calling '*Smart Computing and smart communication.*' Smart computing and communication provide intelligent solutions to the problems which are more complex as compared to general problems. We can say that the main objective of *Ambient Computing and Communication Sciences* is to make software, techniques, computing and communication devices which can be used effectively and efficiently.

Computational and computer science have a wide scope of implementation in engineering sciences through networking and communication as a backbone. Keeping this ideology in preference, this book includes the insights that reflect the immediate surroundings' developments in the field of communication and computer sciences from upcoming researchers and leading academicians across the globe. It contains the high-quality peer-reviewed papers of '*International Conference on Recent Advancements in Computer, Communication and Computational Sciences (RACCCS 2019)*, held at Aryabhatta College of Engineering and Research Center, Ajmer, India, during August 16–17, 2019. These papers are arranged in the form of chapters. The contents of this book cover three areas: *Intelligent Hardware and Software Design, Advanced Communications, Intelligent Computing Technologies, Web and Informatics, Intelligent Image Processing.* This book helps the perspective readers from computer and communication industry and academia to derive the immediate surroundings developments in the field of communication and computer sciences and shape them into real-life applications.

Contents

Advanced Communications

Topology-Aware Low-Cost Video Streaming for Video Data Over Heterogeneous Network 3
Garv Modwel, Anu Mehra, Nitin Rakesh and K. K. Mishra

Comparative Analysis of Graph Clustering Algorithms for Detecting Communities in Social Networks 15
Menta Sai Vineeth, Krishnappa RamKarthik, M. Shiva Phaneendra Reddy, Namala Surya and L. R. Deepthi

Energy-Efficient Heterogeneous WCEP for Enhancing Coverage Lifetime in WSNs 25
Amandeep Kaur Sohal, Ajay K. Sharma and Neetu Sood

Middleware Frameworks for Mobile Cloud Computing, Internet of Things and Cloud of Things: A Review 37
Tribid Debbarma and K. Chandrasekaran

Blockchain Security Threats, Attacks and Countermeasures 51
Shams Tabrez Siddiqui, Riaz Ahmad, Mohammed Shuaib and Shadab Alam

Intelligent Computing Techniques

Implementation of Handoff System to Improve the Performance of a Network by Using Type-2 Fuzzy Inference System 65
Ritu, Hardeep Singh Saini, Dinesh Arora and Rajesh Kumar

Test Suite Optimization Using Lion Search Algorithm 77
Manish Asthana, Kapil Dev Gupta and Arvind Kumar

Advanced Twitter Sentiment Analysis Using Supervised Techniques and Minimalistic Features 91
Sai Srihitha Yadlapalli, R. Rakesh Reddy and T. Sasikala

**Multi-objective Faculty Course Timeslot Assignment Problem
with Result- and Feedback-Based Preferences** 105
Sunil B. Bhoi and Jayesh M. Dhodiya

**Simulation of Learning Logical Functions Using Single-Layer
Perceptron** 121
Mohd Vasim Ahamad, Rashid Ali, Falak Naz and Sabih Fatima

**Applicability of Financial System Using Deep
Learning Techniques** 135
Neeraj Kumar, Ritu Chauhan and Gaurav Dubey

Intelligent Hardware and Software Design

Novel Reversible ALU Architecture Using DSG Gate 149
Shaveta Thakral and Dipali Bansal

**An Appraisal on Microstrip Array Antenna for Various Wireless
Communications and Their Applications** 157
Gurudev Zalki, Mohammed Bakhar, Ayesha Salma and Umme Hani

**Workload Aware Dynamic Scheduling Algorithm for Multi-core
Systems** 171
Savita Gautam and Abdus Samad

Almost Self-centered Index of Some Graphs 181
Priyanka Singh and Pratima Panigrahi

**Computationally Efficient Scheme for Simulation of Ring
Oscillator Model** 193
Satyavir Singh, Mohammad Abid Bazaz and Shahkar Ahmad Nahvi

Web and Informatics

**Comparative Analysis of Consensus Algorithms of Blockchain
Technology** 205
Ashok Kumar Yadav and Karan Singh

**A Roadmap to Realization Approaches in Natural Language
Generation** 219
Lakshmi Kurup and Meera Narvekar

**A Case Study of Structural Health Monitoring by Using Wireless
Sensor Networks** 231
Der-Cherng Liaw, Chiung-Ren Huang, Hung-Tse Lee and Sushil Kumar

TBSAC: Token-Based Secured Access Control for Cloud Data 243
Pankaj Upadhyay and Rupa G. Mehta

2DBSCAN with Local Outlier Detection 255
Urja Pandya, Vidhi Mistry, Anjana Rathwa, Himani Kachroo
and Anjali Jivani

**Classification of Documents based on Local Binary Pattern Features
through Age Analysis** 265
Pushpalata Gonasagi, Rajmohan Pardeshi and Mallikarjun Hangarge

**An Efficient Task Scheduling Strategy for DAG in Cloud Computing
Environment** 273
Nidhi Rajak and Diwakar Shukla

Web-Based Movie Recommender System 291
Mala Saraswat, Anil Dubey, Satyam Naidu, Rohit Vashisht
and Abhishek Singh

**A Two-Step Dimensionality Reduction Scheme for Dark Web
Text Classification** 303
Mohd Faizan and Raees Ahmad Khan

TLC Algorithm in IoT Network for Visually Challenged Persons 313
Nitesh Kumar Gaurav, Rahul Johari, Samridhi Seth, Sapna Chaudhary,
Riya Bhatia, Shubhi Bansal and Kalpana Gupta

Educational App for Android-Specific Users—EA-ASU 325
S. Siji Rani and S. Krishnanunni

Personality Prediction System Through CV Analysis 337
Alakh Arora and N. K. Arora

Intelligent Image Processing

Image Enhancement: A Review 347
Prem Kumari Verma, Nagendra Pratap Singh and Divakar Yadav

**Diagnosis of Cough and Cancer Using Image Compression
and Decompression Techniques** 357
Ashish Tripathi, Ratnesh Prasad Srivastava, Arun Kumar Singh,
Pushpa Choudhary and Prem Chand Vashist

**3D Lung Segmentation Using Thresholding and Active
Contour Method** 369
Satya Prakash Sahu, Bhawana Kamble and Rajesh Doriya

DWT-LBP Descriptors for Chest X-Ray View Classification 381
Rajmohan Pardeshi, Rita Patil, Nirupama Ansingkar, Prapti D. Deshmukh
and Somnath Biradar

**Comparative Study of Latent Fingerprint Image Segmentation
Techniques Based on Literature Review** 391
Neha Chaudhary, Harivans Pratap Singh and Priti Dimri

**Hybrid Domain Feature-Based Image Super-resolution Using Fusion
of APVT and DWT** 401
Prathibha Kiran and Fathima Jabeen

**A Novel BPSO-Based Optimal Features for Handwritten Character
Recognition Using SVM Classifier** 415
Sonali Bali, Shekhar Sharma and Preeti Trivedi

About the Editors

Prof. Yu-Chen Hu received his PhD. degree in computer science and information engineering from the Department of Computer Science and Information Engineering, National Chung Cheng University, Chiayi, Taiwan in 1999. Currently, Dr. Hu is a professor in the Department of Computer Science and Information Management, Providence University, Sha-Lu, Taiwan. He is a member of ACM and IEEE. He is also a member of Computer Vision, Graphics, and Image Processing (CVGIP), Chinese Cryptology and Information Security Association, and Phi Tau Phi Society of the Republic of China.

He also serves as the Editor-in-Chief of International Journal of Image Processing since 2009. In addition, he is the managing editor of Journal of Information Assurance & Security. He is a member of the editorial boards of several other journals.

His areas of interest are: image and signal processing, data compression, information hiding, and data engineering.

Dr. Shailesh Tiwari currently works as a Professor in Computer Science and Engineering Department, ABES Engineering College, Ghaziabad, India. He is an alumnus of Motilal Nehru National Institute of Technology Allahabad, India. His primary areas of research are software testing, implementation of optimization algorithms and machine learning techniques in various problems. He has published more than 50 publications in International Journals and in Proceedings of International Conferences of repute. He is edited Scopus, SCI and E-SCI-indexed journals. He has also edited several books published by Springer. He has organized several international conferences under the banner of IEEE and Springer. He is a Senior Member of IEEE, member of IEEE Computer Society, Fellow of Institution of Engineers (FIE).

Dr. Munesh C. Trivedi currently works as an Associate Professor in Department of Computer Science and Engineering, NIT Agartala, Tripura, India. He has published 20 text books and 80 research publications in different International Journals and Proceedings of International Conferences of repute. He has received Young

Scientist and numerous awards from different national as well international forum. He has organized several international conferences technically sponsored by IEEE, ACM and Springer. He is on the review panel of IEEE Computer Society, International Journal of Network Security, Pattern Recognition Letter and Computer & Education (Elsevier's Journal). He is Executive Committee Member of IEEE UP Section, IEEE India Council and also IEEE Asia Pacific Region 10.

Dr. K. K. Mishra is currently working as Assistant Professor, Department of Computer Science and Engineering, Motilal Nehru National Institute of Technology Allahabad, India. His primary area of research includes evolutionary algorithms, optimization techniques and design, and analysis of algorithms. He has also published more than 50 publications in International Journals and in Proceedings of Internal Conferences of repute. He is serving as a program committee member of several conferences and also edited Scopus and SCI-indexed journals. He is also member of reviewer board of Applied Intelligence Journal, Springer.

Advanced Communications

Topology-Aware Low-Cost Video Streaming for Video Data Over Heterogeneous Network

Garv Modwel, Anu Mehra, Nitin Rakesh and K. K. Mishra

Abstract The wireless security surveillance system is one of the widely implemented applications of the Internet of things technology. These systems, enabled with security cameras, are connected to the Internet and provide complete security access to critical infrastructure. Recently, the usage of unmanned aerial vehicles for surveillance, especially in defense setups and areas beyond physical reach, is on the rise. These systems feed the base station with continuous video stream content and therefore need strong network support for meeting the set performance requirements. It can be achieved by offering better network streaming capabilities to the base station. In this paper, we address the problem of streaming over a network of UAV systems that are designed to handle video stream content by providing a cost-effective path to base station. Contrary to a lot of existing works in this direction, we do not rely on GPS information to select streaming routes. We use topology information and a graph-theoretic approach to select paths. The proposed streaming route selection mechanism is shown to be more stable, preserves the quality of video data, and is energy efficient compared to other related schemes.

Keywords Object detection · Image processing · Object comparison · Image extraction elementary streaming

G. Modwel (✉) · A. Mehra
Amity University, Sector 125, Noida, Uttar Pradesh 201313, India

A. Mehra
e-mail: amehra@amity.edu

N. Rakesh
Sharda University, Knowledge Park 3, Greater Noida, Uttar Pradesh 201310, India

K. K. Mishra
MNNIT, Prayagraj, Uttar Pradesh 201313, India
e-mail: kkm@mnnit.com

© Springer Nature Singapore Pte Ltd. 2020
Y.-C. Hu et al. (eds.), *Ambient Communications and Computer Systems*, Advances in Intelligent Systems and Computing 1097,
https://doi.org/10.1007/978-981-15-1518-7_1

1 Introduction

With the rapid proliferation of embedded and wireless technologies, everyday devices are getting connected to the Internet. As more and more devices are getting connected to the Internet, a number of applications of this technology have come to the fore. These devices that are connected to the Internet generate data, which is then be used for effective decision making. One such application of IoT technology is a surveillance system. A typical surveillance system involves a group of wireless devices usually equipped with cameras constantly monitoring a predefined area of Interest and streaming video stream content to a base station. By enabling Internet connectivity to such a system, user(s) can remotely monitor and alert applicable stakeholders, with or without the involvement of the user. An example of such a surveillance system is a home security solution. However, the same security surveillance system for a home cannot be employed for defense/border areas monitoring. The main reason is being the hostile environment.

A surveillance system for military/defense establishments typically operates in hostile and extreme conditions that are non-prevalent in a typical home security solution. Therefore, the usage of unmanned aerial vehicles equipped with cameras can offer a reliable and robust solution to monitor sensitive areas. A group of UAVs may monitor a predefined region of interest that constantly engages with the base station by streaming video stream data that can be used to detect any form of intrusion at the earliest.

Connectivity in dynamic networks similar to UAV whose operating conditions match is vehicular ad hoc networks. Connectivity in these networks can be managed with MAC layer schemes and network layer protocols. At the MAC layer, nodes aim to maintain reliable one-hop links using acknowledgments. Alternately, devices also employ multiple radios to boost the chances of delivery. At the network layer, routes are preferred that are short (low latency) and offer more reliability, that is, routes whose underlying links that do not suffer frequent disconnections.

To increase reliability in dynamic networks where nodes are always mobile, one of the straightforward ways to build reliable routes is by employing location information. Location information that is obtained from geographical positioning systems suffers from different limitations that include rapid energy drain, failure to determine location information in certain weather conditions. Apart from energy concerns and additional complex hardware involved to obtain GPS coordinates, failure to obtain location information may severely affect route performance or may result in non-selection of routes. To avoid relying on GPS systems and thereby increasing the network lifetime of UAV networks, we propose a topology information-based route selection algorithm that provides high-reliability routes suited to transfer multimedia content.

In this paper, we propose a topology information-based route selection mechanism for the network of UAVs. The proposed route selection mechanism uses stored topology information to select routes. The rest of the paper is organized as follows. The related work in this direction is presented in Sect. 1. Section 2 presents the proposed route selection algorithm along with the considered network model. Section 3 presents the experimental results along with some discussions. Finally, the conclusion and future challenges are presented in Sect. 4.

1.1 Related Work

A set of UAV devices equipped with security cameras can be entrusted to monitor activities of interest forming a surveillance system. The group of such devices frequently sends video stream feed to the base station that can be used for analysis. In such a network, maintaining connectivity is very important to have continuous monitoring capability. In the network stack, MAC and network layer protocols are responsible for maintaining connectivity among UAV devices. At the MAC level, link instabilities are mainly caused due to frequent movements. Nodes changing their positions lead to varying link characteristics [1, 2]. We here are ignoring the changes in link status due to external factors.

In [3], separate radios were used to coordinate transmissions. Omnidirectional radios were used for performing management, and control operations and directional antennas were employed for data transmission. In contrast, [4] goes a step further and uses three antennas. The first ones are used for discovering neighbors the second one for exchanging control messages and last for data transmissions. As expected, multiple radios result in very complex network setup and further their results in higher energy consumption. In [5], a routing scheme has been proposed to improve node links by using geographical information aiming to transmit video content over UAV networks. It makes use of a link status array to make better path selection decisions by each UAV.

In [6], quality of experience, center to video stream applications, is focused on primarily. Opportunistic routing strategies are employed to disseminate video content. It involves choosing a set of forwarders that can probably transfer content reliably toward the base station. The following set of parameters is used to make a forwarding choice like residual energy, location, delay, quality of wireless link along with the remaining buffer space. This is important as each node needs to store and forward, where storing the time of data might be a little higher than non-opportunistic schemes.

In [7], the authors have highlighted the need for proper selection of next-hop UAVs that offer support to ground wireless mesh networks. Specifically, they consider the post-disaster scenario where burst increase in VoIP and video traffic is considered. They aim to increase network throughput with the help of UAV-assisted ground wireless mesh networks, where there are high fluctuations in network traffic. In the proposed work, we aim to design a simple and low-complexity routing protocol that can improve connectivity among cooperating UAVs using topology information. We show that with reasonable assumptions, it is possible to maintain high connectivity to boost throughput in transmitting video stream content.

1.2   Problem Formulation

Unmanned aerial vehicles (UAVs) with surveillance capabilities are a desirable solution to monitor hostile environments where physical monitoring may not be possible or feasible. In this regard, a group of such UAVs can offer a reliable solution to monitor sensitive areas. Applications of such solutions are innumerable ranging from defense, emergency rescue, disaster management, relief, and rescue, etc. However, the major problem with UAV systems for surveillance is maintaining connectivity and meeting the latency and performance requirements of a surveillance system. Existing solutions aim to address this problem by considering the dynamic nature of these devices. Most of the existing solutions rely on position on GPS-based information to establish connectivity and maintain it. However, such solutions are complex and need precise geographical information. In addition, it has been shown that GPS systems are energy draining and are not feasible when such information is lacked. Therefore, there is a need for a low-cost implementable solution that provides connectivity even when GPS information is not available.

2   Methodology

2.1   System Network Model and Assumptions

We consider a group of moving UAVs acting in coordination to survey a region of interest. The set of UAVs forms a surveillance system that systematically moves around and sends out data, usually video stream content, to an on-ground base station. The movement of each UAV is restricted. That is a UAV move/covers only a specified region and does not travel beyond it. This region is usually specified in terms of square meters of the area that it is supposed to monitor. By restricting the movement of UAVs, their location prediction is simplified. This assumption is valid and in line with any real-time monitoring systems where a group of UAVs usually are designated to cover a certain distance and do not overlap with the rest of the monitoring area of other UAVs. For example, a UAV-enabled security system for a home may have one UAVs based on location. They usually do not crossover to avoid sending redundant video stream feed.

The set of UAVs is represented as a graph for simplicity. Even though the network model is modeled as a graph, the aerial vehicles are dynamic and under constant movement. However, the movement of each aerial vehicle is within a specific area and at any given time instant can be mapped to a node in the graph. Thus, it can be safely concluded that the snapshot of these vehicles at any instant of time translates to a graph. Moreover, since these vehicles usually operate at a specified altitude, their positions translate to a 2D graph with a fixed set of nodes and edges between them represent communication links.

2.2 Proposed Streaming Protocol

The proposed streaming protocol works in multiple phases. It begins with the initial network setup phase and builds over the induction of every new UAVs (herein referred to as node for simplicity) in the network. Once the network setup finishes, the streaming protocol can be invoked anytime to reach the base station. All traffic generated by each UAV is assumed to be the base station. Changes in topology do not need to necessitate repetition of the complete process and can be handled with the help of neighboring UAVs of a newly inducted UAV.

As specified in the network model, the very first node is provided a virtual coordinate with respect to the base station. All further nodes that become part of the network compute their virtual coordinates using a localization algorithm like MDS [8]. After computing their relative positions, each node obtains the set of neighboring nodes' positions. Each node repeats this process, thereby establishing a topology map of the entire network. To frequently update the status of links, each node may compute the quality of the link (LQI) by sending probe packets and counting the number of acknowledgments. The following phases constitute the route building process as listed from part I to part III (Fig. 1).

Depending on the vision capacity of the UAV, we have considered the following criteria for the creation of individual node which are the UAV location. According to the height, the field of view of each UAV is assumed to be having Gaussian distribution as shown in Fig. 2. This calculation is according to the height and the range of effective coverage. According to the height, the field of view of each UAV is assumed having Gaussian distribution mentioned in Eq. 1. Through this calculation

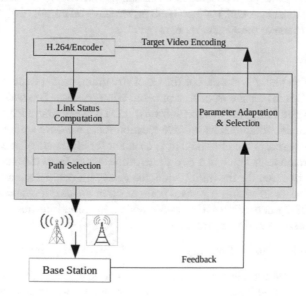

Fig. 1 Overall system architecture [8]

Fig. 2 Coverage area at the
height of 50 m above the
found. The bluer the region
is the less effective the vision
of the UAV

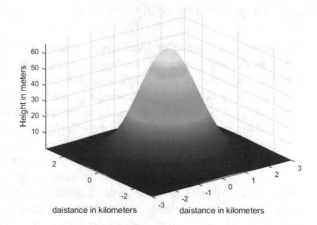

according to the height the range of coverage

$$R = \frac{1}{\sigma\sqrt{2\pi}} e^{-\frac{(x-\mu)^2}{2\sigma^2}} \tag{1}$$

where μ is mean of the distribution, σ is the variance of the distribution, and R
is Gaussian distribution. As the height of the drone increases, the field of view
also increases according to the distribution. Meanwhile, the clarity of the captured
video also decreases if a certain height exceeds. So in the conducted experiment is
conducted at a height of 50 m as a standard height. Figure 2 shows the distribution of
the field of view of the area covered at 50-m height, but even if the area of coverage
at this height is clear around 1000×1000 m, we have taken our area of interest as
200×200 m to make the system more accurate.

Phase I: Setup

Even though this phase is not part of the exact streaming process, the importance of
initial setup steps makes it part of the route selection process. The nodes that are set
up begin with obtaining the location information. The very first node that becomes
part of the network computes its relative location from the base station as a 3-tuple
field (x, y, z) where x is the altitude that is constant for all the next UAVs, and y and
z specify the relative distance in a two-dimensional plane from the base station. On
localizing the position for the initial node, the newly inducted node at this position
is ready to serve as an anchor (or base) point of other nodes. New nodes that aim to
become part of the network use the anchor node's location information to compute
their virtual position in the network.

Phase II: Neighborhood discovery and Link Status Computation

In this phase, a node completes the localization process; that is, as soon as it computes
its relative position with respect to the root node, it tries to discover its neighbors.
This step is to discover and establish multiple links toward the root and further to the

base station. Each node that completes, phase I, begins phase II either by sending out beacon-like advertisement packets or by receiving such packets from neighboring nodes. Each such packet contains a node's identity, its virtual position information, residual energy, and the distance to the root/base station. On receiving such a packet, a node can populate the topology graph by using the position information supplied by that node and computing the distance and cost to reach the root node if it uses the route to the base station through the received node. Since, each node maintains a route of communication towards its root node, obtaining the cost to any such neighboring node will facilitate it to efficiently transmit towards the base station. Each node maintains a list of alternate routes, their link quality, and their energy levels to enable robust and reliable route selection [9, 10].

Link quality is measured by each node with the help of a periodic link quality determination process. This process involves transmitting LQI probe packets and counting the number of acknowledgments (ack). A neighboring node can piggyback multiple probe packets with a single ack packet to lower control overhead. A node maintains the latest link status to facilitate the selection of low-latency quality links, which are a need for video and video stream content.

Phase III: Path Selection and Maintenance

In this phase, a node selects or chooses the route to be followed to reach the base station. Based on the selection criterion adopted by the network and requirements, a node selects the next hop toward the network. The choice of next hop is only changed when the current next hop moves farther and the LQI indicates loss of packets beyond the permitted threshold established by the application as part of the application requirements. That is, as part of the frequent movement by a UAV, if a node that computes the LQI as in phase II anticipates that if the current link is taken to reach, the base station will result in link losses that are beyond the permitted values of the network. In such a situation, the node refers to the route selection table and chooses the next best route toward the base station. This usually prevents frequent fluctuation of routes. Any node that also detects the failure of adjacent nodes will update their route selection table and also the topology map to reflect the changes in topology. Similarly, the joining of new nodes is incorporated after they duly follow the above phases. The simplicity of the proposed route selection algorithm is in the non-dependency on absolute location information and but also on the relative positional information that allows each node to maintain an independent topology map. This distributed way of maintaining the topology information allows moving UAVs to reasonably approximate their position and also reach the base station reliably. The assumption that any UAV does not move away from its designated region and only moves around the region contributes to the correctness of the algorithm. This is because any UAV moving out of its designated violates the constructed topology and resulting in the formation of loops in the network.

3 Result

The performance of the proposed protocol was evaluated in the NS-3 network simulator version 3.17. We considered a UAV network with nodes employed in surveillance of an urban neighborhood. We considered a Gauss–Markov 3D mobility model [11] as in [5]. However, to meet the requirements of our network, the altitude of the nodes is fixed to a constant value of 50 m, representing the altitude. The complete region of interest is fixed to 1000 × 1000 m. Each UAV is restricted to cover a 200 × 200 m of non-overlapping area in the network. Each UAV is designed to transmit a video stream as in [5]. The number of UAVs was fixed to 24 for all the experiments, and the speed of movement is varied between 1 and 5 m/s. As specified, the height is fixed at 50 m. The UAVs employ IEEE 802.11b wireless radios operating in an ad hoc mode that can transmit data at a maximum rate of 11 Mbps. The transmission range of each UAV is set to 300 m. This will enable a moving UAV to receive and transmit to its neighbors if the adjacent node is at least 100 m into its region of interest or terrain. Beyond that, the LQI computation process captures the non-availability of an adjacent link. We consider that UAVs have sufficient energy, to begin with and throughout the simulation process; energy is drained by nodes at a constant rate. The relation between energy losses with speed in meter per second for the drone while deciding the suitable route for movement is shown in Fig. 4. The comparison between the proposed algorithm with GPSR scheme is represented in Fig. 4. Two of the total 24 nodes are randomly chosen to transmit video data encoded in H.264 at 300 kbps as considered in [5].

Network throughput as shown in Fig. 3 allows us to measure the packet delivery ratio within a time duration that is one of the important criteria that need to be satisfied with video traffic. Next is the average end-to-end delay that measures the time taken for the transmitted packet to reach the destination; in addition, the received

Fig. 3 Average network throughput

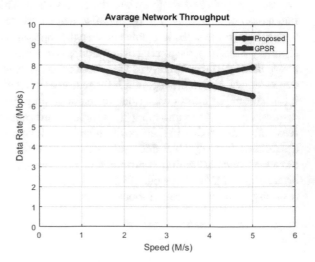

Fig. 4 Route selection overhead

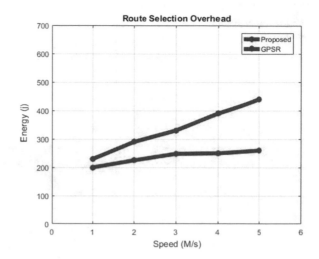

video data is assessed frame by frame to measure the correlation between transmitted and received content to measure the quality of video received. High correlation is an indicator of better quality video content. We consider the performance metrics of average network throughput and end-to-end delay in measuring the performance of the proposed scheme. The results are compared with GPSR [12]—geographical positional-based routing. Network throughput allows us to measure the packet delivery ratio within a time duration that is one of the important criteria that need to be satisfied with video traffic. Next is the average end-to-end delay that measures the time taken for the transmitted packet to reach the destination. Also, the received video data is assessed frame by frame to measure the correlation between transmitted and received content to measure the quality of video received. High correlation is an indicator of better quality video content (Fig. 4).

Lastly, we aim to compare the average end-to-end delay of the two schemes. Figure 5 shows the end-to-end delay comparison in terms of milliseconds in both the schemes for varying speeds of UAVs transmitting at a constant rate of 300 kbps. It can be seen that GPSR has higher end-to-end delay compared to the proposed scheme due to the complexity involved in the route selection process. As the proposed route selection scheme has alternate route options available in case of failures or requirement of low-latency links, alternate options are chosen quickly and lower end-to-end delays are achieved. This makes it possible to meet the video stream requirements of the network. In the experiment the simulation parameters considered are the height of individual node, energy loss while movement from one place to another and quality of the video streaming.

The proposed protocol simplifies the route selection process in UAV networks with the help of some simple assumptions. These assumptions pave the way to build stable and robust paths without relying on location-based systems, like GPS. This simplifies the overall system and allows a much simpler and low-memory footprint on the UAV. However, these assumptions also impose some restrictions on the system.

Fig. 5 Average end-to-end delay

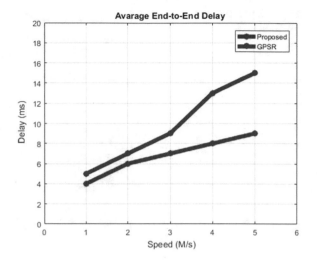

This restriction, however, is validated by the real-world applications setup where the moving pattern of cooperating UAV devices is more or less pre-determined as multiple devices cooperate with the intention to monitor a large geographical area. Overlapping or movement of UAVs into adjoining UAV areas defeats the purpose of the monitoring system as duplicate video stream would be reported by those overlapping UAVs. In this way, we simplify the route selection problem by translation topology information to route selection data.

4 Conclusion

In this paper, we presented a simple yet efficient route selection mechanism for highly dynamic unmanned aerial vehicle-based networks. These systems are designed to stream continuous video data to a designated base station. To meet the performance expectations of such a system, we employed topology information-based graph-theoretic approach to find and select routes in the network. Each UAV node computes partial topology information with the help of an anchor node and further discovers alternate routes using the neighborhood discovery process. The proposed algorithm operates in three distinct phases where phases 2 and 3 are carried out repeatedly on-demand basis. Contrary to many existing protocols in this direction, we do not rely on GPS information to select routes. We use topology information and a graph-theoretic approach to select routes. The proposed route selection mechanism is shown to be more stable and offers reliable routes and energy-efficient compared to other related schemes. Implementing the technology mentioned in the manuscript, we can live stream a larger area. For the same if we place cameras in some particular locations in the surveillance area, it will cost more for the same area of coverage. This way a lot of cost can be saved for the surveillance with improved flexibility.

References

1. Flushing, E.F., L.M. Gambardella, and G.A. Di Caro. 2016. On using mobile robotic relays for adaptive communication in search and rescue missions. In *2016 IEEE international symposium on Safety, Security, and Rescue Robotics (SSRR)*, October 2016, 370–377.
2. Oubbati, Omar Sami, Abderrahmane Lakas, Fen Zhou, Mesut Güneş, and Mohamed Bachir Yagoubi. 2017. A survey on position-based streaming protocols for Flying Ad hoc Networks (FANETs), *Vehicular Communications* 10: 29–56.
3. Alshbatat, Abdel Ilah, and Liang Dong. 2010. Adaptive MAC protocol for UAV communication networks using directional antennas. In *2010 International Conference on Networking, Sensing and Control (ICNSC)*. IEEE.
4. Temel, S., and I. Bekmezci. 2013. On the performance of Flying Ad Hoc Networks (FANETs) utilizing near space high altitude platforms (HAPs). In *Proceedings of the 6th international conference on Recent Advances in Space Technologies (RAST' 13)*, vol. 10, 461–465.
5. Costa, R., D. Rosario, E. Cerqueira, and A. Santos. 2014. Enhanced connectivity for robust videostream transmission in UAV networks. In *2014 IFIP Wireless Days (WD)*, 1–6.
6. Pimentel, Larissa, Denis Rosario, Marcos Seruffo, Zhongliang Zhao, and Torsten Braun. 2015. Adaptive beaconless opportunistic streaming for multimedia distribution. In *13th international conference on Wired/Wireless Internet Communication (WWIC)*, May 2015, Malaga, Spain.
7. Zhao, W., W. Xin, X. Zheng, and T. Hara. 2018. Comparison study on UAV movement for adapting to videostream burst in post-disaster networks. In *2018 IEEE International Conference on Smart Computing (SMARTCOMP)*, 339–343.
8. Raushan, A., and R. Matam. 2018. Fast localization and topology discovery scheme to handle topology changes in industrial IoT networks. In *2018 International Conference on Advances in Computing, Communications and Informatics (ICACCI)*, September 2018, 179–184, Bangalore, India.
9. Zaouche, Lotfi, Natalizio Enrico, and Bouabdallah Abdelmadjid. 2015. ETTAF: Efficient target tracking and filming with a flying ad hoc network. In *Proceedings of the 1st international workshop on experiences with the design and implementation of smart objects*, 49–54, Paris, France.
10. Ho, D., and H. Song 2013. networking cost effective video streaming system over heterogeneous wireless networks. In *2013 IEEE 24th annual international symposium on Personal, Indoor, and Mobile Radio Communications (PIMRC)*, London, 3589–3593.
11. Broyles, Dan, Abdul Jabbar, and James P.G. Sterbenz. 2010. Design and analysis of a 3-D gauss-markov mobility model for highly dynamic airborne networks.
12. Karp, Brad, and H.T. Kung. 2000. GPSR: Greedy perimeter stateless streaming for wireless networks. In *Proceedings of the 6th annual international conference on mobile computing and networking*, 243–254, Boston, USA.

Comparative Analysis of Graph Clustering Algorithms for Detecting Communities in Social Networks

Menta Sai Vineeth, Krishnappa RamKarthik, M. Shiva Phaneendra Reddy, Namala Surya and L. R. Deepthi

Abstract Community detection in social networks is often thought of a challenged domain that has not been explored completely. In today's digital world, it is forever laborious to make a relationship between people or objects. Community detection helps us to find such relationships or build such relationships. It also can facilitate bound organizations to induce the opinion of their product from certain people. Many algorithms have emerged over the years which detect communities in the social networks. We performed a comparative analysis between six completely different bunch of algorithms for detecting communities in social network by taking into account parameters like run-time, cluster size, normalized mutual data , adjusted random score and average score.

Keywords Community Detection · Clustering · Agglomerative · Divisive · NMI · ARS

1 Introduction

Social network analysis (SNA) plays a vital role in today's world. It helps us to find the mapping, relationships and information flow between people, groups, organizations or computers. Generally, a social network is represented by a graph which consists of a set of nodes which represents the individuals and the interactions between the individuals which are represented by edges [1]. A community or a cluster in a social network can be considered as a group of individuals who have similar tastes, preferences or choices. Community detection can be used in applications like, in research where we have to find a common research area for collaboration, for recommendations, in which group of people with similar tastes can be identified, in biological networks, to find protein interaction networks [2]. We have used parameters like run-time, cluster size, adjusted random score, average score, normalized

M. S. Vineeth (✉) · K. RamKarthik · M. S. Phaneendra Reddy · N. Surya · L. R. Deepthi
Department of Computer Science Engineering, Amrita Vishwa Vidyapeetham, Amritapuri, India

© Springer Nature Singapore Pte Ltd. 2020
Y.-C. Hu et al. (eds.), *Ambient Communications and Computer Systems*, Advances in
Intelligent Systems and Computing 1097,
https://doi.org/10.1007/978-981-15-1518-7_2

15

mutual data to search out the simplest rule which will be won't to sight the input from the community.

We provide the data to completely different algorithms for community detection, and from that we get some communities as their output [3]. This is compared to those obtained once applying pre-existing algorithms to identical knowledge. We have used completely different parameters to check these rules and realize the optimum algorithm. This paper has different sections in which Sect. 2 explains related work; Sect. 3 explains proposed work; experimental results are dealt in Sect. 5.

2 Related Works

The literature survey lists totally different graph partitioning and agglomeration ways employed in community detection. It additionally discusses the comparison of community detection algorithms with different parameters [4]. The literature survey lists numerous ways employed in community detection for graph partitioning and agglomeration.

The idea of graph partitioning and ranked bunch is borrowed from classic approaches to finding communities in networks. This study is conducted by Andy Alamsyah et al. Graph partitioning approaches require to grasp data regarding the network's international structure and confirm the quantity and size of the subgroup they need to induce before. Hierarchical partitioning could be a technique of cluster analysis within which the interest network is split into many subgroups [5]. The division is somewhat natural because it depends on node relationships at intervals the network similarly as node properties themselves. Node relationships are measured by metric similarity, for instance, the similarity of the vertex and also the edges between them [6].

Amit Aylani This article proposes a new community detection strategy based on user social activity, interest and interaction. This process utilizes activity as Number of Tags, Common Interest and TagComment. A fresh parameter is obtained using the values of activities Interest, TagLike and TagComment. A clustering algorithm is implemented using this fresh parameter and the number of tags that results in suggested communities for seed users. Hence, communities that are highly interactive and that can be a useful source of recent updates and advertising for fresh goods of comparable concern to the community participants are proposed on the grounds of social activities. This article outlines the fundamental ideas of social networks, community structure and techniques of grouping similar items. The future semantic strategy can be implemented that can more accurately provide user areas of concern. Sutaria et al. [7]: An Adaptive Approximation Algorithm for Community Detection in Social Network Data Mining is a method for collecting significant data from a wide range of data resources. It offers a way to extract important data using suitable algorithms from this big quantity. They suggested a technique that detects the vertex and the modularity of the vertex. They make our technique more accurate on the basis of a dynamic modular calculation. They given inputs without the target vertex as a series of edges with the origin vertex. New vertex is added or updated to the community index during each phase of pre-processing. At the last phase of this process, we receive a list of vertexes with the community in which the partitions are located and modular.

A Review of Community Detection Algorithms in Signed Social Networks: They provided an outline of the algorithms for community detection of unsigned social networks. Some current methods have the primary concentrate on entry parameters that are utilized to detect the communities and others are automated methods that do not need input parameter. This is an overall analysis of all community detection algorithms available for signed social networks.

Survey on Efficient Community Detection in Social Networks: In this research, several algorithms for both the directed, un-directed networks, community detection performs a significant part in the management of complicated or large networks. Mostly, algorithms work on small networks, but we contrasted several techniques of community detection used in social networks in this study. The primary difficulty in this study is to find the finest technique or algorithm. It also mentioned in detail about the time complexity of each method. For many other scientists who research community detection on both real-time networks and social networks, all methods mentioned in the study will be of excellent use.

3 Proposed Work

Here in this paper, we make a comparative analysis of the clustering algorithms which are used to detect communities and find the efficient one using some well-known parameters. We have taken data set from snap repository of Stanford University which is a Facebook data set to work on. This data set contains information about users and their corresponding linkages [8]. The clustering algorithms we took into account are:

A. Agglomerative nesting
B. Affinity propagation
C. DBSCAN
D. Spectral clustering
E. K-means.

3.1 Agglomerative Clustering

Agglomerative clustering operates "bottom-up". That is, every object is at first thought of as a cluster (leaf) of single parts. At each step of the formula, the two clusters that are the foremost similar are combined into a replacement larger cluster (nodes). This procedure is iterated till all points are member of only one single huge cluster (root). The opposite of collective clustering is a discordant bunch, conjointly referred to as divisive analysis (DIANA) and operating in a very top-down manner. It starts with the basics, which has all objects in a very single cluster. The foremost heterogeneous cluster is split into two at every stage of iteration.

Data Transformations Choice depends on data set!

– Centre and standardize

  1. Centre: subtract from each vector its mean
  2. Standardize: divide by standard deviation

$$\Rightarrow Mean = 0 \text{ and } STDEV = 1$$

– Centre and scale with the `scale()` function

  1. Centre: subtract from each vector its mean
  2. Scale: divide centred vector by their root mean square (rms)

$$x_{rms} = \sqrt{\frac{1}{n-1} \sum_{i=1}^{n} x_i{}^2}$$

$$\Rightarrow Mean = 0 \text{ and } STDEV = 1$$

Hierarchical Clustering Steps

1. Identify clusters (items) with closest distance
2. Join them to new clusters
3. Compute distance between clusters (items)
4. Return to step 1.

3.2 Affinity Propagation

Refianti et al. [9] in affinity propagation, supported similarities between pairs of knowledge points are considered, and at the same time, it considers all data points as a possible cluster. AP finds clusters by traversing recursively through associate degree unvarying method, and the real-valued messages are transmitted or changed between information points till top quality set as exemplars, and the corresponding cluster emerges [10].

Jyothisha and Nair [11] Affinity propagation incorporates a range of appreciated similarity between sources of data, wherever the similarity S(i, j) tells us how well the j coefficient data point is suitable for data purposes i. The AP objective is to reduce the mistake in a square, and each similitude is therefore prepared for the negative square error (Euclidean distance). In affinity propagation, there are two varieties of information interchanged among information points, and they are responsibility and handiness.

3.3   DBSCAN

Density-based spatial clustering of applications with noise (DBSCAN) could be a well-known knowledge clustering rule that is usually employed in data processing and machine learning.

Based on a collection of points, DBSCAN teams along with points that are near one another supported a distance measuring (usually geometrician distance) and a minimum range of points. It additionally marks outliers as the points that are in low-density regions.

Labelling convention:

- 0: Unlabelled
- −1: Noise
- 1: Cluster number 1
  ...
- k: Cluster number k

Procedure:

1. Index all points, and label all as '0'
2. for each point:

   (a) If it is labelled already, goto next point.
   (b) Get points (neighbours) within 'ϵ' distance from the chosen point.
   (c) If # of neighbours < 'minPts', label as NOISE(-1) and goto next point.
   (d) Else, label it as a new cluster (c_{new}).
   (e) Select the neighbouring points as a new set 'S', for each point in 'S':
       i. If labelled as −1, relabel to new cluster number (c_{new}), and go to next point in the set.
       ii. If point is already labelled, skip it.
       iii. If it is unlabelled, then label it (c_{new}),
            Get neighbours in ϵ boundary and if count > minPts add to the set 'S'
       iv. Continue to next point in the set, if set is empty, the go back to step (a).

3.4   Spectral Clustering

Spectral clump has become more and more well-liked because of its easy implementation and promising performance in several graph-based clusterings. It is often resolved expeditiously by normal algebra package, and really typically outperforms ancient algorithmic rules like the k-means algorithm.

3.5 K-means

K-means' cluster is one in all the foremost wide used unsupervised machine learning algorithms that form clusters of information supported the similarity between data instances. The amount of clusters should be outlined beforehand. The K within the K-means refers to the amount of clusters.

The K-means' rule starts by arbitrarily selecting a centre of mass price for every cluster. The rule iteratively performs three steps:

(i) Realize the Euclidean distance between all data and centre of the clusters.
(ii) Assign the information objects to the cluster's centroid with the nearest distance.
(iii) Calculate new centroid values that supported the mean values of the coordinates of all the data from the corresponding cluster.

1. Start with initial guesses for cluster centres (centroids).
2. For each data point, find the closest cluster centre (partitioning step).
3. Replace each centroid by an average of data points in its partition.
4. Iterate $1+2$ until convergence

Write $x_i = (x_{i1}, \ldots x_{ip})$:

If centroids are $m_1, m_2, \ldots m_k$, and partitions are

$c_1, c_2, \ldots c_k$, then one can show that K-means converges to a *local* minimum of

$$\sum_{k=1}^{K}\sum_{i \in c_k} ||x_i - m_k||^2 \quad \text{Euclidean distance}$$

(within cluster sum of squares)

– Try many random starting centroids (observations) and choose solution with smallest of squares.

4 Simulated Results

Figure 1 is the result obtained by agglomerative clustering. It took a time of 0.31 s, and 6 clusters are formed. NMI is 1; ARS is 1; and Avg is 1. Each colour represents a cluster in the figure.

Figure 2 is the result obtained by affinity propagation. It took a time of 10.24 s, and 8 clusters are formed. NMI is 0.53; ARS is 0.25; and Avg is 0.39. Each colour represents a cluster in the figure.

Fig. 1 Clusters formed by
Spectral clustering

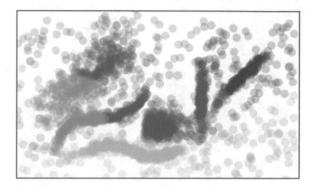

Fig. 2 Clusters formed by
K-means

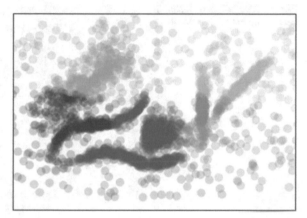

Figure 3 is the result obtained by DBSCAN. It took a time of 0.05 s, and 8 clusters are formed. Each colour represents a cluster in the figure.

Figure 4 is the result obtained by spectral clustering. It took a time of 0.84 s, and 6 clusters are formed. NMI is 0.36; ARS is 0.26; and Avg is 0.31. Each colour represents a cluster in the figure.

Figure 5 is the result obtained by K-means. It took a time of 0.12 s, and 6 clusters are formed. NMI is 1; ARS is 1; and Avg is 1. Each colour represents a cluster in the figure.

5   Experimental Results

The parameters that are taken into account for comparing the graph clustering algorithm are: time taken, number of clusters, NMI, ARS and Avg. Normalized Mutual Information(NMI): If the value of NMI is close to 1, then the communities that are found out by the algorithm are more close to real communities Adjusted Random Score: ARS is a measure of the similarity between two data clusters. Larger the value, higher is the similarity between the two clusters [12].

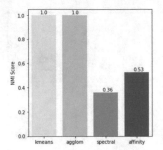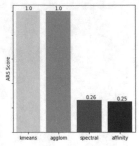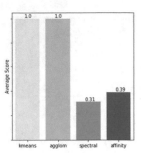

Fig. 3 Parameters versus algorithms

Fig. 4 Time taken to cluster versus number of data points

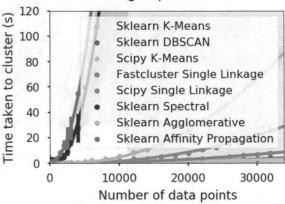

Figure 3 shows the result of plotting NMI, ARS and average score against the algorithms: K-means, agglomerative, spectral clustering and affinity propagation. Figure 4 depicts the number of clusters formed for a set of data points for the various algorithms. Time taken for execution by each of these algorithms is given in Fig. 5.

Table 1 provides a comparative analysis of the various graph clustering algorithms against prominent parameters. As shown in K-means, agglomerative clustering provides better results, since the NMI and ARS values are 1. Even though the time taken by the DBSCAN is low, NMI and ARS values, which are the most important parameters, are low.

6   Conclusion

Our paper aimed at detecting communities in real social networks using various clustering algorithms, and then we made a comparative analysis of the graph clustering algorithms. For comparison, the well-known parameters are taken into consideration,

Fig. 5 Number of clusters
versus data points

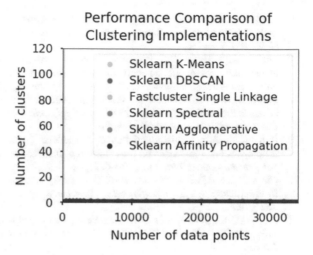

Table 1 Comparison of various parameters

Algorithm	Time taken	No of clusters	NMI	ARS	Avg
K-means	0.12	6	1	1	1
Agglomerative	0.31	6	1	1	1
Spectral	0.84	6	0.36	0.26	0.31
Affinity propagation	10.24	8	0.53	0.25	0.39
DBSCAN	0.05	8	–	–	–

and from the experimental results and the background of those real social networks, we can conclude that K-means and agglomerative clustering find the communities that are very close to real communities.

References

1. Despalatovi, L., T. Vojkovi, and D. Vukicevic. 2014. Community structure in networks: Girvan-newman algorithm improvement. In *2014 37th international convention on information and communication technology, electronics and microelectronics (MIPRO)*, 997–1002.
2. Archana J.S., and M.R. Kaimal. 2014. Community detection in complex networks using randomisation. In *2014 ACM international conference proceeding series*, 10.
3. Jayasree, N., K.S. Krishnavajjala, and G.P. Sajeev. 2018. A novel community detection for colloborative networks. In *2018 international conference on advances in computing, communications and informatics, ICACCI*.
4. Suryateja, G., and S. Palani. 2017. Survey on efficient community detection in social networks. In *2017 international conference on intelligent sustainable systems (ICISS)*, 93–97.
5. Aylani, A., and N. Goyal. 2017. Community detection in social network based on useras social activities. In *2017 international conference on I-SMAC (IoT in social, mobile, analytics and cloud) (I-SMAC)*, 625–628.

6. Ruby, and I. Kaur. 2017. A review of community detection algorithms in signed social networks. In *2017 international conference on energy, communication, data analytics and soft computing (ICECDS)*, 413–416.
7. Sutaria, K., D. Joshi, C.K. Bhensdadia, and K. Khalpada. 2015. An adaptive approximation algorithm for community detection in social network. In *2015 IEEE international conference on computational intelligence & communication technology*, 785–788.
8. Jyothisha, A.C., and J. Nair. 2019. Influencing community detection using overlapping communities. In Proceedings, 2018. In *4th international conference on computing, 2019. ICCUBEA: Communication control and automation.*
9. Refianti, R., A.B. Mutiara, and A.A. Syamsudduha. 2016. Performance evaluation of affinity propagation approaches on data clustering. *International Journal of Advanced Computer Science and Applications* 7 (3): 420–429.
10. Girvan, M., and M.E.J. Newman. 2002. Community structure in social and biological networks. In *Proceedings of the national academy of sciences of the United States of America*, vol. 99, 7821–7826.
11. Jyothisha, A.C., and J. Nair. 2018. Uncovering relationships among overlapping communities. In *Proceedings - 2018 IEEE international conference on communication and signal processing, ICCSP.*
12. Yang, Z., R. Algesheimer, and C.J. Tessone. 2016. A comparative analysis of community detection algorithms on artificial networks. In *Scientific reports.*

Energy-Efficient Heterogeneous WCEP for Enhancing Coverage Lifetime in WSNs

Amandeep Kaur Sohal, Ajay K. Sharma and Neetu Sood

Abstract The coverage of the monitoring field is the most important issue in geographical regions, traffic, and battlefields. Although the coverage-enhancing clustering algorithms for homogeneous WSNs have been investigated, heterogeneity in context to energy, sensing range, pre-determined deployment scenario, etc., has not been explored. In this paper, heterogeneous weight-based coverage-enhancing protocol (H-WCEP) is proposed that incorporates heterogeneity of sensors in terms of initial energy for load balancing at geographically far away sensors from the sink. The drop in the remaining energy and subsequent decrease in coverage has been overcome by deploying a certain number of higher initial energy sensors in specific areas. Thus, the H-WCEP achieves an increase in the lifetime of the network and full coverage of the monitoring area for a longer time. The simulation results show that H-WCEP outperforms as compare to existing coverage-enhancing clustering algorithms.

Keywords Clustering · Wireless sensor networks · Energy efficient · Coverage · Network lifetime

1 Introduction

Presently, wireless sensor networks (WSNs) are motivating with great potential for healthcare monitoring, environmental/earth sensing, transportation, agricultural monitoring, military surveillance, home automation, industrial monitoring, and many

A. K. Sohal (✉)
Amandeep Kaur Sohal, GNDEC, Ludhiana, Punjab 141006, India

A. K. Sharma
I. K. Gujral Punjab Technical University, Kapurthala, Punjab, India

N. Sood
B.R.Ambedkar NIT, Jalandhar, Punjab 144011, India
e-mail: soodneetu@nitj.ac.in

© Springer Nature Singapore Pte Ltd. 2020　　　　　　　　　　　　　　　25
Y.-C. Hu et al. (eds.), *Ambient Communications and Computer Systems*, Advances in Intelligent Systems and Computing 1097,
https://doi.org/10.1007/978-981-15-1518-7_3

more [1]. Moreover, advancements in micro-electro-mechanical systems and wireless communications have fueled up the development of tiny inexpensive and low-power sensors. These sensors are connected wirelessly with each other and are used to monitor the surrounding for measuring temperature, humidity, vibration, pressure, position, sound, etc. These tiny devices (sensors) performs sensing, processing of data with small communication range and can be deployed in harsh environmental conditions to collect useful information [2]. Such sensors are employed in numerous real-time applications such as smart detecting in home automation, identification of neighboring node in vehicular ad hoc networks, target (wild animal) tracking, node localization, and effective routing between the sink and sensors [3].

In military and disaster management applications, large numbers of sensors are densely deployed because of the failure-prone area and unable to replace the batteries. Although the cost of sensors is less still, there is frequently a need to encourage the coverage of application area for longer network lifetime with limited energy resources. The clustering can contribute to network scalability, resource sharing, decreases of overall energy consumption, and increase in network lifetime [4]. The thousands of sensors are densely deployed over the application-specific region; further grouping the sensors into clusters has been widely adopted to overcome the above-said challenges of WSNs. The sensor monitors or senses periodically and transmits local information to cluster head (CH). The CH processes the information collected from sensors and forwards it to sink or base station (BS) [4]. The intra-cluster communication refers to communication within a cluster, whereas the communication between CHs and sink is called inter-cluster communication.

In this paper, we propose a concept based on weighted clustering scheme for heterogeneous WSN and compared with two other clustering approaches, weight-based coverage-enhancing protocol (WCEP) [5] and coverage-preserving clustering protocol-minimum weight (Cmw) (CPCP) [6]. These three approaches follow different categories of clustering methods. We analyzed all these protocols on the basis of coverage lifetime, the average remaining energy of cluster heads, total network remaining energy, number of dead nodes, and average cluster size in terms of cluster heads and cluster members.

The remaining paper is arranged in the following manner. Section 2 describes a brief description of recent work, WSN device, and limitations of related coverage-based clustering protocols. Section 3 explains the procedure of proposed work, assumptions, and energy model. Section 4 discusses the comparative results of the implemented proposed approach and another state of arts. Finally, Sect. 4 presents the conclusion and future scope of the presented work.

2 Background Details and Related Work

The main purpose of clustering-based (hierarchical) protocols in WSNs is an appropriate method for the optimal use of limited available energy [2]. Moreover, in coverage-based clustering algorithms, reduction in energy consumption helps in longer network lifetime of sensors which maintains the coverage of monitoring field. Generally, this is done by clustering and sleep scheduling mechanism in homogeneous WSNs [5, 6]. In sleep scheduling, sensors are periodically subjected to power down state (sleep state) and power up state (active state) to minimize the energy consumption during unnecessary information sensing [7]. In addition to clustering, an optimal number of CMs and CHs are selected to obtain the QoS parameter. The cluster-based WCEP is one of the clustering protocols for maintaining coverage for a longer time for homogeneous WSNs in past years.

The key components that make up a typical WSN device are as follows: (i) low-power embedded processor (processing of both locally sensed information as well as information communicated by other sensors), (ii) memory/storage (program and data memory), (ii) radio transceiver (low-rate, short-range wireless radio), (iv) sensors (low-data-rate sensing), (v) geo-positioning system (satellite-based GPS), and (vi) power source (battery powered) [8]. The deployments of sensor and network structure are dependent on the applications and environmental conditions. Generally, the deployment of sensors is based on two main objectives: coverage and connectivity. The term coverage indicates sensing of information from all sensors in the application-specific region, and connectivity refers to the arrangement of sensors over which information is routed [9]. Other issues, such as establishment cost, energy limitations, and the need for robustness, should also be taken into account.

The various energy-efficient coverage-aware protocols are focused on balancing of energy consumption and load balancing of sensor nodes to achieve coverage lifetime along with longer network lifetime. There are three types of coverage considered in WSNs [10]. The types are point coverage, area coverage, and barrier coverage. The point coverage is defined as a set of points/targets (wild animals, vehicle, etc.) which are monitored by sensor nodes which are distributed over an application-specific region [11–13]. The area coverage indicates the given area for application is sensed by sensors distributed over a geographical region [14]. The barrier coverage specifies the detection of movement of targets across a barrier/boundary of sensor that is used to prevent the intruders [15]. The coverage-aware clustering protocols are classified based on clustering characteristics, network model, and type of coverage. The work is motivated from idea to enhance area coverage from a specific region to complete area of the geographical region [5]. For this, heterogeneity is incorporated with weight-based coverage-enhancing protocol (WCEP) to enhance the lifetime of full area coverage in WSNs. The energy and coverage-aware distributed clustering (ECDC) protocol was introduced based on the heterogeneity of energy to improve the point and area coverage of WSN [16] but does not discuss full area coverage. The WCEP is an energy-efficient full-coverage approach for WSNs [5]. The WCEP achieves longer full coverage, improved scalability with minimum energy consumption for homogeneous WSNs. It minimizes overhead messages of sensors during protocol operation. The WCEP is based on four controlling parameters that are remaining energy, overlapping degree, node density, and degree of the sensor node. The different weight values are assigned to these controlling parameters which are responsible for selecting active nodes as CMs and CHs. The periodic selection of CMs and CHs w.r.t. coverage in each complete round makes the WCEP efficient.

The CH selection technique called coverage-preserving clustering protocol (CPCP) was investigated with the aim of coverage preservation [6]. The CPCP is based on four coverage aware cost metrics i.e. energy-aware (Cea), minimum-weight (Cmw), weighted sum (Cws), and coverage redundancy (Ccc) respectively. Out of these cost metrics, CPCP (Cmw) provides best result for 100% coverage of the monitoring field. In CPCP, densely deployed sensors act as CH, active nodes and routing nodes. The CPCP integrates problem of network coverage and energy consumption. But these cost metrics cause a large overhead over sensors. The CPCP is also unable to minimize the redundantly overlapped area in each complete round.

3 Proposed Approach

The main drawback of WCEP algorithm is sensors which depletes their energy with a faster rate which is deployed at edges (especially at corners of square monitoring area) and near to sink (due to hot spot problem). At the edges, sensors (as CMs/CHs) consume more energy to transmit information to sink due to distance. The sensors near to sink also depletes its remaining energy due to forwarding information to sink. To overcome this problem, two concepts are introduced in the proposed work. Firstly, heterogeneous sensors (energy of 1/4 sensors gets increased to double) are used. Secondly, these sensors are deployed at critical zones (corners and around the sink). Therefore, the proposed algorithm enhances the full-coverage lifetime significantly by using heterogeneity in context to the initial energy of sensors. The proposed algorithm is named as heterogeneous WCEP (H-WCEP).

3.1 Procedure

The stepwise procedure of proposed work, i.e., H-WCEP, is discussed as follows:

Step 1

At the beginning of the algorithm, 'N' numbers of sensors are deployed in a random uniform manner in the square monitoring area. These sensors fully covered the monitoring area.

Step 2

Among the total number of sensors, one-fourth of sensors are initialized with 2 J and remaining with 1 J. One-fourth of sensors with double initial energy is deployed in critical zone area of the monitoring field.

Step 3

As the sensors are well aware of their coordinates, computation of all the distances between each other is done using the Euclidian formula. The H-WCEP algorithm is based on four coverage controlling parameters remaining energy, overlapping degree, node density, and degree of the sensor. The remaining energy of the sensor is updated after each complete round. One complete round indicates the time taken by sensors to transmit the sensed information to sink which includes sensing from the surrounding, transmits to CHs, and then CHs to sink. The overlapping degree of each sensor can be determined by a number of sensors which are generating overlapping area. The node density is the total count r of sensors within twice of its sensor range (i.e., cluster range). The number of covered points in the monitoring area is termed as the degree of the sensor. The above-said controlling parameters are updated after the end of every complete round.

Step 4

In the clustering process, a group of sensors forms a cluster. The members of the cluster are called cluster members (CMs), and head of the cluster is called cluster head (CH). In the H-WCEP, clustering is performed along with the sleep scheduling mechanism. As a result, optimal numbers of sensors are selected for 100% coverage of monitoring area and turned into power up (as alive/active sensors). The remaining sensors are turned into power down (sleep state). The appropriate numbers of alive sensors are computed using the weighted sum method. The H-WCEP works with 0.7, 0.2, 0.05, and 0.05 weight values for the above-said controlling parameters.

Step 5

The selected active/alive members became CMs and broadcast an updated packet within its sensing range. The selection of CHs is based on weighted sum value and the remaining energy of the sensor. The CH could be active/alive or sleep state sensor set which turned the power up accordingly. The selected CHs broadcast can update packet within its cluster radius.

Step 6

The CMs join their CHs according to their cluster range. The CMs transmit its sensed information to CHs and CHs forward collected information to sink via multi-hop communication and completes a single complete round. At the end of each complete round, four controlling parameters are computed, and if the remaining energy of the sensor is less than threshold energy, then the sensor is declared as dead. The threshold energy of sensor is 10^{-4} J for H-WCEP.

The H-WCEP procedure from steps 1–6 is persisted until all sensors will become dead.

3.2 Assumptions

- In H-WCEP, the following assumptions are taken in implementation:

- All the sensors are randomly uniformly distributed.
- All the sensors and sink are fixed in position after deployment.
- The BS is placed at the center of the application-specific region.
- All sensors are well aware of their coordinates and sink.
- The identification of sensors is unique.
- The initial energy level of sensors is defined.
- All sensors have a fixed sensing range.
- The sensor is taken at the center of the disk in the circular disk sensing model in 2D space [8].
- All the sensors can communicate within its transmission range that is considered as a circular disk sensing model in 2D space [8].
- The relation between transmission range and sensor range is
- Transmission range ≥ 2 times sensor range [8].
- The link between sensors is wireless and symmetric.

3.3   Energy Model

The proposed approach refers simple radio model similar to WCEP and CPCP (Cmw) [5, 6]. The simple radio model has transmitter, receiver, and power amplifier. The energy consumed by the sensor to transmit 'L' number of bits to the next sensor at distance 'd' is expressed in Eq. (1). The two propagation models are (i) free space model (d^2 power loss) and (ii) multipath propagation model (d^4 power loss). These propagation models can be used according to the distance between the source sensor and destination sensor. When the distance 'd' is less than a threshold distance (do) first model; i.e., free space (fs) model is used. Otherwise, the multipath (mp) model is used.

$$E_T = L E_{\text{elec}} + L \in d^\alpha \tag{1}$$

$$L E_{\text{elec}} + L \in_{\text{fs}} d^2 \, d < d_0$$
$$L E_{\text{elec}} + L \in_{\text{fmp}} d^4 \, d \geq d_0$$

where $\alpha = 2$ or 4

The d_0 is calculated as Eq. (2)

$$\text{do} = \sqrt{\frac{\in_{\text{fs}}}{\in_{\text{fmp}}}} \tag{2}$$

The energy consumption in receiving 'L' number of bits is determined by Eq. (3) [17].

$$E_{Rx}(l) = L E_{\text{elec}} \tag{3}$$

The electronics energy E_{elec} is based on digital coding, modulation, filtering, and spreading of the signal. The amplifier energies ($\in_{\text{fs}} d^2$, and $\in_{\text{fmp}} d^4$) rely on the distance to the receiver.

4 Simulation Setup and Results

This section explains the performance of H-WCEP. The implementation of H-WCEP is done with MATLAB software. The performance of H-WCEP (energy level of 1/4 sensors is doubled) is compared with WCEP [5] and CPCP (Cmw) [6]. The WCEP is based on the weighted sum method, and CPCP is based on four coverage-aware cost metrics. These are energy-aware (Cea), minimum weight (Cmw), weighted sum (Cws), and coverage redundancy (Ccc). Among these four cost metrics, CPCP (Cmw) is responsible for complete coverage of the monitoring region. Therefore, CPCP (Cmw) is considered for comparison in our simulations. Table 1 shows the simulation parameters used in the implementation of H-WCEP.

The performance of H-WCEP is evaluated with various performance parameters such as network lifetime, coverage lifetime, total network remaining energy, number of dead nodes, average CHs, and CMs are used and results are discussed in the following sections.

4.1 Comparison of Coverage Lifetime

Figure 1 shows the results of coverage lifetime with time. The comparison has been made between H-WCEP, WCEP, and CPCP (Cmw) in case of random deployment. While clusters are formed, H-WCEP opts efficient set of CMs and CHs. Because

Table 1 Simulation parameters

Parameters	Value
Monitoring area	$200 \times 200 \ (\text{m}^2)$
Total sensors	100
Initial energy for 75 sensors	1 J
Initial energy of 25 sensors	2 J
Sink	At center (at 0,0)
Transmission/receiving energy constant	50 (nJ/bit)
Amplifier constant	10 (pJ/bit/m^2)
Packet size	4000 bits
Sensor range	15 m

Fig. 1 Coverage lifetime
with time

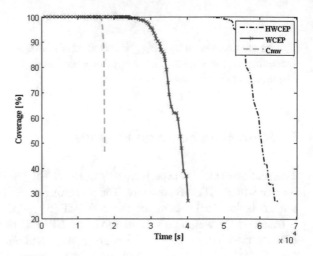

of higher energy level at critical zone sensors, H-WCEP maintains 25 and 58% prolonged full coverage with longer network lifetime in comparison to WCEP and CPCP (Cmw), respectively.

4.2 Comparison of Average Remaining Energy of CHs

Figure 2 depicts the performance of the average remaining energy of CHs with network lifetime. The unit of remaining energy of CHs is joules and unit of network lifetime in seconds. The average remaining energy of CHs is defined as follows

Fig. 2 Average remaining
energy of CHs with time

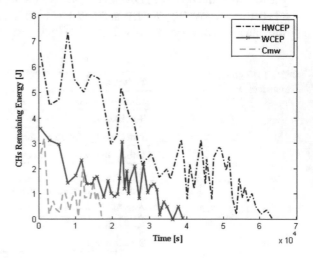

Fig. 3 Total network
remaining energy with time

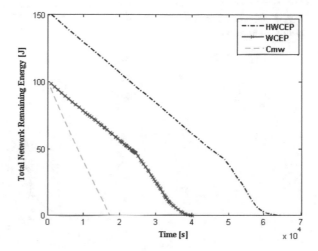

$$\text{Remaining } E_{(CHs)} = \left(\frac{\text{The remaining energy of CHs}}{\text{total number of CHs}} \right) \text{ at particular time t.}$$

In Fig. 2, simulation results indicate overall average CHs remaining energy of H-WCEP is more as compared to WCEP, and CPCP (Cmw) due to high-energy sensors performs their tasks for a longer time.

4.3 Comparison of Total Network Remaining Energy with Time

Total network remaining energy is defined as follows

$$\text{Remaining } E_{(sensor)} = \left(\frac{\text{The remaining energy of sensors}}{\text{total number of sensors}} \right) \text{ at particular time t.}$$

As in H-WCEP, heterogeneous sensors live longer at critical zones to maintain coverage. So, it achieves efficient load balancing in H-WCEP. Figure 3 indicates the overall network remaining energy of H-WCEP is higher in comparison to WCEP and CPCP (Cmw).

4.4 Comparison of Number of Dead Nodes with Time

As the H-WCEP uses two different levels of energy sensors (1/4 of total sensors have double initial energy), the rate of decay of dead nodes with time is much slower as compared to WCEP and CPCP (Cmw). Therefore, H-WCEP provides much longer

Fig. 4 Number of dead
nodes' time with time

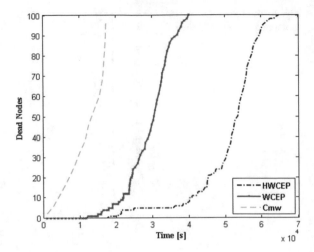

network lifetime with prolonged full coverage of the monitoring region. Figure 4
shows a decay of dead nodes w.r.t. time for H-WCEP, WCEP, and CPCP (Cmw).

4.5 Comparison of Cluster Size in Terms of Number of CHs and CMs Per Cluster

Figure 5 indicates the number of CHs with time. The number of CHs in H-WCEP at
complete coverage of the monitoring area is 20 and 36% less as compared to WCEP
and CPCP (Cmw), respectively. Similarly, in network operation, an average number
of alive sensors of H-WCEP is less as compared to WCEP and CPCP (Cmw). Figure 5
shows the trend of an average number of CMs along with time for H-WCEP, WCE,
and CPCP (Cmw).

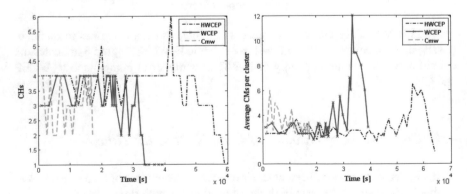

Fig. 5 Number of CHs and CMs per cluster with time

5 Conclusions

The coverage of the monitoring field is the most challenging issue in wireless sensor networks (WSNs) real-time applications. In this paper, we have proposed heterogeneous-based WCEP that improves full-coverage lifetime as compared to homogeneous WCEP in WSNs. The energy heterogeneity for sparse sensor helps to prevent hot spots and holes in the monitoring field which enhances the full-coverage and network lifetime. So, H-WCEP increased full-coverage lifetime by 25% from the original version of WCEP. Simulation results show that our proposed method is capable to enhance full coverage in comparison to WCEP and CPCP (Cmw). In the future, work may be extended with an increasing level of heterogeneity, mobile sensors, and multiple sinks in WSNs.

References

1. Siqueira, I.G., Ruiz, L.B., Loureiro, A.A.F. 2014. Coverage area management for wireless sensor networks. *International Journal of Network Management* 17–31. https://doi.org/10.1002/nem.

2. Verdone, R., D. Dardari, G. Mazzini, and A. Conti. 2009. Applications of WSANs. In *Wireless sensor and actuator networks: technologies, analysis, and design*, ed. T. Pitts, 16–21. Hoboken, New Jersey: Wiley.

3. Dargie, W., and P. Christian. 2010. Motivation for a network of wireless sensor nodes. In *Fundamentals of wireless sensor networks theory and practice*, ed. Xuemin (Sherman), Yi Pan, 17–32. Willey Series on Wireless Communications and Mobile Computing.

4. Mamalis, B., D. Gavalas, C. Konstantopoulos, and G. Pantziou. 2009. Clustering in Wireless Sensor Networks. In *RFID and sensor networks: architectures, protocols, security, and integrations*, ed. Y. Zhang, L.T. Yang, and J. Chen, 323–354. New York, USA: CRC Press, Taylor and Francis Group.

5. Sohal, A.K., A.K. Sharma, and N. Sood. 2017. enhancing coverage using weight based clustering in wireless sensor networks. *Wireless Personal Communications* 98: 3505–3526. https://doi.org/10.1007/s11277-017-5026-1.

6. Soro, S., and W.B. Heinzelman. 2009. Cluster head election techniques for coverage preservation in wireless sensor networks. *Ad Hoc Networks* 7: 955–972. https://doi.org/10.1016/j.adhoc.2008.08.006.

7. Ben-Othman, J., K. Bessaoud, A. Bui, and L. Pilard. 2013. Self-stabilizing algorithm for efficient topology control in wireless sensor networks. *Journal of Computer Science* 4: 199–208. https://doi.org/10.1016/j.jocs.2012.01.003.

8. Agrawal, D.P. 2011. Designing wireless sensor networks: from theory to applications. *Central European Journal of Computer Science* 1: 2–18. https://doi.org/10.2478/s13537-011-0007-z.

9. Patnaik, S., X. Li, and Y.-M. Yang. 2014. *Recent development in wireless sensor and ad-hoc networks*, vol. 233. https://doi.org/10.1007/978-81-322-2129-6.

10. Wang, Yun, L.J. Zhang Yanping, and R. Bhandari. 2015. Coverage, connectivity, and deployment in wireless sensor networks. *Recent Development in Wireless Sensor and Ad-hoc Networks, Signals and Communication Technology* 25–44. https://doi.org/10.1007/978-81-322-2129-6_2.

11. Sohal, A.K., A.K. Sharma, and N. Sood. 2019. A study on energy efficient routing protocols in wireless sensor networks. *International Journal of Distributed and Parallel Systems (IJDPS)* 6: 9–17. https://doi.org/10.5121/ijdps.2012.3326.

12. Özdemir, S., B.A. Attea, Ö.A. Khalil. 2012. Multi-objective evolutionary algorithm based on decomposition for energy efficient coverage in wireless sensor networks. *Wireless Personal Communications* 195–215. https://doi.org/10.1007/s11277-012-0811-3.
13. Zeyu, S., W. Huanzhao, W. Weiguo, and X. Xiaofei. 2015. ECAPM : An enhanced coverage algorithm in wireless sensor network based on probability model. 2015. https://doi.org/10.1155/2015/203502.
14. Amgoth, T., and P.K. Jana. 2015. Energy and coverage-aware routing algorithm for wireless sensor networks. *Wireless Personal Communications* 81: 531–545. https://doi.org/10.1007/s11277-014-2143-y.
15. Mostafaei, H., and M.R. Meybodi. 2014. An energy-efficient barrier coverage algorithm for wireless sensor networks. *Wireless Personal Communications* 77: 2099–2115. https://doi.org/10.1007/s11277-014-1626-1.
16. Gu, X., J. Yu, D. Yu, G. Wang, and Y. Lv. 2014. ECDC: An energy and coverage-aware distributed clustering protocol for wireless sensor networks. *Computers & Electrical Engineering* 40: 384–398. https://doi.org/10.1016/j.compeleceng.2013.08.003.
17. Heinzelman, W.B., A.P. Chandrakasan, and H. Balakrishnan. 2002. An Application-specific protocol architecture for wireless microsensor networks. *IEEE Transactions on Wireless Communications* 1: 660–670. https://doi.org/10.1109/TWC.2002.804190.

Middleware Frameworks for Mobile Cloud Computing, Internet of Things and Cloud of Things: A Review

Tribid Debbarma and K. Chandrasekaran

Abstract Mobile cloud computing (MCC) is an extension of cloud computing (CC) technologies. It provides seamless access of different cloud services to smart mobile devices (SMDs). There is no denying that CC can be scaled to a great extent in terms of computing, storage and other services, but the SMDs used for accessing those services are limited on battery capacity, storage and computing power due to their small form factors. The limitations of SMDs can be minimised/resolved by using MCC platforms. Though MCC is advantageous in many ways, it has its own inherent challenges and issues due to the heterogeneous hardware and software platforms used by SMDs and CC platforms, which makes it difficult to have interoperable services and the development of applications for those devices. This paper studied recently (from 2012 onwards) proposed/developed middlewares for the Internet of things (IoT), cloud of things (CoT), context-aware middlewares (CaMs) and mobile cloud middlewares (MCMs). Different middleware architectures are chosen, as in many cases, these technologies converge in terms of features, functions and services they provide. The study finds that the present middlewares lack in providing an integrated solution that complies with interoperability, portability, adaptability, context awareness, security and privacy, service discovery, fault tolerance requirements. At the end of the paper, the challenges pertaining to achieve portability, interoperability, context awareness, security are discussed and identify the gaps in the existing approaches in MCC interoperability and context adaptability.

Keywords Mobile cloud computing · Middlewares · Interoperability · Context awareness

T. Debbarma (✉)
Computer Science and Engineering, NIT Agartala, Agartala, Tripura, India
e-mail: tribid@ieee.org

K. Chandrasekaran
Computer Science and Engineering, NIT Karnataka, Mangalore, Karnataka, India
e-mail: kchnitk@ieee.org

© Springer Nature Singapore Pte Ltd. 2020   37
Y.-C. Hu et al. (eds.), *Ambient Communications and Computer Systems*, Advances in
Intelligent Systems and Computing 1097,
https://doi.org/10.1007/978-981-15-1518-7_4

1 Introduction

Cloud computing (CC) is expanding in many forms in the current ICT world, viz. mobile cloud (MC), cloud of things (CoT), Internet of things (IoT) with CC, fog computing (FC), edge computing (EC). The new forms of cloud computing are growing at a very rapid rate. The International Data Corporation (IDC) [1] predicts the expenditure on IoT which will reach $1.2 Tn in 2022. The similar trend is seen in case of CoT, FC, EC fields also.

In conventional cloud computing systems, the client and cloud servers are well connected to each other by high-speed and reliable communication technologies. But it is becoming more complicated in the new forms of cloud computing platforms, as these systems mostly rely on wireless and ad hoc communication systems to allow mobility and reduce the need for wired communications.

Among the extended CC services, the MCC has become one of the most widely used and an essential part of human life in the current age. Further, the MCC can help in achieving higher acceptance of CC services and ease of use by acting as an intermediator between CC and the sensor systems. MC services provide ubiquitous computing and other services to the general end users. End users are using various cloud services through SMDs without knowing the presence of cloud services that are taking place in the back end.

But with all these rapid transformation and growth, these new technologies have brought complex challenges due to the high level of heterogeneity in the hardware and software present in the platforms such as MC, EC and IoTs. The differences in these platforms make it difficult to achieve interoperability, portability, security, privacy, scalability, discovery of services and adaptability with the changing contexts. Moreover, the industry still lacks any standards for addressing these issues. For meeting these issues, different types of middleware technologies have been proposed with various features and functions.

In this paper, 16 carefully selected articles and sources that proposed middlewares for MCC, IoT, CoT and CaM platforms published from 2012 onwards are critically discussed and summarised. There are few studies which have viewed the different forms of mobile cloud computing and tried to address a unified solution in a generalised framework.

After the study of different features, functions and methodologies, the gaps in those systems are identified for further work to develop a middleware which shall be able to converge the various MCC forms and able to cater the MCC interoperability to the heterogeneous cloud platforms and mobile technologies with context adaptability.

The rest of the paper is organised in the following manner—Sect. 1—introduction, Sect. 2—background and related works, Sect. 3—different middlewares, Sect. 4—the issues and challenges, Sect. 5—summary of gaps and challenges found in Sects. 3 and 4, and Sect. 6—the conclusion and future direction of works.

2   Background and Related Works

Different types of middlewares are used for integrating different applications, devices and services. [2] lists the following middlewares which are commonly used in ICT platforms.

Message-Oriented Middleware: This type of middlewares allows distributed applications to communicate through messages.

Procedure Call (RPC) Middleware: This type of middlewares calls remote procedures. Synchronous and asynchronous communications can be performed through this.

Object Middleware: This type of middlewares communicates through object-oriented system.

Portals: Enterprise portal servers are the front-end services which connect to the back-end systems.

Content-Centric Middleware: Similar to publish/subscribe middlewares that are used in many Web applications.

Transaction Middleware: Web application servers and transaction processing monitors are an example of such middlewares.

Based on the implementation, architecture style middlewares can be classified as follows:

Layered: In this architecture, each of the layers integrates different functions and they are transferred to the higher layers.

Distributed: In this architecture, the integration is done in distributed components. This gives better performance than layered architecture.

Hybrid: In this, both distributed and layered structures are combined.

Though there are several comparative studies on middleware of MCC, IoT, edge computing, cloud computing, etc., there are only a few studies which have studied the different forms of mobile cloud computing middleware architectures in an integrated approach. The technologies of MCC, IoT, CoT and EC are in many ways part of the broad cloud computing platform. These new forms of technologies when connect to the conventional cloud and mobile devices with dynamic- and context-aware decision-making mechanisms, would perform on a larger scale and bring better user experience to the end users.

There are few survey works which studied different middlewares proposed for MCC, IoT and CoT platforms. Very few survey studies are available on exclusively MCC middleware architectures. The survey conducted on CaM, IoT, CoT middlewares also considered some of the MCC middlewares. In [3], the authors studied the computation offloading issues and challenges in MCC offloading. It pointed out security and privacy, fault tolerance and uninterrupted connectivity, platform diversity, automatic mechanisms, division of offloading, and offloading overheads as the common issues in MCC.

Farahzadi et al. [4] by Farhazadi et al. studies different middlewares for cloud of things (CoT) which identifies and discusses some of the open challenges and issues in the CoT platforms. Such as security and privacy enhancement, supporting of different protocols for e-healthcare, QoS, needs for standardisation, better resource discovery and utilisation, provisioning in a timely manner are some of the challenges and issues discussed in this work.

The survey work [5] studied several IoT middlewares to find the implementation of security issues in those middlewares. Their work identifies that most of the middlewares do not satisfy the various requirements to be a secure system. Further, they propose some points to address those gaps by considering privacy and security in the initial stage of designing the model and architecture.

Context awareness or adapting the system services based on current context can make a big difference in terms of user experience, service discovery, QoS, energy consumption and overall system performance in cloud computing platform. There are several middlewares which implement/propose context awareness in the middlewares. This middleware does different jobs of integrating context awareness of applications in the heterogeneous MC by means of different context management solutions. Xin Li et al. [6] is one of the comprehensive surveys on different context-aware middlewares (CaMs). This survey discusses different features of the CaMs, proposed by different researchers in several articles. In this study, it identifies the gaps and challenges in the field of context-aware middleware solutions such as degree of context awareness, security and privacy issues that needs to be incorporated and further extensive study that needs to be carried out in this field.

3   Reviewed Middleware Frameworks

In the following subsections, different middlewares proposed/developed for MCC, CoT, IoT and CaM architectures are discussed.

3.1   CoCaMAAL

Cloud-oriented context-aware middleware in ambient assisted living (CoCaMAAL) [7], proposed by Forkan et al., is a middleware to integrate different biomedical sensors for monitoring and data aggregation purpose. CoCaMAAL is a context-aware middleware for providing ambient assisted living services using sensors for monitoring the health condition of patients from the comfort of homes. This architecture has five main modules which are designed to work with the cloud architecture. The main components of the proposed architecture are (i) ambient assisted living (AAL) systems; (ii) context aggregator and provider (CAP); (iii) service providers; (iv) context-aware middleware; (v) context data visualisation.

3.2 SeCoMan

Semantic Web-based context management [8]—It uses semantic Web for data description, data modelling and reasoning of things in the IoT context. A layered architecture is used in this framework. The three main layers of SeCoMan framework are application, context management and plugin. Context management layer is the main module which is responsible for context-aware location-related services. The context awareness of this framework is to provide the location of the users, and by using semantic rules, it maintains the privacy of user's location with restriction to its access.

3.3 BDCaM

Big data for context-aware monitoring (BDCaM) by Forkan et al. [9] proposes a context-aware-based monitoring system based on knowledge discovery approach. This architecture has the following components:

(i) AAL systems—ambient assisted living; (ii) personal cloud servers(PCS); (iii) data collector and forwarder (DCF); (iv) context aggregator (CA); (v) context providers (CP) (vi) context management system (CMS); (vii) service providers (SP); (viii) remote monitoring system (RMS).

This architecture uses ontologies for processing of context data. This proposed system, however, does not consider the integration of application, services and interoperability issues.

3.4 FlexRFID

FlexRFID [10] is a policy-based middleware solution that supports context-aware application development and devices' integration. It has the following layers—application extraction layer, business event and data processing layer, policy-based business rules layer, device abstraction layer. These layers further consist of several other layers and components within them. But it is not designed for cloud platform and not interoperable.

3.5 DropLock

Vinh et al. [11] presented a middleware to integrate mobile devices, sensors and cloud computing. It integrates MCs and IoTs and uses representational state transfer (REST) and bluetooth low-energy (BLE) technologies. This proposed system has the following components: (i) network management services; (ii) data management services; (iii) system management services; (iv) security and authentication services—this component handles privacy and security of data in the cloud by a trusted service manager. The mobile devices and sensors may be provided such services through SIM cards, trusted execution environment and secure elements.

3.6 Mobile Cloud Middleware

Flores et al. [12] proposed a mobile cloud middleware (MCM) to address multi-cloud interoperability, dynamic allocation of cloud resources and mobile tasks' delegation using asynchronous communication. It uses interoperability API engine for multi-cloud communication. In this proposed architecture, mobile application performs distributed computing using tasks through delegation approach. The eclipse modelling framework is used for plugin developments in this proposed framework. The following are the main components of MCM—(i) TP handler; (ii) MCM manager; (iii) session; (iv) interoperability engine; (v) composition engine; (vi) asynchronous notifier.

3.7 VIRTUS

Conzon et al. [13] is e-health solution proposed for IoT platforms. Extensible message-passing protocol (XMPP) is used for communication in this middleware to provide communication among heterogeneous devices. This system is aimed to provide preventive measures to patients by collecting and monitoring the status of his/her health condition. Virtus use distributed architecture, and its main application is in e-health care. It is publish/subscribe type of middleware without any cloud-based service support.

3.8 OpenIoT

OpenIoT [14] is a service-based architecture aimed to enrich smart city projects by employing semantic interoperability among heterogeneous systems in the cloud. The main elements of this middleware are scheduler, service delivery and utility manager, request definition, request presentation, configuration and monitoring, cloud data storage and a sensor middleware.

3.9 Carriots

Carriots [15] is designed to work as a PaaS for IoTs in the cloud. In this middleware, sensor data is collected through Web connectivity and stored in the clouds. Message Queuing Telemetry Transport (MQTT) IoT protocols are used, and transmission is carried out in JSON or XML format. The main USP of this platform is scalability and reliability along with cost–benefit and low development effort.

3.10 mCloud

Zhou et al. [16] is a context-aware offloading framework proposed for heterogeneous mobile cloud. It proposes code offloading of applications to the nearby cloudlets and public clouds. Based on the wireless networks and cloud resources' availability, the context-aware algorithm takes the decision of offloading of codes in the runtime.

3.11 CHOReOS

Ben Hamida et al. [17] collection of software components are organised in this middleware to provide assistance in the development and executing expansive web service compositions. IoT, Internet of services and CC technologies are converged in this middleware to extend the heterogeneous service support. The architecture of this middleware is organised mainly in four sections: a executable service composition (XSC), extensible service access (XSA), extensible service discover (XSD) and cloud and grid middleware. XSC deals with the composition of services and things, XSA deals with the connectivity of heterogeneous services and things, and it is handled by extensible service bus (XSB), client–server, publish–subscribe or tuple paradigms which are used for interconnection of heterogeneous services. The XSD module provides discovery services as well as the organisation of services and things. Cloud and grid middleware takes care of scalability and distribution needs in the proposed system.

3.12 C-MOSDEN

Context-aware mobile sensor data engine [18] is a middleware for mobile devices and used as a plugin. This functions as an activity and location-aware mobile plat-form. On-demand distributed crowd sensing-based architecture is proposed to avoid overheads in processing, storage which is based on users' request and location. It provides management of costs and resources by collecting only the required data. Interoperability and scalability are also considered in the proposed system.

3.13 Thingsonomy

Hasan and Curry [19] designed this framework to tackle a variety of IoT events happening in IoT middlewares and application layers. It proposes the use of thematic tagging for explaining types, attributes and values to represent their meanings. The article gave the name thingsonomies to these tags done on the things and taxonomies. The authors explained a loosely coupled producers and consumers approach that is adapted to work with the approximate semantic matching of events. It is a publish/subscribe type of middleware.

3.14 M-Hub

Mobile-Hub [20] proposes a general-purpose middleware for the Internet of mobile things (IoMT). This middleware aims to provide mobility to mobile smart objects in the IoT platform and also works towards providing scalable data communication with enhanced reliability. However, this work did not consider security and privacy issues.

3.15 NERD

No Effort Rapid Development (NERD) [21] is a middleware framework that tries to address the human–machine interface (HMI) challenges such as discovery, provisioning and co-evolution with associated industrial control systems (ICSs). It proposes the use of data-driven domain-specific language (DSL) to address the transmission speed, the low data capacity of device markers and BLEs and for on-device storage capability. The DSL gives specific importance to the data source and sinks between the two major components, i.e. HMIs and ICSs.

3.16 CloudAware

Orsini et al. [22] proposed a middleware for mobile cloud computing (MCC) and mobile edge computing (MEC) with context adaptability. It uses active components concept that is build along with Jadex.

Table 1 summarises the proposed middlewares' important properties.

Table 1 Summary of the studied frameworks' major properties (abbreviations: Dist.—distributed, Layd.—layered, ND—not defined)

Properties middleware	Interoperability	Security and privacy	Service discovery	Scalability	Context awareness	Architecture	Cloud service support
CoCaMAAL	Yes	ND	Yes	Yes	Yes	Dist.	Yes
SeCoMan	Yes	Yes	No	No	Yes	Layd.	No
MCM	Yes	No	Yes	Yes	No	Dist.	Yes
FlexRFID	No	Yes	No	No	Yes	Hybrid	No
DropLock	No	Yes	No	Yes	Yes	Service oriented	Yes
VIRTUS	Yes	Yes	Yes	Yes	Yes	Dist.	No
OpenIoT	Yes	Yes	Yes	Yes	Yes	Service-based	Yes
mCloud	Yes	No	Yes	Yes	Yes	Dist.	Yes
BDCaM	Yes	ND	Yes	Yes	Yes	Dist.	Yes
CHOReOS	Yes	Yes	Yes	Yes	Yes	Hybrid	Yes
C-MOSDEN	Yes	ND	Yes	Yes	Yes	Dist., component	Yes
Carriots	Yes	Yes	Yes	Yes	No	Dist.	yes
Thingsonomy	Yes	ND	Yes	Yes	Yes	Publish/subscribe	Yes
M-Hub	Yes	ND	Yes	Yes	Yes	Dist.	Yes
NERD	Yes	Yes	Yes	Yes	ND	Hybrid	No
CloudAware	Yes	ND	Yes	Yes	No	Dist.	Yes

4 Issues and Challenges

4.1 Interoperability

It is one of the most important features of middleware due to the heterogeneous systems in mobile computing (MC) and cloud computing (CC). Most of the middlewares are intended to provide interoperability among the different hardware and software platforms.

Debbarma and Chandrasekaran [23] discussed the interoperability issues and challenges in MCC along with privacy and security. It suggests the use of ReSTful services with combination to other message-passing technologies for the transmission of abstract data to achieve interoperability and data portability.

4.2 Portability

Easy portability of services from one platform to another platform makes a useful feature to the end user as well as developers. But due to heterogeneity in the existing software, hardware platform and different data formats make it a difficult task to fulfil this requirement.

4.3 Scalability

Scalability of middleware architectures provides the advantage to reduce or increase the number of associated entities easily, and it ensures that there is no need to change or minimal change is required to it to accommodate with changing requirements of the user demands. This feature is very much desired for cloud-based systems as it is very likely the amount of data transaction and processing requirements will grow at different times, and it should be scalable to meet these requirements.

4.4 Adaptability

The adaptability of the middleware services according to changing platform makes it a convenient feature to end users. To meet this challenge, the systems can be designed to identify the changes in services and platforms where it is being executed.

4.5 Context Awareness

Context management feature of middlewares can provide users with information with respect to the current scenario and make necessary decisions making an effortless one. Current SMDs are equipped with many sensors. Raw data from these sensors is of little use. The middleware should utilise the sensor data and provide necessary context awareness using different context management systems.

4.6 Security and Privacy

Every technology in the information and communication technology (ICT) must be dealt with security and privacy issues with the utmost importance. Current SMDs carry enormous personal data of the user, and mobile devices are equipped with low-security measures unlike the stationary server or PC systems. Different wireless technologies used by the SMDs are susceptible to various types of attacks.

4.7 Service Discovery

Service discovery can enhance the performance of the mobile cloud services, and the end user will get better and more accurate services. The availability of services discovery may be improved by combining and orchestrating services. Pieces of information about different services with details may be stored in one or multiple cloud repositories.

5 Gaps in Existing Approaches Towards Interoperability

It can be summarised from the summary of the different middlewares study that Web API methods alone cannot solve the interoperability and the integration of context-aware mobile cloud application and services. Web APIs need further support from other components. Also due to version changes and non-updating, it may become non-operational over a period of time. Web APIs can also create storage and operational overheads in the resource constraint mobile devices. Several researchers used the middleware approaches for integration of context-aware services. Though many studies proposed middleware for IoTs, CoTs with context-aware application and services, there is less number of middleware frameworks that support context-aware applications for integrated mobile cloud computing services within a single framework, and they lack the level of awareness and self adaptability to counter the ever-changing contextual environment.

Some of the proposals [7–9, 12–20] do support interoperability, but many of them are not designed to support cloud services, and some of them did not consider the context awareness issues.

Forkan et al. [7] proposed a cloud-oriented context-aware middleware in ambient assisted living (CoCaMAAL) is one of the works that feature a context awareness services integration using middlewares for assisted living. It uses distributed architecture and semantic technologies for context awareness. But this work did not consider the mobile cloud computing services in their scope.

The other work called mobile cloud middleware [12] by Flores et al. addressed important issues of integration of different mobile cloud platforms and application services. However, in this work, the context awareness issues of mobile computing services are not considered.

Most of the works reviewed are specific to a particular service and platform such as IoT, MC or EC. Many of the works used different semantic technologies for achieving context awareness. Ontology-based context acquisition and modelling are the most preferred by different researchers. However, these works fail to provide interoperability of mobile cloud platforms and applications with context awareness that supports scalability, adaptive context awareness, privacy and higher autonomy in a single framework. The authors are motivated to work further on these issues to address these gaps and aim to achieve higher context awareness and adaptability through machine learning algorithms. MCC middleware with support for XMPP, XML, Javascript Object Notation (JSON) [24], HTTP/HTTPS REST would help system interoperability and data portable architecture. Also, protocols such as message queuing telemetry transport (MQTT) and constrained application protocol (CoAP) [25] would help IoT interoperability and lightweight data transfer between sensors and SMDs.

6 Conclusion

The different categories of middlewares discussed in this paper provide us with an idea about the existing systems that try to incorporate various features into a middleware. The middlewares of MCC, IoT and CoT converge in many features and functions, but current systems lack in providing an integrated solution for these platforms.

The gaps in the studied works of literature give us directions to focus on those specific issues and challenges which need to be further worked on to address them. Most of the proposed solutions in most cases do not provide much importance to interoperability, and only a few efforts are seen that tried to resolve the issue. In the absence of any standard in the interoperability of heterogeneous MCs, middlewares developed with careful integration of API management systems and context management modules with machine learning capabilities along with essential cloud features shall be able to provide the required solutions.

References

1. IDC Report. https://www.idc.com/getdoc.jsp?containerId=prUS43994118.
2. Troy. 2017. Types of Middleware—Apprenda. https://apprenda.com/library/architecture/types-of-middleware.
3. Khadija, A., M. Gerndt, and H. Harroud. 2017. Mobile cloud computing for computation offloading: issues and challenges. *Applied Computing and Informatics*. https://doi.org/10.1016/j.aci.2016.11.002.
4. Farahzadi, A., P. Shams, J. Rezazadeh, and R. Farahbakhsh. 2018. Middleware technologies for cloud of things: a survey. *Digital Communications and Networks* 4: 176–188. https://doi.org/10.1016/j.dcan.2017.04.005.
5. Fremantle, P., and P. Scott. 2017. A survey of secure middleware for the internet of things. *PeerJ Computer Science* 3: e114. https://doi.org/10.7717/peerj-cs.114.
6. Li, X., M. Eckert, J.F. Martinez, and G. Rubio. 2015. Context aware middleware architectures: Survey and challenges. *Sensors* 20570–20607.
7. Forkan, A., I. Khalil, and Tari Z. Cocamaal. 2014. A cloud-oriented context-aware middle-ware in ambient assisted living. *Future Generation Computer Systems* 35: 114–127.
8. Huertas Celdran, A., F. Garcia Clemente, M. Gil Perez, and G. Martinez Perez. 2013. Seco-man: A semantic-aware policy framework for developing privacy-preserving and context-aware smart applications. *IEEE Systems Journal*.
9. Forkan, A., I. Khalil, A. Ibaida, and Z. Tari. 2015. Bdcam big data for context-aware monitoring—A personalized knowledge discovery framework for assisted healthcare. *IEEE Transaction on Cloud Computing*.
10. Khaddar, M., M. Chraibi, and H. Harroud. 2015. A policy-based middleware for context-aware pervasive computing. *International Journal of Pervasive Computing and Communications* 43–68.
11. Vinh, T.L, S. Bouzefrane, J.M. Farinone, A. Attar, B.P. Kennedy. 2015. Middleware to integrate mobile devices, sensors and cloud computing. In *The 6th international conference on ambient systems, networks and technologies*, 234–243. Procedia Computer Science.
12. Flores, H., and S.N. Srirama. 2015. Mobile Cloud Middleware. *Journal of Systems and Software*.
13. Conzon, D., T. Bolognesi, P. Brizzi, A. Lotito, R. Tomasi, and M. Spirito. 2012. The virtus middleware: An xmpp based architecture for secure IoT communications. In 21st International Conference on Computer Communications and Networks (ICCCN), 1–6. IEEE, Piscataway.
14. OpenIoT. http://www.openiot.eu/.
15. Carriots. https://www.carriots.com/.
16. Zhou, B., A.V. Dastjerdi, R.N. Calheiros, S.N. Srirama, and R. Buyya. 2017. mCloud: A context-aware offloading framework for heterogeneous mobile Cloud. *IEEE Transactions on Services Computing* 10 (5). https://doi.org/10.1109/tsc.2015.2511002.
17. Ben Hamida, A., F. Kon, N. Lago, A. Zarras, and D. Athanasopoulos et al. 2013. Integrated CHOReOS middleware—Enabling large-scale, QoS-aware adaptive choreographies. Hal-00912882.
18. Perera, C., D.S. Talagala, C.H. Liu, and J.C. Estrella. 2015. Energy-efficient location and activity-aware on-demand mobile distributed sensing platform for sensing as a service in IoT clouds. *IEEE Transactions on Computational Social Systems* 2 (4): 171–181.
19. Hasan, S., and E. Curry. 2015. Thingsonomy: Tackling variety in internet of things events. *Internet Computing. IEEE* 19 (2): 10–18. https://doi.org/10.1109/mic.2015.26.
20. Talavera, L.E., M. Endler, I. Vasconcelos, R. Vasconcelos, M. Cunha, and F.J.D.S. De Silva. 2015. The mobile hub concept: Enabling applications for the internet of mobile things. In *IEEE International Conference on Pervasive Computing and Communication Workshops (PerCom Workshops)*, 123–128. Piscataway: IEEE.
21. Czauski, T., J. White, Y. Sun, H. Turner, and S. Eade. 2016. NERD—Middleware for IoT human machine interfaces. *Annals of Telecommunications* 71 (3–4): 109–119. https://doi.org/10.1007/s12243-015-0486-3.

22. Orsini, G., D. Bade, and W. Lamersdorf. 2015. Cloudaware: Towards context-adaptive mobile cloud computing. In *IFIP/IEEE IM 2015 Workshop: 7th International Workshop on Management of the Future Internet (ManFI)*, 1190–1195. IEEE.
23. Debbarma, T., and K. Chandrasekaran. 2019. Springer International Conference on Inventive Communication and Computational Technologies (ICICCT), Tamilnadu (2019).
24. JSON. https://www.json.org/.
25. Eclipse. https://www.eclipse.org/community/eclipse_newsletter/2014/february/article2.php.

Blockchain Security Threats, Attacks and Countermeasures

Shams Tabrez Siddiqui, Riaz Ahmad, Mohammed Shuaib and Shadab Alam

Abstract This paper is about the information security and solution to it by the use of blockchain implementation in data security. Blockchain security implementations mostly witnessed in sensitive information systems, like the financial database. Furthermore, in this research paper, threats to computers and information systems are highlighted and analyzed as well as solutions to them are elaborated. The paper majorly based on the step-by-step solution to the threats to the information systems by blockchain technology. This paper systematically covers various risks aspects, demonstrates different attacks and discusses the challenges faced while incorporating blockchain technology for securing the system.

Keywords Blockchain · Threats · Attacks · Security

1 Introduction

The term blockchain was mainly used to refer to the applications that were new of the distributed blockchain database. The blockchain technology is one of the significant trends in the economy, whereby the contracts and wills are excited on their own [1]. The blockchain technology gives the allowance to the user to access the blockchain transactions whenever they have the internet.

The blockchain systems are very resistant to any form of data manipulation of any kind. Furthermore, the new blocks validated before incorporation into the blockchain. This feature enables the blockchain systems to be used as the distributed ledger [2]. Blockchain technology acts as the open ledger that is distributed all over and can take a record of transactions between two ends without straining [3]. The transactions between the two parties can be verified as the record of the transactions is permanent. The management of the blockchain by peer-to-peer network and the protocol for internode communication had to be adhered to when validating new blocks.

S. T. Siddiqui (✉) · M. Shuaib · S. Alam
Department of Computer Science, Jazan University, Jizan, Saudi Arabia

R. Ahmad
Department of Computer Science, Aligarh Muslim University, Aligarh, India

© Springer Nature Singapore Pte Ltd. 2020
Y.-C. Hu et al. (eds.), *Ambient Communications and Computer Systems*, Advances in Intelligent Systems and Computing 1097,
https://doi.org/10.1007/978-981-15-1518-7_5

51

The blockchain technology had some attributes that attracted the many users to employ and embrace the technology. The blockchain embraces the decentralization which allows the users to be linked together in one market place whereby the transactions and even the transfer of ownership of assets are conducted among the users in an open way without any mediator. The specific time of the transaction in the block can be verified [2]. Along the chain, every block has the value of the previous transactions, and thus, the blocks automatically authenticated and inference to the information stored usually prohibited. The blockchain is designed with the resilience properties whereby each node stores a copy of the information on the chain.

2 Literature Review

The blockchain technology is a technology with very little interference. The economic transactions ledger was digitally incorruptible and was programmed to store the financial transactions information as well as record the value of almost everything. The data held on the blockchain existed as a continually merged dataset. The blockchain database was not stored in any location, rendering the records stored on public chain easy verification [4]. The information in the blockchain is not centralized; thus, hackers and crackers had no chance to modify or steal the stored data. The accessibility of the information in the blockchain system was for anyone on the internet. The blockchain could host millions of computers at the same time.

The digital information on the internet distributed without the allowance of copying the information. These helped the blockchain technology to create a new type of the internet from the original purpose [5]. The blockchain technology was majorly used as the digital currency by use of bitcoin. Blockchain technology has inbuilt robustness, in that blocks with similar data distributively stored in the network. Some issues accompanied the use of bitcoin [4]. The issues were due to the hacking or mishandling the system.

3 Structure of Blockchain

The block has the main data. In this section, the major information of the blockchain is recorded. For instance, when the blockchain is used in the banking systems, bank clearing records will be stored in the main data [6]. The transaction data and even the contract records are stored and found in the main data. More than 500 transactions contained by the single block [7]. The block height determined by the number of transactions in every block.

Hash values generated when the transaction executed and hashed to a code, which broadcasted to every node [8]. Since there were many transaction records in each block node, the Merkle tree function was utilized to produce the final hash value. A data transmission resource reduced by the use of the Merkle tree function.

Timestamps in the blocks give or show the time taken to generate a block, and in the other information section, additional features of the block are generated. The user may define what data to be in the additional section [1]. The signature of the block and even nonce value are in most cases present in the section of the other information.

The hash functions in the block were used to verify the transaction information stored in the system. The credibility of the transactions maintained as the information was not tampered with, because the hash functions were coded with, the creation of the new blocks done by the hash functions [8]. The values in hash replicated onto the other hash values in the different block; hence, the manipulation of the information hindered. The current hash value in block 2 linked to the previous has value in block 1. The connection blocks create the dependability of information, and thus, the security of the blockchain system enhanced.

3.1 Types of Blockchain

There are three types of blockchain

Public Blockchain: In public blockchain, anyone who accessed the internet was able to use blockchain technology [9]. The transaction information and verification of same were publicly available for everyone. The users in the public blockchain can participate in the consensus process. The example of the systems that employ the public blockchain technology is bitcoin and the ethereum (Fig. 1).

Hybrid Blockchain: The information in the hybrid blockchain is usually either public or private. The hybrid blockchain systems are partly decentralized. The main examples of the methods that do use the hybrid blockchain systems are a Hyperledger.

Private Blockchain: In the private blockchain, the write permissions of the blocks monitored by certain authority, whereas the read permission is public or it is either

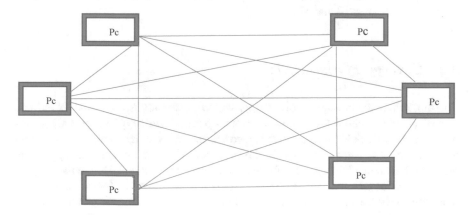

Fig. 1 Public blockchain

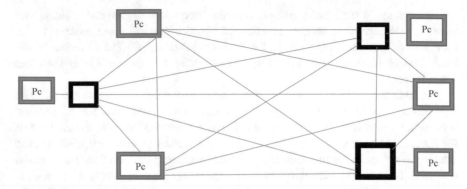

Fig. 2 Hybrid blockchain

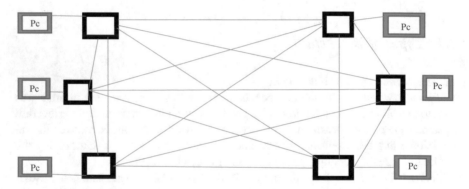

Fig. 3 Private blockchain

restricted. The nodes that participate in this blockchain are restricted [5]. Not every node can access the transaction information in the blockchain. This blockchain is highly secured (Figs. 2 and 3).

3.2 Concepts of Blockchain

The blockchain technology involves other various technologies for proper functions. The blockchain technology in most cases incorporated with cryptography, algorithm and economic models. Mathematics is also useful in blockchain technologies. Peer-to-peer networks were combined in the blockchain, and the consensus algorithms were used to solve distributed database synchronize problem [1]. Due to the cooperation of various technologies into the blockchain technology, the blockchain system was made to be an integration of multi-field infrastructure. Some elements make the blockchain technologies applicable in various fields.

Decentralization: The major characteristic of the blockchain network is the decentralization of the data. The centralized nodes are not used in blockchain [4]. The transactions in the blockchain systems were carried out between two peers without the approval from a centralized system. The data recorded, stored and even updated in the database distributively. By doing so, the server cost was low.

Transparent: There is transparency in the information records in the blockchain system. Each and every node in the blockchain accessed the clear data record. During the information update in the system, the process was transparent and thus built trust among the blockchain system users. Blockchain systems, in most cases, are public, whereby anyone accessed the transaction information in the system [10]. Every node can check the recorded data; the users had the opportunity of using the blockchain technology of the applications they wanted.

Autonomy: The information in the blockchain system between the blocks is interconnected. Due to the linkage, the data between the blocks, every node is able to update the information without any hindrance [9]. Whenever there is trust between an individual and system, the person is able to transfer the information in the blockchain system without any inventions.

Immutable: The transaction information record in the blockchain system is reserved permanently. The modifications on the records were not easy unless an individual controlled 51% and more nodes at the same time [7]. It was next to impossible for a person to control 51% of the nodes at the same time. Hence, the security of the data in the blockchain system was high.

Anonymity: The users interacted with blockchain network using the generated address. Some users generated many addresses to hide the exposure of their identity. The trust issue between the nodes solved, and hence, the transfer of data was anonymous [4]. The mechanism created some privacy for the transactions in the blockchain.

Auditability: The transactions on the blockchain system were validated using a timestamp. The previous records were easily verified and traced by the user accessing any node in the blockchain distributed network. The traceability of the information stored in the blockchain system improved through this feature.

3.3 Design

A network of the computing nodes makes up the blockchain system. The information on blockchain network copied automatically in the nodes. Every node in the blockchain system as administrator and the user joined the network voluntarily; thus, the blockchain system decentralized [9]. Nodes do compete by solving computational puzzles to win the bitcoin. Different nodes interconnected to each other.

4 Challenges of Blockchain

Blockchain systems do have challenges during the implementation process or after the implementation process [11].

Privacy: The privacy of the transaction data and confidentiality was still a challenge in the blockchain network [4]. Each node in the blockchain network can access the transaction data of the other nodes. An individual who took a look at the blockchain can also see the information record; thus, there is no privacy of the data stored in the blockchain.

Decentralized nature of the Blockchain: Decentralization of the blockchain system creates a protection to the system users from being attacked at the same time [12]. The blockchain network maybe disrupted, by a developer who is fully aware of the layout of the network, when it is an inside job.

Software Vulnerability: The bugs do exist in the software code and some exist in the poorly designed software. The software of this kind is usually vulnerable to the hacks. Software that uses bitcoin is prone to the malfunctioning [7]. As the software gets more complicated and interconnected to various nodes as in the blockchain network, the reliability of the system goes down and the software malfunctioning goes up.

Redundancy: The blockchain system faced challenges of redundancy in the network. The copies of every transaction are in the nodes of the blockchain network, with an aim of prohibiting intermediation [13]. Thus, it is very illogical for the system to have the redundancy and intermediation at the same time.

Distributed and Replicated: The blockchain system is a distributed network, which entirely relies on the replication and duplication of the data, to ensure the consistency of the information among its users [5]. The replication and duplication processes posed significant challenges to the implementation of the blockchain, in that they were complicated, time-consuming, and computationally intensive.

Initial implementation cost: The initial cost of implementing the blockchain system is very high. The mining of the blockchain requires a very high computational aid. Thus, it is a great challenge for the firms to adopt the blockchain network.

Regulatory Compliance: The application of the blockchain technology in the finance field, with currencies not in the form of bitcoin, faces regulatory challenges. Thus, the regulatory that applies to the infrastructure is not similar to that applied on the blockchain.

4.1 Risks of Blockchain

51% Vulnerability: In proof of stake-based blockchain system, the system is vulnerable when the single miner owns 50% and more coins. The attacker, with 51% of the coins, was able to manipulate and interfere with the blockchain data.

Private Key Security: The private key is generated and maintained by the participants in the blockchain network. The system is vulnerable during the signature production process; there is no assurance of the randomness production of signatures [4]. The attackers are able to recover the private key, and hence, the data in blockchain is tampered with.

Criminal Smart Contracts: Smart contracts in cooperated in the blockchain network suffer critical risks. A criminal smart contract leaks out the confidential information, steals the cryptographic keys and even facilitates crimes.

Under-Optimized Smart Contract: Some contract development and deployment in the blockchain were under-optimized [10]. This risk resulted to the additional operation to be run plus the additional cost on the users.

4.2 Attacks of Blockchain

Selfish Mining Attack: The blocks are mined secretively on the private branch, while the legit miner mines the blocks on the public chain [14]. The attacker tends to develop a long chain of blocks and then honest miner who end up wasting computer resources.

DAO Attack: The decentralized autonomous organization is a smart contract that is used in the ethereum, which used in the implementation of the crowd-funding platform [8]. The attacker publishes the malicious contract, which ignites the callback function for malicious gain. Through this way, the malicious attacker was able to snatch the ether from the decentralized autonomous organization.

Border gateway protocol attack: BGP is a routing protocol that regulates Internet Protocol packets to their destination [15]. The attacker manipulates BGP routing hence, gain control of the blockchain network, causing the delay in the network message.

Eclipse Attack: The attacker controls all connections of the blockchain network users and ends up causing the unnecessary computing power cost to the users [10]. Liveness attack delays confirmation time of the transaction of the targeted blockchain users (Table 1).

5 Recommendation

The blockchain systems are very much reliable for the use, but the security mechanism has to be deployed in every point of the network, to ensure the use of bitcoin technology in the blockchain [20]. Multisig which ensures multiple signatures for the transaction data has to be utilized [13].

Multisig secures the bitcoin. In the blockchain, the bitcoin model has to be modified so that they can verify the transaction script. The use of the blockchain technology should be thoroughly analyzed, before setting up the system [21]. The private key

Table 1 Countermeasures to blockchain in security

Countermeasures to the blockchain in security

Method	Application	The impact of the method
Quantitative framework	The quantitative framework is composed of two sections, one section has blockchain stimulator, and another section has security model design [1]. The stimulator resembles the working of the blockchain systems. The parameters are consensus protocol and the network, which is the inputs	The quantitative framework outputs a high common strategy to curb the attacks. By doing so, the framework helps in building security for the blockchain system
Oyente	The Oyente is created in a way that it is able to sense bugs in ethereum–based contracts. The technology is designed to evaluate the byte codes of the blockchain smart contracts in ethereum [4]. The ethereum blockchain systems store the EVM byte code of the smart contract	Oyente becomes very much useful when deployed into the system. Oyente detects any bug that may be present in the system
Hawk	The framework is used to develop the privacy of smart contracts. In the use of the hawk, the developers were able to write the private smart contract that had no codes; thus, the system was more secure	By using hawk, the financial transactions are not explicitly stored in the blockchain network system because the developers do separate the system into two major portions [1]. The private portion stores the data that is not for the public; the financial transaction information is stored in the private portion. The code or the information that does not require privacy is written on the public portion [16]. Hawk provided protection for the private information record in the blockchain system

(continued)

Table 1 (continued)

Countermeasures to the blockchain in security

Method	Application	The impact of the method
Town crier	Town crier functions by obtaining information request from the user and also gather the data from the HTTP websites [9]. A digitally signed blockchain message is returned to the user contract by the town crier	The town crier provided security during the data request by the users. The robust security model for the blockchain smart contracts is provided by the town crier
Segwit	Segwit is among the features of the side chain, which runs parallel to the main blockchain network [17]. The signature data move on from the main blockchain to the extended side chain	By use of the side chain, a lot of blockchain space is freed-up, and hence, more transactions are carried out [8]. The signature data is arranged in the form of the Merkle tree in the parallel side chain. By this arrangement, the overall block size limit increased with the no interference with the block size. By diversification of data, the security of the network improves
Off-chain state channels	The channel is always a two-way interaction of participants while sharing the information. This normally occurs in the blockchain network	Through the use of the off-chain channel, the transaction time is reduced exponentially, because there is no dependency on the third party like the miners [13]. This reduces the fraud that may occur on exposure of the data to the many participants

(continued)

Table 1 (continued)

Countermeasures to the blockchain in security

Method	Application	The impact of the method
Lightning network	Lightning network creates the double-signed transaction receipts. The transaction is termed to be valid after the parties involved in a transaction, signs for the acceptance of the new check [9]	The network helps two individuals to carry out a transaction between themselves, without the interference by the miner, who is the third party. The double signing ensures the security of the transactions for the involved parties
Permissioned blockchain	The companies and the organizations deal with the confidential information, in that not to allow anybody to have access to the information. They have to make sure that the insiders do not corrupt system network, and the outsiders do not have access to the information [18]	The permissioned blockchain network prevents the fraud as there is a restriction on who has to participate and in what capacity [8]. The member in the permissioned blockchain network has to be validated upon invitation before they participate. The identity key address of the participants is controlled and managed in the permissioned network. Hyperledger fabric is a blockchain framework, whereby the users have the membership card, which grants them access to their transaction information [19]. The participants cannot tamper with blockchain data without consensus
Integration of blockchain with artificial intelligence (AI)	Artificial intelligence is whereby the machine is built in a way that is able to perform tasks that demand intelligence	By use of the machine learning, the personnel in charge of the security can detect any unusual behavior in network and hence prevent any attacks to the system [9]

address for the blockchain user should be highly coded to increase the safety of the logging in the information. The designer of the blockchain network should figure out the possible attacks to the network before the implementation [14]. The self-detecting software for the attacks should be incorporated in the systems [22]. This ensures that any attack is dealt with, at it is emergence.

6 Conclusion

Blockchain eliminated the risks which come along with data centralization, by storing information across the network. By use of the encryption technology in the blockchain security systems, the security is enhanced. The public and private key addresses are very much useful. The private key acts like passwords which give the user the allowance to access their bitcoin or other digital assets available on the blockchain network. Blockchain, the foundation of bitcoin, has become one of the utmost prevalent technologies to create and manage safe digital transactions. The attackers who control all connections of the blockchain network users and end up causing the unnecessary computing power to the users are very dangerous attackers that should be avoided at any cost.

This paper explores the depth of a comprehensive survey on blockchain. Structure and types of blockchain, namely public, hybrid and private blockchain are discussed in detail. Furthermore, numerous risks listed that may muddle the implementation of blockchain technology. Selfish mining and the eclipse attack are the worst attacks. The privacy of the transaction data and confidentiality is a challenge in the blockchain network. The private encryption of data is employed to safeguard the information that required privacy. The deployment of artificial intelligence in blockchain is the way to go, to detect and curb the attacks. The use of permissioned blockchain system tends to prevent the attacks at early stages, hence most effective for the countermeasure to many blockchain attacks.

References

1. Latifa, E.R., and A. Omar. 2017. Blockchain: Bitcoin wallet cryptography security, challenges and countermeasures. *Journal of Internet Banking and Commerce* 22 (3): 1–29.
2. Karame, G. 2016. On the security and scalability of bitcoin's blockchain. In *Proceedings of the 2016 ACM SIGSAC conference on computer and communications security*, 1861–1862. ACM.
3. Zyskind, G., and O. Nathan. 2015. Decentralizing privacy: Using blockchain to protect personal data. In *Security and Privacy Workshops (SPW), 2015 IEEE*, 180–184.
4. Karame, G.O., and E. Androulaki. *Bitcoin and blockchain security*. Artech House.
5. Pilkington, M. 2016. 11 Blockchain technology: principles and applications. *Research Handbook on Digital Transformations* 225.
6. Shrier, D., W. Wu, and A. Pentland. 2016. Blockchain & infrastructure (identity, data security). *MIT Connection Science* 1–18.
7. Weimer, J., an R. Fox. 2017. U.S. Patent Application No. 15/626,054.

8. Cachin, C. 2017. Blockchain and consensus protocols: Snake oil warning. In *2017 IEEE 13th European Dependable Computing Conference (EDCC)*, 1–2. IEEE.

9. Idelberger, F., G. Governatori, R. Riveret, and G. Sartor. 2016. Evaluation of logic-based smart contracts for blockchain systems. In International symposium on rules and rule markup languages for the semantic web, 167–183. Springer, Cham.

10. Heilman, E., A. Kendler, A. Zohar, and S. Goldberg. 2015. Eclipse attacks on bitcoin's peer-to-peer network. In *USENIX security symposium*, 129–144.

11. Alam, S., S.T. Siddiqui, F. Masoodi, & M. Shuaib. 2018. Threats to information security on cloud: Implementing blockchain. In *3rd International conference on SMART computing and Informatics (SCI), Kalinga Institute of Industrial Technology*, Odisha. Springer.

12. Tosh, D.K., S. Shetty, X. Liang, C.A. Kamhoua, K.A. Kwiat, and L. Njilla. 2017. Security implications of blockchain cloud with analysis of block withholding attack. In *Proceedings of the 17th IEEE/ACM international symposium on cluster, cloud and grid computing*, 458–467. IEEE Press.

13. Crosby, M., P. Pattanayak, S. Verma, and V. Kalyanaraman. 2016. Blockchain technology: Beyond bitcoin. *Applied Innovation* 2: 6–10.

14. Eyal, I., and E.G. Sirer. 2018. Majority is not enough: bitcoin mining is vulnerable. *Communications of the ACM*, 61 (7): 95–102.

15. Wright, A., and P. De Filippi. 2015. Decentralized blockchain technology and the rise of lex cryptography.

16. Aitzhan, N.Z., and D. Svetinovic. 2016. Security and privacy in decentralized energy trading through multi-signatures, blockchain and anonymous messaging streams. *IEEE Transactions on Dependable and Secure Computing*.

17. Kiayias, A., and G. Panagiotakos. 2015. Speed-security tradeoffs in blockchain protocols. *IACR Cryptology ePrint Archive*.

18. Lim, I.K., Y.H. Kim, J.G. Lee, J.P. Lee, H. Nam-Gung, and J.K. Lee. The analysis and counter-measures on security breach of bitcoin. In *International conference on computational science and its applications*, 20–732. Springer, Cham.

19. Biswas, K., and V. Muthukkumarasamy. 2016. Securing smart cities using blockchain technology. In *2016 IEEE 18th International Conference on High Performance Computing and Communications; IEEE 14th International Conference on Smart City; IEEE 2nd International Conference on Data Science and Systems (HPCC/SmartCity/DSS)*, 1392–1393. IEEE.

20. Yli-Huumo, J., D. Ko, S. Choi, S. Park, and K. Smolander. 2016. Where is current research on blockchain technology? A systematic review. *PLoS One* 11 (10): e0163477.

21. Gupta, A.K., S.T. Siddiqui, S. Alam, and M. Shuaib. 2019. Cloud computing security using blockchain. *Journal of Emerging Technologies and Innovative Research*.

22. Lin, I.C., and T.C. Liao. 2017. A survey of blockchain security issues and challenges. *IJ Network Security* 19 (5): 653–659.

Intelligent Computing Techniques

Implementation of Handoff System to Improve the Performance of a Network by Using Type-2 Fuzzy Inference System

Ritu, Hardeep Singh Saini, Dinesh Arora and Rajesh Kumar

Abstract Wireless networks comprise of different networks such as cellular, WiFi, WIMAX, etc. Mobility-based networks are increasing at a rapid rate and similarly, the demands of the users to access Internet are also increasing at any time and at any place. Thus, to fulfill the demands of the users, an effective handover is required. However, with the advancement in technology, the methods used earlier were not seemingly more effective. In the traditional approaches, the limited numbers of parameters are used. This paper consists of a scheme of advancement in the fuzzy system of inference, i.e. fuzzy type-2, as it provides an additional design for systems used in situations with large amount of uncertainties. It is concluded with the simulation results that the proposed system outperformed the traditional approach by increasing the number of parameters to be utilized in achieving the handoff.

Keywords Wireless network · Handoff · Fuzzy system · Crisp values · Membership functions

1 Introduction

The transfer of information from one node to other wirelessly without their coupling is done with the help of electrical conductors. Radio communication is one of the examples of wireless technologies. Due to the development of such a system, it becomes easy to send packets to distance; also it reduces the overall size of the packet which makes distance communication of up to hundreds of kilometers possible. Many portable applications are fixed as well as mobile applications are supported by this technology [1]. Many networking algorithms are introduced into this technique. Apart from these applications, radio wireless technology consists of PDAs, cordless telephone keyboards, wireless computers, GPS units, cellular telephones, garage door opener and broadcast television.

Ritu · H. S. Saini (✉) · R. Kumar
Indo Global College of Engineering, Abhipur, Mohali, Punjab 140109, India

D. Arora
Chandigarh Engineering College, Landran, Mohali, Punjab 140307, India

© Springer Nature Singapore Pte Ltd. 2020   65
Y.-C. Hu et al. (eds.), *Ambient Communications and Computer Systems*, Advances in Intelligent Systems and Computing 1097,
https://doi.org/10.1007/978-981-15-1518-7_6

Moreover, these applications comprise of many electro-magnetic wireless technologies by taking into account light, use of sound or electric field. Wireless networking has been used in several applications and especially to fulfill the requirements. For instance, laptop users who need to travel location to location, to connect them wireless networking has used [2]. With the help of satellites, it becomes possible to communicate mobile phones and to provide network connection. The local area network mostly consists of wireless communication. Various sensors are buried in many areas to test the physical as well as environmental conditions like sound, temperature as well as pressure. Moreover, the network is used to pass the data from one node to another or the main location. The more advanced networks are two-way, thus enabling the sensor activity to be controlled [3–5]. Creating WSNs depends upon several applications specifically military applications, i.e. battlefield surveillance. With the changing technology, people started adopting new techniques of communications in various fields, such as hospitals and offices.

The types of communication where mobile phones are able to communicate with each other are considered as cellular communication. As mobile phones are easily to carry, so the users do not stay at a particular location. Similarly, in this communication mobile phone users do not stay at a single position and freely moves from one to another place [6]. As communication is done through mobile phones, therefore, it is essential to offer effective and noiseless connection. Considering this responsibility, cellular systems initiated a concept of handoff in the network [7].

2   Fuzzy Inference System

The fuzzy logic is like our brain that takes in data or information such as 'mango is ripe,' 'weather is cloudy' and then responds with accurate actions. With having no knowledge of mathematical analysis, a method which controls the complex system is called as 'Fuzzy Logic. ' It is a system which uses fuzzy logic to bring output from input. Fuzzy inference is the process in which an output is mapped from input through implementation of fuzzy logic as shown in Fig. 1. This mapping of output and input allowed the system to differ with the patterns and to make decisions on the basis of that. Both the FIS are written as follows:

Sugeno Fuzzy Model: A proposed model for fuzzy logic was advised by Tomohiro and Sugeno [8] in 1985. With the help of datasets for input and output, it becomes possible to develop the rules for fuzzy logic. Following is the syntax of this proposed technique:

$$\text{If } x \text{ is } A \text{ and } y \text{ is } B \text{ then } z = f(x, y) \tag{1}$$

A and B are the predecessors in a crisp function such as $z = f(x, y)$. $f(x, y)$ defines the overall output of the system as well as describes the performance of the system. For a zero-order Sugeno fuzzy model, the interference level depends upon the polynomial one used in his analysis. It will be known as first-order Sugeno fuzzy

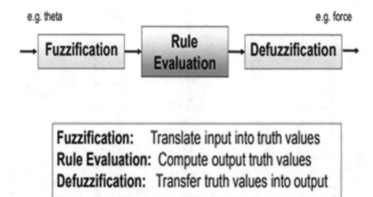

Fig. 1 Fuzzy inference process [8]

model if $f(x, y)$ is in the form of polynomial with the first order. Zero-order fuzzy used Mamdani fuzzy inference system. Also to identify consequents of rules, fuzzy singleton is required. Mamdani and Assilian [9] use fuzzy set concepts to develop control of fuzzy logic. To execute, FLC required the above steps in order to make decisions. The task of decision making in the medical field is very difficult and as there is no such model provided by the traditional analytical approaches to find the modes of reasoning that may be encountered in the process of decision making in the medical field. Mamdani and Assilian [9] proposed a model of combination of boiler and steam engine by using the setoff linguistic control rules in fuzzy logic. In Fig. 2, the demonstration of obtaining the overall output z from the two-rule fuzzy inference system (FIS) of Mamdani type with the use of min and max for fuzzy AND and OR operator if x and y are used. After obtaining the output, defuzzification process is required for translating the fuzzy controller's output to numerical value. These obtained values are known as crisp values.

In general, centroid of the area is the defuzzification approach, that is mostly used is expressed as:

$$z_{COA} = \frac{\int_z \mu_{C'}(z)z\,dz}{\int_z \mu_{C'}(z)z\,dz} \tag{2}$$

where, $\mu_c(z)$ is the membership of combined output functions. Many applications contain the maximum of smallest, largest of the maximum, the bisector of area, mean of the maximum which are used to apply the approaches of defuzzification [10]. Figure 3 shows the process of defuzzification technique.

It might be possible that certain reasoning mechanisms used by FIS are not according to the inference's compositional rule. User can use it according to his requirement, for example, qualified rule outputs or firing strengths or both can be calculated either using minor product. The different example is rather than using max, using pointwise summation (sum) in standard fuzzy reasoning. However, sum is not the fuzzy

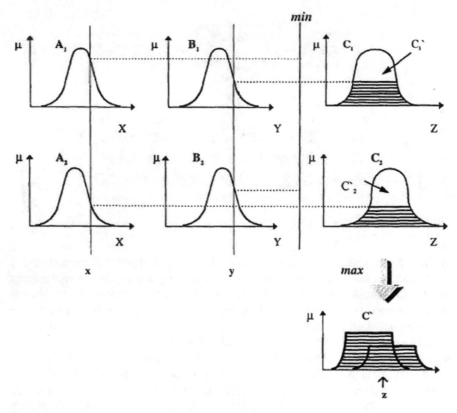

Fig. 2 Fuzzification process by using AND or OR operators [8]

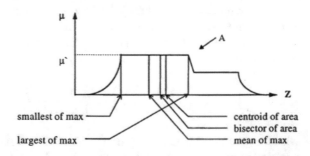

Fig. 3 Defuzzification process for obtaining crisp output [8]

OR operator. Kosko [11] projected a work which was beneficial as composition of sum–product and the final crisp output made by the author using centroid defuzzification is equivalent to the weighted average of each value of crisp output. The weighting factor is equal to the product of its firing strength to area of the output MF of the rule. The crisp output is equivalent to the centroid defuzzification value of its output MF. By attaining the centroid and the area of each output MF function in advance, the burden of calculation is reduced.

2.1 Type-2 Fuzzy Systems

With type-2 fuzzy sets, it become possible to handle linguistic uncertainties easily. These are expressed as:

'For different people, there is different meaning of same words.'

- A fuzzy type-2 relation has been considered as one way to enhance the relation's fuzziness. According to Hisdal, 'enhanced fuzziness is the increased ability for tackling vague information in a plausibly right manner.' Zadeh and Kacpyrzy [12]; Zadeh [13] introduced the concept of type-2 fuzzy sets as an extended version of the theory of an ordinary fuzzy set, for example, a type-1 fuzzy set.
- Fuzzy sets with type 2 consist of membership grades that act as fuzzy. A type-2 membership grade could be any of the subsets in [0, 1]—the primary membership—and there is a secondary membership for each primary which delineates all the possible situations for the primary membership.
- A Type-1 fuzzy set is a unique situation in a fuzzy set with type-2; its derived affiliation feature is a subset containing only single element.
- Type-2 fuzzy logic sets are computer-intensive due to the intense nature of the type-reduction.
- Things are simplified if secondary MF's are interval sets. In this scenario, secondary memberships are known as interval type-2 sets.

Inspired from these properties of the fuzzy type-2 system, the proposed model in this paper is focused on the fuzzy type system in the handoff decision making. The simulation of the proposed model is done MATLAB and analyzed in this paper.

3 Literature Survey

A quality of service-aware fuzzy rule-based vertical handoff mechanism which was used to make the multi-criteria based decision was proposed by Kantubukta et al. [14]. It is an effective approach to the fulfillment of the requirements of various applications in a heterogeneous networking environment. PresilaIsrat et al. [15] developed an adaptive handoff management protocol on the basis of the fuzzy logic and collaborate it with existing cross-layer handoff protocol. Yaw and Johnson [16] considered multi-criteria vertical handoff decision algorithm's design as well as implementation. This algorithm of vertical transfer of decision was demonstrated in a heterogeneous wide-range wireless network such as LAN and WAN environment. This technique was used in IP centric field workforce automation that provided free movement while connected with IP mode between networks and use of a single device on multiple networks. Orazio et al. [17] explained the concept of handoff where the condition occurred that the mobile node lost its temporary connection at the time of transferring from one wireless connection to another.

4 Problem Formulation

A handover is often regarded as the procedure by which the received signal is moved to the nearest base station in order to maintain the connection at the coverage limits of the original base station. However, a transfer between heterogeneous networks is not simply a way to maintain a connection but also to select the optimal network at a particular point. In the traditional work, a handoff system was prepared on the basis of the fuzzy system. The fuzzy system is somewhere useful for handoff technique as the parameters are limited; therefore, the fuzzy system is good for the limited number of parameters. But in a network, a number of nodes occur, whereas every node has its individual dependency. In addition, the other flaw of using fuzzy is that if the data is increased, then to define the rules as well as defining the range of the membership function are difficult. Also in the present system, only two factors are considered for the decision model, whereas more factors are responsible for handoff in the network. So there is need to increase dependent factors and also up gradation of fuzzy model. The major objectives of the proposed work are illustrated as:

- To enhance the existing handoff model with extending dependency factors, i.e. coverage area, speed and traffic load.
- To replace the existing fuzzy model with the type-2 fuzzy system for more effective decision capability.
- To perform the comparison analysis over the network performance parameters with the traditional fuzzy system.

5 Proposed Work

As discussed in the above section, traditional work includes some limitations, however, to conquer these limitations, a new mechanism is offered in this work. It is seen in the existing work that there may be an issue with the limited number of parameters. In addition, the other flaw of using fuzzy is that if the data is increased, then to define the rules as well as defining the range of the membership function are difficult. The proposed model will have the additional factors on which the handoff will be dependent these factors are: coverage area, speed and traffic load. Also, in addition to type-1 fuzzy system, traditional fuzzy is replaced with type 2 fuzzy system. Figure 4 shows the flow diagram of the proposed framework.

5.1 Methodology

The methodology of proposed work has mentioned in the following steps which are followed to attain the handover decision.

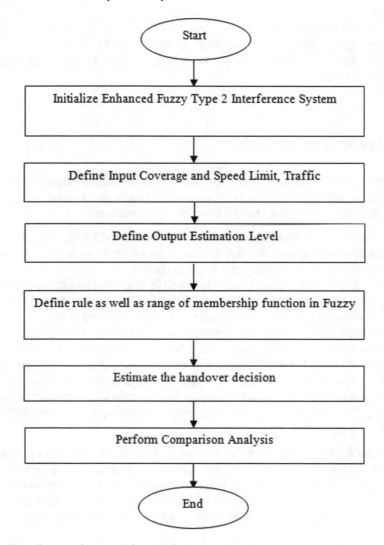

Fig. 4 Flow diagram of proposed framework

Step 1: Initialization of enhanced fuzzy interference system is done by using the type-2 fuzzy interface to increase the efficiency and defuzzification approach.

Step 2: The input parameters as proposed, coverage, speed limit and traffic load are used as input in the proposed model as an increment to traditional factors.

Step 3: The rules on which the estimation level are defined, i.e. the condition according to which the estimation is decided.

Step 4: After that, the fuzzy rules are clubbed along with the membership function in FIS system so that input-based decision can be calculated.

Step 5: After this, handover decision is taken based upon the factors that are input to fuzzy type 2, on behalf of rules defined an estimation level is obtained.
Step 6: Finally, the comparison analysis of the proposed technique with the traditional technique is performed.

6 Experimental Results

A fuzzy system has been designed to increase the ability to make decisions for hand-offs while increasing the number of parameters used to make a decision effectively. Following parameters are taken into account to generate output:

- **Coverage**: The coverage area of the network is one of the main factors. A small area in coverage would create excessive handoffs within the same network amongst access points that may cause a great loss to packets.
- **Speed:** This parameter determines the speed at which the mobile terminal (MT) is moving. The speed of a node in the network is used as an input parameter to make a decision of the handoff initiation with RSS threshold value.
- **Load:** During efficient handoff, network load must be taken into consideration. In order to prevent inefficient service performance, it is imperative to maintain the balance of network load. Changes in traffic loads will decrease the capacity for traffic. To provide mobile users with a high-quality communication service and to increase their traffic capacity when traffic changes take place, attention should be given to the network load. This parameter was, therefore, regarded as initial phase of making handoff decision.

The designed fuzzy system generates a single output on the basis of above parameters. Figure 5 explains that the network includes three inputs and produces the single output with regard to the handoff decision based on rules following the involvement of the Sugeno type-2 interfaces. Each input comprises of three membership functions. An estimation level is delineated on the basis of these membership functions.

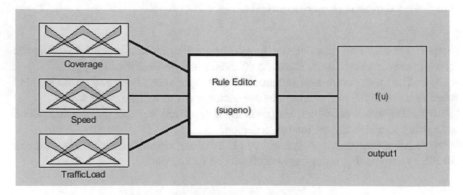

Fig. 5 Fuzzy interface of the multi-layered fuzzy interface's first layer

Fig. 6 Membership function
of input parameter
'Coverage'

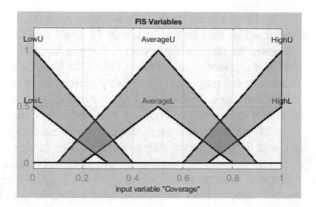

When the parameters have been entered in the fuzzy interface, a decision is made
based on predefined rules. Figures 6, 7 and 8 explain the membership function that
works on the fuzzy system. The range is between 0 and 1.

Fig. 7 Membership function
for 'Load'

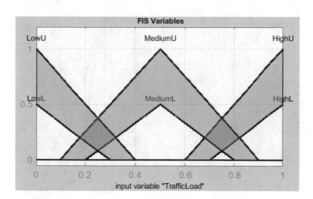

Fig. 8 Membership function
of input parameter 'Speed'

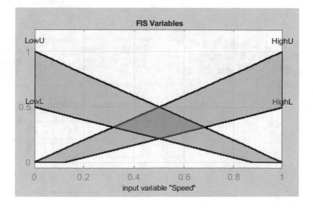

The handover comparison in terms of straight and random motion method is shown in Fig. 9. It is observed from the evaluation that the type-2 fuzzy algorithm surpasses the conventional fuzzy technique. In conventional technology, the probability discontinuity at a different distance from RSS was varying from 0.5 to 0, whereas in the proposed type-2 fuzzy system, the results are improved as the probability is going to be decreased and the variation become between 0.47 and 0. This is providing an approach toward an effective handoff in the proposed mode using type-2 fuzzy interface system. Table 1 shows the comparison between the proposed and traditional system at different distance levels in meters.

Figure 9 shows the likelihood of disconnection in terms of distance. Traditional techniques were compared with the proposed type-2 fuzzy techniques for simulation evaluation. It has been obtained from the acquired results that proposed technique has less likelihood of disconnection with regards to other techniques. The likelihood of disconnection has reduced since the range has risen but in perspective of the various conventional methods, the proposed method is proved to be effective.

Fig. 9 Probability of disconnection with regard to distance

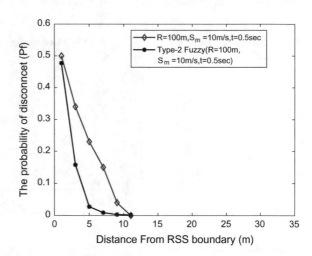

Table 1 Comparison of proposed and traditional system at different distance from RSS

Distance from RSS	Traditional probability of discontinuity	Proposed probability of discontinuity
1	0.5	0.47
3	0.34	0.158
5	0.23	0.0275
7	0.15	0.0090
9	0.04	0.03
11	0	0

7 Conclusion

This work introduced a method of taking handover decision. The coverage, speed and loading method are involved in this method. These parameters influence the communication process. These parameters are taken as input variables for the type-2 FIS, according to which the estimate level has been assessed and, therefore, the decision made. Moreover, the rules for reasoning are based on parameters along with their logical relationships in order to take decisions for handover. For reasons which are based on non-linear data (i.e. terminal and network information), the type-2 FIS is determined as an appropriate system. In the work proposed, the enhanced type-2 fuzzy system was replaced by FIS for improved predictions. In view of traditional techniques, analysis of the results is done with the MATLAB software. The proposed technique outperforms than traditional fuzzy inference systems. In comparison with other techniques at different distances, the proposed model has likelihood of disconnection.

References

1. Urmi, N.M., K. Sunera, and A.T. Shweta. 2017. Improving QoS in 4G network during handoff by using fuzzy logic based more precision handover algorithm (FMPHA). In *International Conference on Energy, Communication, Data Analytics and Soft Computing (ICECDS), IEEE*, 955–962, Chennai, India. https://doi.org/10.1109/icecds.2017.8389578.
2. Dinesh, A., and S.S. Hardeep. 2019. A split network based routing approach in wireless sensor network to enhance network stability. *International Journal of Sensors, Wireless Communications and Control* 9 (1). https://doi.org/10.2174/2210327909666190208152955.
3. Manu, M.T., K. Thomas, A. Joanna, and M.G. Elisabeta. 2017. Echo: A large display interactive visualization of ICU data for effective care handoffs. In *IEEE Workshop on Visual Analytics in Healthcare (VAHC), IEEE*, 47–54, Phoenix, AZ, USA. https://doi.org/10.1109/vahc.2017.8387500.
4. Preeti, S., and S.S. Hardeep. 2014. Localization algorithms for mobile wireless sensor networks-a review and future scope. *International Journal of Electronics & Communication Technology (IJECT)* 4 (Spl 3), 90–92. http://www.iject.org/vol4/spl3/c0115.pdf.
5. Preeti, S., and S.S. Hardeep. 2013. Enhanced event triggered localization algorithm with a parameter of energy transmission and energy received in mobile wireless sensors networks. *International Journal of Computer Applications (IJCA)* 81 (14). https://doi.org/10.5120/14181-1531.
6. Kiichi, T., and M. Naoto. 2018. Data rate and handoff rate analysis for user mobility in cellular networks. In *IEEE Wireless Communications and Networking Conference (WCNC), IEEE*, 1–6, Barcelona, Spain. https://doi.org/10.1109/wcnc.2018.8377167.
7. Charu, C., A. Dinesh, and S. Hardeep. Hand-off techniques for cellular mobile network-a review. In *International Conference on Recent Trends in Electronics, Data and Communication Computing (ICRTEDC-2014)*, Gurukul Vidyapeeth Institute of Engineering and Technology, Banur, Patiala, India. (Published in IJEEE 1(Spl.2), 189–191).
8. Tomohiro, T., and M. Sugeno. 1985. Fuzzy identification system and its applications to modelling and control. *IEEE Transaction on Systems Man and Cybernetics* SMC-15 (1), 116–132. https://doi.org/10.1109/tsmc.1985.6313399.

9. Mamdani, E.H., and S. Assilian. 1975. An experiment in linguistic synthesis with a fuzzy logic controller. *International Journal of Man-Machine Studies* 7 (1), 1–13. 10.1016/S0020-7373(75)80002-2.

10. Stegmaie, P.A., J.X. Brunner, N.N. Tschichold, T.P. Laubscher, and W. Liebert. 1994. Fuzzy logic cough detection: A first step towards clinical application. In *Proceedings of 1994 IEEE 3rd international fuzzy systems conference*, 1000–1005, Orlando, FL, USA . https://doi.org/10.1109/fuzzy.1994.343872.

11. Kosko, B. 1992. *Neural network and fuzzy systems: A dynamical systems approach*. Englewood Ciffs, NJ: Prentice Hall.

12. Zadeh, L.A., and K. Kacpyrzy. 1992. *Fuzzy logic for the management of uncertainty*, vol. 217. Willey.

13. Zadeh, L.A. 1975. The concept of a linguistic variable and its application to approximate reasoning-I. *Information Sciences* 8 (3): 199–249. https://doi.org/10.1016/0020-0255(75)90036-5.

14. Kantubukta,V., M. Sumit, M. Sudipta, and S.K. Cheruvu. 2012. QoS-aware fuzzy rule-based vertical handoff decision algorithm incorporating a new evaluation model for wireless heterogeneous network. *EURASIP Journal on Wireless Communications and Networking* 322. https://doi.org/10.1186/1687-1499-2012-322.

15. PresilaIsrat, N.C., and M.M.A. Hashem. 2008. A fuzzy logic-based adaptive handoff management protocol for next-generation wireless systems. In *11th international conference on computer and information technology,* IEEE, Khulna, Bangladesh. https://doi.org/10.1109/iccitechn.2008.4802978.

16. Yaw, N.G., and A. Johnson. 2006. Vertical handoff decision algorithms using fuzzy logic. In *International conference on wireless broadband and ultra wideband communications*, 1–5.

17. Orazio, M., R. Antonino, B. Michele, L.B. Lucia. 2007. Fast handoff for mobile wireless process control. In *IEEE conference on Emerging Technologies and Factory Automation (EFTA 2007)*, Patras, Greece. https://doi.org/10.1109/efta.2007.4416771.

Test Suite Optimization Using Lion Search Algorithm

Manish Asthana, Kapil Dev Gupta and Arvind Kumar

Abstract Testing is a continuous activity since visualization of the product. Regression testing is a type of testing which is performed to make sure that change in code has not impacted any already working functionality of the software. This is an unavoidable and expensive activity. Running all the test cases of regression test suite takes a lot of time and is expensive too. At the same time with the evolution of software, the software test suite size also increases, which increases the cost of test case execution. It is not feasible to rerun each test case. One of the most efficient ways to improve regression testing and reduce the cost is test case prioritization for regression test suite. This is a technique to prioritized regression test suite according to some specific criteria and execute the test cases according to the prioritized list, i.e. higher priority test case first and then the lower priority test cases. But the challenge is how to optimize the test cases order according to criterion. To optimize test suite, in this paper Lion optimization algorithm (LOA) has been proposed. LOA is a population-based metaheuristic algorithm. This approach utilized the historical execution data of the regression cycles, which will generate the list of prioritized test cases. At last, the optimized outcome has been compared by fault detection matrix.

1 Introduction

Software engineering not only about the writing code for software applications but it is a field that comprises many different activities. Success of a software application depends upon each and every activity involved in software development cycle (SDLC). In SDLC software testing, is one of the most important and time-taking activities. While the development of a software application, software testing is an

M. Asthana · K. D. Gupta
Amity University, Noida, India

A. Kumar (✉)
Bennett University, TechZone II, Greater Noida 201310, India

© Springer Nature Singapore Pte Ltd. 2020   77
Y.-C. Hu et al. (eds.), *Ambient Communications and Computer Systems*, Advances in Intelligent Systems and Computing 1097,
https://doi.org/10.1007/978-981-15-1518-7_7

activity which has to be executed repeatedly, with time and resource's constraint environment. In general, testing activity stops only because of time or finical limitations, which could cause some complications such as problems with the quality of software and customer contract. To reduce this complication, test team has to effectively utilize their time and resource. Effective testing directly affects the cost of the project, because the cost of the fixing a defect grows exponentially according to the phase in which defect is been found [1]. Industry has focused on how to control this cost. Defect found in the later stage of development which is not only about the cost but also matter of confidence on the testing team. It has been seen in real time that if the trivial defect has been uncovered in the later stage of SDLC, entire engineering team will lose confidence on quality of testing. Also, if defect found in a project after release to the customer, cost to fix the defect is more than the defect found in laboratory environment Fig. 1. This also impacts on the image of company and its reputation. At the same time, the involvement of different teams like service team and product support team also has to happen which is an additional cost to the project. In such case, development and test team have to rework on the same area and release a hotfix or patch which attributes to an additional cost to the project.

In software testing activities, regression testing is one of the important activities. Regression testing is a testing activity that ensures that modification or change in already working code or addition of new code in already working code, to support new functionality, has not introduced any bug in already working functionality. This is an unavoidable and expensive activity. Also, this is the most time-consuming activity. This is the reason that there are so much research works going on in this area.

With the evolution of software, the software test suite size also increases which increases the cost of test case execution also. The main purpose of software test suite optimization is to find and remove the redundant test cases that do not provide the required value to the overall testing effort and are not able to find the defects. Different regression testing methods were analysed to understand the importance of

Fig. 1 Cost Of Defect

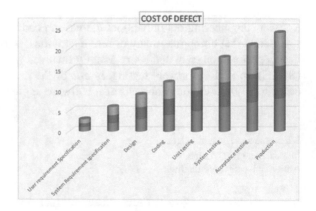

the best test suite in regression testing. These studies further categorized approaches in three different groups [2]:-

- Test suite minimization (TSM)
- Test case selection (TCS)
- Test case prioritization (TCP).

Intention of TSM is to identify repetitive effort and test case from test suite and then eliminate from test suite, with goals to reduce the number of test case from test suite [2]. Intention of TCS is also to reduce the total count of test cases in a test suite to be executed for regression testing. Approach for test case selection is around the modification done in software [2]. TCP objective is to reorder a group of test cases to achieve an early optimization based on specific criteria. It works on approach that highly significant test case should run first according to some criteria and provide the required result such as uncover fault early than the older ordered test case execution. It also helps to find the perfect combination of test case in a test suite and should be executed according to priority [2].

This is not possible to always run entire regression test suite to ensure the impact of change in software. So it is very much required to apply some technique to select specific test cases from test suite, according to some criteria like test case selection and test case prioritization techniques and order the set of test cases for improved and effective testing [3].

2 Literature Review

Nature is the best programmer, scientist and mathematician, that is why we could see such a complex yet beautifully evolved and maintained the world around us. Each species has a significance and a specific place in the ecology. There has been interference from the developments made by nature that the ecological balance would have been perfect. Looking up to nature for solutions of complex problems has received a lot of attention due to this. For solving different types of problems, inspiration has been drawn from different phenomena of nature.

It has been observed that many optimization problems in engineering are difficult to resolve. In such problem, search space grows exponentially and several applications have to face such complex problems. For such problems, traditional optimization methods do not provide an appropriate solution. In past few years, to provide an appropriate solution for such problems, metaheuristic algorithms have been designed. To solve different software engineering problems with the time, many of metaheuristics algorithms have been used. Researchers have illustrated excellent performance of metaheuristic algorithms in a wide range of complex problems such as data clustering, scheduling problems, tuning of neural networks, image and video processing and pattern recognition [4].

With time, it is been observed that human has unitized nature guidance to find the best appropriate solution for the complex problem. Therefore, there has been a significant attempt in developing algorithms inspired by nature in the last decades [4]. For example, on the basis of Darwinian's evolution concepts, genetic algorithm was proposed by Holland [5]. By inspiring the behaviour of ants foraging for food, ant colony optimization algorithm introduced [6]. Group of migrating birds trying to reach an unknown destination is the social behaviour of the birds. Based on this, particle swarm optimization simulates this behaviour [7]. Cuckoo optimization algorithm is based on the cuckoos' reproduction strategy [8].

Senthil et al. have proposed method in which test cases are optimized by using particle swarm optimization algorithm (PSO). Selected optimized test cases further prioritized using improved cuckoo search algorithm (ICSA) [9]. Reetika et al. have proposed cuckoo search algorithm for test case selection and prioritization. The objective of the algorithm is to provide the prioritized list of the test case set which can detect the fault in spending minimum time. To do the same, algorithm selects and prioritizes the relevant test cases. They have used random test data to demonstrate the proficiency of the algorithm [10]. Indumathi et al. have proposed prioritization of test cases using genetic algorithm (GA) and measured efficiency of the proposed algorithm by utilizing average percentage of faults detected (APFD) [11]. Then they have illustrated that the prioritized list of test cases is minimized. In order to reduce the cost and improve efficiency, identify and eliminate the redundant test case from test suite which will generate a reduce set of test cases for regression and help to reduce the execution effort also. Gurinder Singh et al. have proposed method to reduce the regression testing cost. By reduction of test suite, their proposed approach is based on concepts of ant colony optimization (ACO) and genetic algorithms (GA) [12]. The method chooses the group of test case from the already available test suite which will cover all the faults that identified previously in minimum execution time. To identified the minimum set of test case which can discover all the defects in application, ants work as agent here. K.K. Mishra et al. have proposed improved environmental adaption method (IEAM) to improved version of environmental adaption method (EAM) algorithm, which provides ways to select the best solution on the basis of current environmental fitness and its own fitness. Proper tuning of parameters involved in algorithm has been done to find the optimal solution in minimal time. In this study, establish the application of IEAM in generating the test suites especially for white box testing that achieves maximum branch coverage [13]. A hybrid of PSO and GA algorithm is also proposed in [14].

For all the optimization problems, there is no specific algorithm to achieve the most suitable solution. Few of the algorithms deliver improved solution for few specific problems compared with other algorithm. Thus, the researching for new optimization methods is an open problem.

3 Proposed Algorithm

Aim of this paper is to optimize the effectiveness of regression testing by selecting the list of test cases and need to perform for a successful regression testing for a given application. This can be done by providing input as test case matrices and sort them, according to some specific conditions. So that the effectiveness of the test suite can be optimized [15]. To reduce the regression testing cost by test suite minimization, we proposed Lion optimization algorithm (LOA). LOA is a population-based metaheuristic algorithm. In this research work, the lifestyle of Lions, their behaviour amongst themselves and their cooperation characteristics has been the key inspiration. These characteristics help the lions to survive as well as help them in retaining their status of the King of the jungle.

In this paper, we utilize LOA for test suite optimization. The proposed technique will select the test case from the set of test cases which has been executed earlier and detected all the fault in minimum time. In this technique, Lions will be used as an agent to explore the optimum set of test cases which can cover all the defects in minimum time. If required new test cases will be added in test suite to increases the fault discovering capacity, then adding the new test case in test suite crossover operation will be performed. The process will be repeated till the optimum set of test cases will not be discovered which can cover all the defect in minimum time.

3.1 Lion Optimization Algorithm (LOA)

Metaheuristic algorithms are of particular interest when the problem is complex with the search space growing exponentially with the problem size [2]. One such metaheuristic algorithm, the Lion optimization algorithm (LOA) [16], is utilized here to solve test suite optimization problem. The LOA algorithm is inspired by the characteristics of Lions as nomads or as a member of their group called pride. Various activities of the Lions have been used as the source for the development of these algorithms [4]. LOA may be summarized in Table 1.

Lions particularly hunt in groups with two or more females going for hunting. Rest females stay back and look for safe place suitable for safe stay of cubs, and other group members till the time the hunters return back form hunting. During hunting, the team of female lions surround the prey from different directions and making it very difficult for the prey to escape. The attack positions change in accordance with the changes in the position of the prey. The males roam about for various reasons including better place and mating propositions. Mating is an important criterion for survival and continuation of the species. A nomad male can mate with a lioness in the pride by expelling out the male in the pride. A lioness in the pride can mate with one or more lions in the same pride. In nomads, however, it is different in the sense that a nomad female can mate with only one male selected on random basis. In a pride, when males become mature, they fight between themselves and the male who is defeated or beaten has to leave the pride. Such males become nomads. On the

Table 1 Lion optimization algorithm workflow

Step 1: Pride Generation
    Step 1.1: Generate territorial male and female subject to solution constraints

Step 2: Mating
    Step 2.1: Crossover
    Step 2.2: Mutation
    Step 2.3: Gender grouping
    Step 2.4: Kill sick/ weak cubs
    Step 2.5: Update Pride

Step 3: Territorial Defence
    Step 3.1: Keep the record of cubs age
    Step 3.2: Do
        Step 3.2.1: Generate and Trespass
            Nomadic lion
            Until stronger Nomadic lion trespasses
            Until Cubs get matured

Step 4: Territorial Takeover
    Step 4.1: Selection of best lion and lionesses
    Step 4.2: Go to Step 2 until termination criteria is met

---

other hand, a strong nomad male may defeat a mature male in the pride and enter the pride with expelling the pride male. Thus, defence is an important factor for males in the pride within the pride from mature males as well as from nomads outside the pride [17].

The migratory behaviour in the lifestyle of lions helps in improving the diversity of the pride and thus improving the chances of survival and continuation of the pride. The number of females in a pride would depend on the number of total members in the group. The best females are retained, and the others have to leave the group to become nomads. Amongst them also, the best females are selected to fill the gap of migrated females in other prides. This migratory behaviour also ensures that the best traits are passed on and diversity is maintained in the pride. It also helps in information sharing in different prides.

Lion search algorithm provides the optimum solution for any engineering problems because of two unique behaviours of lions which are territorial defence and territorial takeover. For territorial defence, a war carried out between the pride lion and nomad lion, whereas the territorial takeover is carried out between the old territorial male lion and new territorial male lion in pride [18].

3.2 LOA in Problem Optimization Context

LOA in problem optimization context may be summarized as follows:

1. Here lion represents the solution which needs to determine for a problem and cubs are the solution which derived from already available solutions.
2. Process of evaluating existing and new generated solution, replacing the new solution with older if new is better is called territorial defence.
3. Process of having only the finest male and female solution which is better than the latest solution and removing the present solution from pride is territorial takeover [16].
4. In LOA, pride is dynamically changed set of solutions which change in size, initiated with few random lions and updated by derived better solution and undesired solution is being removed from set of solution.
5. Whereas mating is a process of developing a better solution from already available solutions which include crossover and mutation for generating new set of solution.

Thus, the basic structure of lion's algorithm can be clustered into four main modules based on the nature of its tasks [16], as follows:

   i. Generation of pride, which is mainly responsible for solution generation.
  ii. Mating is the process of deriving new solutions and from existing once.
 iii. Territorial defence, process of evaluating new solution (nomad lion) with already existing solution (territorial lion) and continue with better one.
 iv. Territorial takeover aims to find and change the worst solution by new better solution (cubes).

The recurring process facilitates heuristic search to meet the solution nearer to the target/desired solution.

4 Experimental Evaluation

We used the above-explained LOA, in test case prioritization providing it the historical execution results and compare it with actual results. To evaluate the performance of Lion optimization algorithm, standard APFD (average percentage faults detected) metrics are used, which give better scores for early detection of faults in the test execution.

$$APFD = 1 - \frac{TF_1 + TF_2 + TF_3 \dots TF_M}{n.m} + \frac{1}{2.n}$$

where TFi = Relative ith position of the test case which has exposed a defect in this cycle; m = Total number of faults exposed in this cycle; n = Test case count in test suite. This APFD metrics' value lies between 0 and 1, where higher value indicates that suite is more optimized and identified the faults early in the execution of the test suite.

4.1 Implementation of Proposed Algorithm

We have used the public data set made available by ABB Robotics, Norway. This data set provides historical information about test case executions, and their results in previous cycles or executions, which will be used to evaluate test case prioritization and selection methods, finding test cases most likely to fail during their next execution. Each row in the data set provides their execution duration, last execution time and results of their recent executions.

To implement the algorithm, MATLAB installed on a Windows 10, with Intel 64-x86 and with 4 GB RAM is used. We have loaded the benchmark data set from Excel file. The algorithm is executed for maximum iteration of 1000 for the entire population of test cases in the data set.

4.1.1 Experimental Data

The data set is available on bitbucket1 with features set as mentioned in Table 2. According to the important features mentioned in Table 3, we have selected test case for proposed algorithm.

Table 2 Data set features

Field's	Details
Test cycle	Iteration number in which selected test case executed
Test Id	Iteration Id for every execution
Duration	Execution time for run the test case
Name	Name of test case
Verdict	Test result (Pass: False, Fail: True)
Last results	Previous result list
Last run	Last run of test case in format of timestamp

Table 3 Filtered features

Column	Description
Test execution record	Unique auto increment value
Test case id	Id of test case
Time taken	Execution time (milliseconds)
Result	Test result (Pass: False, Fail: True)

5 Data Selection

This is always a challenging task to find the correct data set for experiment. Used data set is not only regression testing data, this has mixed execution information such as information of feature testing. We have selected five cycles randomly to apply the proposed Lion optimization algorithm.

5.1 Training Procedure

As with any other metaheuristic algorithm, we need to train the algorithm to predict accurately in the future. For training the algorithm, test results for the previous cycle are ingested into the algorithm and validated for the current cycle. For illustration and verification, we selected few cycles as shown in Table 4, using APFD metrics to measure the efficiency of the algorithm in detecting the faults. Here, we take one test cycle as one lion of a pride, and we will calculate the fitness of each cycle and then perform the cross over to find the best set of prioritized test case for regression testing In matrix of Table 5, an overview of sample execution cycle for test cases is shown. The table captures six days of test suite execution results which span from 1 Mar 2019 to 16 Mar 2019.

Table 6 shows the parameters used for the calculation of fitness of a lion for this sample cycle (test execution).

Here, 1 represents the test case failure and 0 when test case did not failed and no fault identified. In the first iteration, all the lions returned home with no crossover

Table 4 Training and validation data selection

S. No	Cycle	Number of executions	Unique test cases
1	52	88	86
2	100	66	65
3	190	177	125
4	205	61	61
5	305	497	251

Table 5 Six days matrix of test case execution

Test case	03/01	03/02	03/03	03/04	03/05	03/06
T1	1	0	1	0	0	1
T2	0	1	0	0	0	0
T3	0	1	0	0	1	0
T4	0	0	0	1	0	1

Table 6 Test case execution time

Test cases	Fault	Duration
T1	3	4
T2	1	6
T3	2	2
T4	2	6

as every lion has only a single test case. Exit criteria for the loop are when any lion has discovered all the faults, i.e. we have ideal condition, where single test case is able to uncovered all the possible faults in the system. Create new cubs by adding test cases from female lion and male lion. If the fitness value for new cubs increases, i.e. it exposed more faults, the new test case is added to the lion, or else the cycle continues till the remaining list is exhausted. New cubs (offspring) are added to the pride only when following criteria of crossover is met:

1. No new offspring should have the same traits as one of the lion in the pride, i.e. same number of 1's.
2. The new lion added in pride should take less execution time than the maximum execution time of any other lion present in pride during execution.

To start the algorithm, Table 7 shows the initial set of lions in the pride going for hunt.

Step 1: Next test case has to be picked up for each test suite as shown in Table 8.

Step 2: Crossover between lions is to create new offspring as shown in Table 9. As lion L3 execution time does not satisfy the crossover criteria, it will not be go to the list. The female lion will again produce offspring for adding a new sets of test case (Fig. 2).

Step 3: Evaluate the fitness of all the lions in the pride using the parameters shown in Table 6. For example, the first iteration will have lions selected as shown in Table 10.

Table 7 Initial assignment

Lion	Test cases	Test case fault detecting capacity	Fitness value	Time
L1	T2	010000	1	6
L2	T3	010010	2	2

Table 8 Test case assignment

Lion	Test case	Test case fault detecting capacity	Fitness value	Time
L1	T2, T1	111001	4	10
L2	T3, T4	010111	4	8

Fig. 2 Crossover

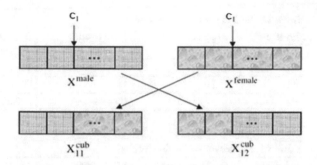

Table 9 Offspring after crossover

Lion	Test case	Fault detection capacity of test case	Value of fitness	Duration
L1	T2, T1	111001	4	10
L2	T3, T4	010111	4	8
L3	T2, T4	010101	3	12
L4	T3, T1	111011	5	6

Table 10 Lion selection

Lion	Test cases	Fault detection capacity of test case	Value of fitness	Duration
L1	T2, T1, T4	111101	5	16
L2	T3, T4, T2	010111	4	14
L3	T3, T1, T4	111111	6	16

The lion L3 has maximum fitness value, and execution time is the same as L1; hence, L3 will be selected for the first prioritized set of test case. Thus the order of prioritized TC will be T3, T1, T4 and T2 in order.

5.2 Experimental Results

Test execution results for each cycle present in Table 4 are divide into groups from P1 to P3. P1 group has utmost priority test cases, and P3 group has the least priority test cases. Each cycle results are evaluated based on their APFD metrics value. As seen in the results, more failures are detected in the group P1, and the most prioritized test cases. Table 10 contains the results from validation for the first cycle (Table 11).

Table 11 First cycle validation results

Priority	Pass count	Fail count	Aggregate discovery (%)	Un-executed
P1	10	12	48.00	286
P2	9	8	80.00	262
P3	43	5	100.00	253

Table 12 110th cycle validation results

Priority	Pass count	Fail count	Aggregate discovery (%)	Un-executed
P1	18	2	50	288
P2	18	1	75	260
P3	26	1	25	274

Table 13 190th cycle validation results

Priority	Pass count	Fail count	Aggregate discovery (%)	Un-executed
P1	44	26	46.42	238
P2	37	19	80.35	223
P3	40	11	100	250

Table 14 205th cycle validation results

Priority	Pass count	Fail count	Aggregate discovery (%)	Un-executed
P1	19	2	67	287
P2	19	1	100	259
P3	20	0	100	281

Above table shows that 80% of the faults are detected by the test cases from groups P1 and P2, cumulatively. So, APFD value for this test suite, i.e. prioritized list of test cases, improved from 0.45 to 0.55. For 52th cycle, total 888 test cases used for training the algorithm, of which 87 were actually executed, the sum of passed and failed test cases in this cycle. Similarly, Tables 12, 13, 14 and 15 show the validation results for rest of the cycles in Table 4.

Graph in Fig. 3 exposes APFD values comparison for all the five cycles selected for validation of the trained prioritization model. Average APFD for all cycles was 0.70 with the default order of test case execution and 0.79 for rest of the cycles where faults were detected later in the execution (Table 16).

Table 15 305th cycle validation results

Priority	Pass count	Fail count	Aggregate discovery %	Un-executed
P1	59	116	36.25	133
P2	59	108	70.00	112
P3	59	96	100.00	146

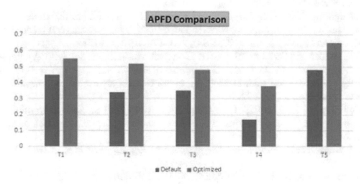

Fig. 3 APFD improvement

Table 16 Final results

	T1	T2	T3	T3	T5
Default	0.45	0.34	0.35	0.17	0.48
Optimized	0.55	0.52	0.48	0.38	0.65

6 Conclusion

Test suite optimization is an important activity of SDLC cycle. It saves not only time but money also for delivering quality products. In this paper, we have used Lion optimization algorithm for the selection and prioritization of the test cases from a given test suite using the historical failure patterns from regression testing. The lifestyle of lions, their behaviour amongst themselves and their cooperation characteristics have been the key inspiration. This approach has been successfully used for numerous problems. One of the scenarios has been described in this research work. After going through the execution results, one can clearly conclude that there is a substantial improvement for detecting the fault by reordering the execution of test cases in the regression suite based on the past cycles before start of the regression testing iteration. This technique takes into consideration the current attributes (previous execution cycles results) and does not have any crucial dependency on other parameter like code coverage analysis, feature risks and requirements and thus allows the automation of test case prioritization reducing the costs and efforts to spend time in preparation and collection of the data for the training. If this method is combined with the legacy cognitive approach of test case prioritization, then it will produce remarkable results.

References

1. Menzies, Tim, William Nichols, Forrest Shull, and Lucas Layman. 2017. Are delayed issues harder to resolve? revisiting cost-to-fix of defects throughout the lifecycle. *Empirical Software Engineering* 22 (4): 1903–1935.
2. Khatibsyarbini, Muhammad, Mohd Adham Isa, Dayang N.A. Jawawi, and Rooster Tumeng. 2018. Test case prioritization approaches in regression testing: A systematic literature review. *Information and Software Technology* 93: 74–93.
3. Gao, Dongdong, Xiangying Guo, and Lei Zhao. 2015. Test case prioritization for regression testing based on ant colony optimization. In 2015 6th IEEE international conference on software engineering and service science (ICSESS), 275–279. IEEE (2015).
4. Yazdani, Maziar, and Fariborz Jolai. 2016. Lion optimization algorithm (loa): A nature-inspired metaheuristic algorithm. *Journal of Computational Design and Engineering* 3 (1): 24–36.
5. Goldberg, David E., and John H. Holland. 1988. Genetic algorithms and machine learning. *Machine Learning* 3 (2): 95–99.
6. Parpinelli, Rafael S., Heitor S. Lopes, and Alex Alves Freitas. 2002. Data mining with an ant colony optimization algorithm. *IEEE Transactions on Evolutionary Computation* 6 (4): 321–332.
7. Kennedy, James. 2010. Particle swarm optimization. *Encyclopedia of Machine Learning*, 760–766.
8. Rajabioun, Ramin. 2011. Cuckoo optimization algorithm. *Applied Soft Computing* 11 (8): 5508–5518.
9. Kumar, K. Senthil, and A. Muthukumaravel. 2017. Optimal test suite selection using improved cuckoo search algorithm based on extensive testing constraints. *International Journal of Applied Engineering Research* 12: 1920–1928.
10. Nagar, Reetika, Arvind Kumar, Gaurav Pratap Singh, and Sachin Kumar. 2015. Test case selection and prioritization using cuckoos search algorithm. In *2015 international conference on futuristic trends on computational analysis and knowledge management (ABLAZE)*, 283–288. IEEE.
11. Indumathi, C.P. and S. Madhumathi. 2017. Cost aware test suite reduction algorithm for regression testing. In *2017 international conference on trends in electronics and informatics (ICEI)*, 869–874. IEEE.
12. Singh, Gurinder, and Dinesh Gupta. 2013. An integrated approach to test suite selection using aco and genetic algorithm. *International Journal of Advanced Research in Computer Science and Software Engineering* 3 (6).
13. Mishra, K.K., Shailesh Tiwari, and Arun Kumar Misra. 2012. Improved environmental adaption method for solving optimization problems. In *International symposium on intelligence computation and applications*, 300–313. Springer.
14. Nagar, Reetika, Arvind Kumar, Sachin Kumar, and Anurag Singh Baghel. 2014. Implementing test case selection and reduction techniques using meta-heuristics. In *2014 5th international conference-confluence the next generation information technology summit (Confluence)*, 837–842. IEEE.
15. Ansari, Ahlam, Anam Khan, Alisha Khan, and Konain Mukadam. 2016. Optimized regression test using test case prioritization. *Procedia Computer Science* 79: 152–160.
16. Rajakumar, B.R. 2012. The lion's algorithm: A new nature-inspired search algorithm. *Procedia Technology* 6: 126–135.
17. Kaveh, A., and S. Mahjoubi. 2018. Lion pride optimization algorithm: A meta-heuristic method for global optimization problems. *Scientia Iranica* 25: 3113–3132.
18. Wang, Bo, XiaoPing Jin, and Bo Cheng. 2012. Lion pride optimizer: An optimization algorithm inspired by lion pride behavior. *Science China Information Sciences* 55 (10): 2369–2389.

Advanced Twitter Sentiment Analysis Using Supervised Techniques and Minimalistic Features

Sai Srihitha Yadlapalli, R. Rakesh Reddy and T. Sasikala

Abstract Social media along with different web forums contribute to a huge amount of data, i.e. opinions, feedbacks, reviews generated on a daily basis. Sentiment analysis is the identification of polarity (positive, negative or neutral) of the data to analyse the opinion of users on various platforms on a range of topics including those that have real business value, i.e. polarity of users on a product. With the advent of platforms like Twitter, people freely express their opinion on almost everything and that generates a huge amount of data which cannot be processed or analysed manually; therefore, we have various NLP and machine learning techniques which can effectively analyse and predict the polarity of the data which enables us to capture the sentiment of the people regarding a particular issue. In this paper, we intend to analyse the different machine learning algorithms for sentiment analysis of Twitter data and compare the algorithms' performance on different datasets with the help of metrics like precision, accuracy, *F*-measure and recall.

Keywords Social media · Twitter · Machine learning · NLP · Pre-processing · Feature extraction · Naïve Bayes · SVM · Maximum entropy · Dictionary-based

1 Introduction

Since the dawn of the social media, people use it as a medium to express their opinion, and this generates a lot of data. One such social media platform is Twitter which is a microblogging social media platform in which where people express their opinions on a wide range of topics like, products they have purchased, movies they have watched, election sentiments and several other ongoing issues in the form of 'tweets' which are

S. S. Yadlapalli (✉) · R. Rakesh Reddy · T. Sasikala
Department of Computer Science & Engineering, Amrita School of Engineering, Amrita Vishwa Vidyapeetham, Bengaluru, India
e-mail: saisrihithay@gmail.com

R. Rakesh Reddy
e-mail: rakeshreddy3743@gmail.com

T. Sasikala
e-mail: t_sasikala@blr.amrita.edu

© Springer Nature Singapore Pte Ltd. 2020
Y.-C. Hu et al. (eds.), *Ambient Communications and Computer Systems*, Advances in Intelligent Systems and Computing 1097,
https://doi.org/10.1007/978-981-15-1518-7_8

restricted to a character count limit of 280. Due to this restriction, people generally tend to use acronyms, emoticons (emojis), hashtags, slang references and so on to express their views instead of using fully well-formed grammatical sentences. This leads to several challenges to find out the true sentiment of these tweets. People rely on this data to make decisions like looking at reviews to buy a product and so on. We can use sentiment analysis to analyse the huge amount of data and give us an overall outlook of user perception since it is extremely hard to go through millions of data points manually to make a decision.

In this paper, we are more focused on the data available on the most popular microblogging platform Twitter. The analysis of the tweets available via Twitter is important for various business organizations to analyse the opinion of users on their products in order to determine the interest of the users at various stages of the product. Before the launch of the product, the industries determine if there is enough buzz on the product in order to have enough inventory based on the expectation of various users, and after the launch of the product, they determine the polarity of the data to know the quantity of product they need to produce and to determine the market reception of the product. So, sentiment analysis is extremely necessary and very widely used by many organizations in some form or the other because it provides the organizations with necessary information about user perception and simplifies the process of making business decisions.

Sentiment analysis can further be classified into document level, sentence level and aspect-based. The document-level sentiment analysis refers to dealing with determining the polarity of the document as a whole, the sentence level determines the sentiment of every sentence in the document, whereas the aspect-based analysis deals with analysing the sentiment of particular features of a given product, i.e. the sentiment of people on a phones battery life, camera, processor and so on. In this paper, we perform document-level sentiment analysis where each tweet is considered as a document. We initially perform dictionary-based sentiment analysis using a corpus of data and a dictionary which tells the polarity of words present in the tweets and compare it to using machine learning techniques to do the same. We compare the performance of each of these algorithms as we cannot blindly say that one algorithm is the clear winner for all types of datasets as each algorithm has its own advantages. Using machine learning techniques also allows us to take into consideration the fast rate of change in the trends of language usage in a social media platform like Twitter as we can train the model using the training dataset consisting of the latest trends to generate accurate results. We will be comparing the performance of several supervised learning algorithms like Naïve Bayes, SVM, gradient boosting and the maximum entropy classifier with the help of several evaluation methods such as accuracy, precision, recall and F-measure.

2 Prior Work

There are a variety of applications of sentiment analysis of Twitter data with the data explosion in the 21st century. Mining opinions is popular to gauge the overall sentiment of the people based on a few metrics about a particular product. The objective of the work was to identify which classifier worked best among a set of classifiers like random kitchen sink algorithm, Naïve Bayes and logistic regression [1]. Pre-processing is an important ritual in opinion mining as you need to eliminate all the unnecessary details from the tweets. This paper talks about the refining data that we extract and analysing the algorithms like SVM, decision trees and Naïve Bayes classifier [2]. Extracting the right features plays an important role in the improving the key metrics of various algorithms after classifying. This work is based on feature extraction, feature selection and feature cleansing based on statistical, hybrid and NLP-based methods [3]. Lexicon-based methods are discussed for opinion mining considering three different lexicons combined with simple strategies. Metrics are discussed based on considering features like negation and intensification [4]. In this, we find out how the patient feedback can be automated for health services and the problems that arose with large and noisy corpora, the language's idiomatic nature and the linguistic problems that arose in that domain [5]. The advantages and disadvantages of sentiment analysis are broadly discussed along with algorithms which perform well in predicting the sentiment based on metrics like F-measure, precision and recall. The findings are then put into how the entire process can be generalized [6]. This work is primarily based on analysing the sentiment in Roman Urdu tweets by first converting them to English. Sentiment analysers like sentiword net, W-WSD and text blob are used to classify the tweets initially. Naïve Bayes and SVM are later used for classification and are compared using metrics like precision and recall [7]. This work is basic review of all the trivial but important methods in opinion analysis. It reviews everything from feature selection to the different methods of classifying sentiment like machine learning methods. It also reviews the different parameters of evaluation [8]. This paper discusses about the use of the Chi square feature clustering method and discusses several feature selection methods which help in reducing the redundancy and increase the overall accuracy irrespective of the data size [9].

This paper incorporates the above understandings obtained for pre-processing, extraction of robust features using POS tagging and NLP methods, using optimal machine learning models that in turn reduce redundancy and increase the overall accuracy irrespective of the size of data and use the fit evaluation metrics to evaluate the model performance.

3 Proposed Work

3.1 Data Collection

In order to mine tweets to suit your requirements, we require access tokens and keys which are parameters required by the Twitter API, so it is necessary to register the application with the developers account on Twitter. Once you register the application and fill in the necessary fields like the loopback address, you get a set of keys which are unique for every application. Using these keys, you can access tweets via the Twitter API. The fields are consumerKey, consumerSecret, accessToken and accessTokenSecret. Once you are able to connect with the Twitter API, we can mine the tweets based on the keywords and apply filters like the language of the tweets, number of tweets and so on. The overall workflow of the model can be seen in Fig. 1.

Fig. 1 Workflow of the sentiment analysis model

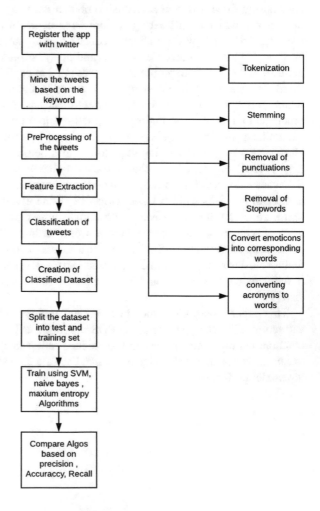

3.2 Data Pre-processing

The tweets when extracted in the raw form contain a lot of unnecessary things like URLs, retweet symbols, names and random tags which interfere with sentiment analysis, so, in order to have the data ready for classification, we follow a few steps like.

Tokenization. Tokenization is the phenomenon where words from a particular text are broken up into small individual units called tokens. These tokens can either be digits, punctuations or words. Tokenization is done by finding boundaries of words. Word boundaries are a word ends and where it begins. Word segmentation is another name for tokenization. Tokenization is quite challenging. Language is a major factor that adds to these challenges. Space-delimited languages like English and French use white spaces to separate most of the words. In languages like Chinese and Thai, the words do not have clear boundaries and are therefore called unsegmented. To tokenize unsegmented languages, we require additional lexical and morphological information. Topographical structure and the writing system are both affected by tokenization.

For example,

Input: I'm not feeling that great
Output: I'm, not, feeling, that, great.

Stemming. Stemming is a process where we produce morphological variants of a base word. A stemmer reduces the words like "clingy", "clinging" to its base form "cling", and "berry" and "berries" reduce to the word "berri". The two errors that occur in stemming are over stemming and under stemming. Over stemming occurs when words belonging to different stems are stemmed to a common root. Under stemming occurs when two words that belong to the same root are stemmed to a different root. Another example of stemming is where we convert exaggerated words having repeating letters like "happyyyyyy" to "happy", "amazingggg" to "amazing", "cooool" to "cool".

Removal of Punctuations. Punctuations when taken individually do not add much relevance to our process, and therefore, regular expressions can be used to remove all alphanumeric characters which remove all the punctuations in the process. The following expression [!"#$%&'()* + ,-./:; <=>?@[\]^_'{|} ~] can be used to remove all the punctuations. For example, wow!!!! is replaced by wow.

Removal of Stop Words. Removing data that is useless and no longer needed, i.e. stop words are an important step in pre-processing. Stop words are very frequently used words (like "a", "an", "the", "am") that search engines usually ignore as they neither add to the sentiment nor express an opinion. These words might take up additional processing time or additional space in our databases and that is totally unnecessary. The stopwords() function removes most of the stop words and is similar to that in the NLTK toolkit.

Emoticons to Words. The emoticons used in the tweets play a major role in determining the sentiment or polarity of the document. We mapped the emoticons to

Table 1 Mapping of emoticons to words

Emoticon	Synonym	Sentiment score
☹	Sadness	−1
☺	Happy	+1
☻	Disappointed	−1
:O	Surprise	+1
<)	Grumpy	−1
:'(	Crying	−1

Table 2 Translation of acronyms to words

Acronyms	Words
LOL	Laughing out loud
NM	Nothing much
ROFL	Rolling on the floor laughing
OMG	Oh my God

words and simultaneously stored it in a vector for every tweet, and each emoticon was assigned a sentiment score which we eventually used to calculate the polarity of the tweet as shown in Table 1.

Acronyms to Words. Acronyms can easily influence the sentiment score just like emoticons can. Every acronym can be translated into the corresponding words like LOL, ROFL, NM, etc., can be expanded to add context to the sentence as shown in Table 2.

3.3   Feature Extraction

Feature Extraction is used to specify the different type of features used in sentiment analysis. These include

- The feature vector is composed of a variety of features like count of positive words, negative words (+1 positive, −1 negative), term frequency, unigrams and bigrams [10].
- The positive, negative words and emoticons are given a weight of +1 and −1, respectively.
- Negation words are represented separately as features.
- Class labels are assigned using dictionary-based approach.
- Emoticons that are converted to words are also considered as a feature.

Table 3 Dataset description

Dataset	Positive	Negative	Total
Training	35,000	25,000	60,000
Testing	12,000	8000	20,000

3.4 Methodology Used

Once the class labels were assigned, we split the dataset into training and testing data as shown in Table 3.

Algorithms like SVM, Naive Bayes, maximum entropy were trained on the data, and the testing set was then used to classify the tweets and compare the models using different metrics like precision, accuracy and recall.

3.5 Dictionary-Based Approach

The sentiment of a tweet can be scored based on the presence of opinion words that could sway the emotion of the tweet to either positive or negative. We compile a list of words that are positive and another list of words that are negative. We now use the pre-processed corpus and match every tweet with the positive opinion words and negative opinion words. We maintain the count of the number of matches of each sentence to the positive and negative words. The sentiment of every tweet can be scored on a scale of five $(-2, -1, 0, 1, 2)$ where:

-2: Extremely Negative
-1: Negative
0: Neutral
1: Positive
2: Extremely Positive.

Sentiment Score = weight of positives words + emoticons – weight of negative words + emoticons.

As observed in Figs. 2 and 3 each tweet, here we have collected tweets based on the keyword "OnePlus6T" which is a popular mobile phone that has been launched in India towards the end of 2018 and hence has flooding reviews. The sentiment has been predicted as positive, neutral and negative or a score of 1, 0, -1, respectively, for each tweet.

As it can be observed in Fig. 4, most of the tweets regarding the launched phone are positive followed by neutral, whereas there are very few negative tweets about it. This can be visualized in the form of a bar graph.

Supervised Machine Learning Techniques for Tweet Classification. Supervised machine learning algorithms are being used for classification of data by using training and test datasets. The input feature vectors and their respective class labels are present in the training dataset. This training set is then used to develop a model that maps

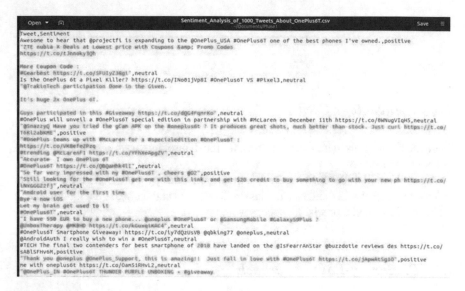

Fig. 2 Output of tweet and sentiment prediction

the new input feature vectors to their corresponding output labels known as classes using a machine learning algorithm. Furthermore, the test set is used to validate the model by prediction of the classification labels of new/unseen data points. Several machine learning algorithms are used to classify the sentiment of the tweet such as gradient boosting algorithm, support vector machines (SVM) and maximum entropy and Naïve Bayes.

Naïve Bayes Classifier. Naïve Bayes classifiers are a collection of classification algorithms based on the Bayes's theorem. It is a collection of algorithms, all of which share the same principle, which is that each and every feature that forms the feature vector for classification is independent of one another. It uses all of the features that are present in the previously obtained feature vector and analyses each of them independently without considering the other features. The following is a representation of Naïve Bayes' conditional probability:

$$(X|yj) = \pi m \, P(xj|yj) \tag{1}$$

In the above Eq. (1), 'X' is the feature vector that is obtained, and it comprises several features such as the total number of positive and negative words, emoticons, and total number of positive and negative hashtags that are utilized efficiently by the Naïve Bayes classifier to perform the classification of tweets into positive sentiment and negative sentiment tweets.

Gradient Boosting Algorithm. Gradient boosting classifier is a technique which is used to perform regression as well as classification problems and is used to generate a predictive model which is a collective set of weak predictive models, usually decision trees. This model is built using stepwise method which is similar to the other boosting

	score	text
1	0	oneplususa four unique oneplus6t features makes d...
2	0	phoneradarblog phoneradar news 27th november 2...
3	0	oneplususa flagship fatigue oneplus6t result consta...
4	0	phoneradarblog oneplus6t mclaren edition expecte...
5	0	simoilblasco1 ♦ widget piscis kwgt haleypredator ♦i...
6	1	amitbhawani new video oneplus6t thunderpurple un...
7	0	oneplus partners mclarenf1 got oneplus6t mclaren ...
8	1	new oneplus 6t month can easily say best androidp...
9	0	phoneradarblog phoneradar news 27th november 2...
10	1	amitbhawani new video oneplus6t thunderpurple un...
11	0	oneplus committed bringing best technology world ...
12	0	simoilblasco1 ♦ widget piscis kwgt haleypredator ♦i...
13	1	amitbhawani new video oneplus6t thunderpurple un...
14	0	checked c4etech oneplus6t thunder purple giveawa...
15	0	oneplus oneplus6t best android phone buy find verge
16	0	oneplus oneplussupport pls add option schedule tex...
17	0	simoilblasco1 ♦ widget piscis kwgt haleypredator ♦i...
18	0	phoneradarblog oneplus6t mclaren edition expecte...
19	0	checked c4etech oneplus6t thunder purple giveawa...
20	0	trakintech oneplus launching oneplus 6t mclaren ed...
21	0	phoneradarblog phoneradar news 27th november 2...
22	0	phoneradarblog phoneradar news 27th november 2...

Fig. 3 Output of score of a given tweet

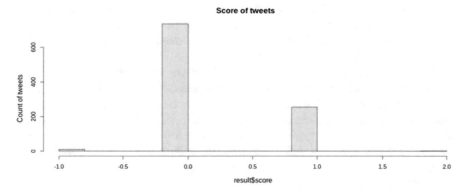

Fig. 4 Representation using ggplot

methods, by enabling the optimization of a random loss function that is differentiable such that it can then be generalized. The making of strong and robust classifiers from a set of weak learners is one of the main characteristics of boosting.

SVM Classifier. The main idea behind SVM is to effectively plot each and every data item present in the dataset as a point with each feature's value as a coordinate in the n-dimensional space where 'n' is the total count of features that are present. It then makes use of a hyperplane in order to differentiate the different classes with the least error.

SVM works using the concept of a discriminative function as shown in Eq. (2) where 'X' represents the feature vector, 'w' being the vector of weights and 'b' as the bias. 'b' and 'w' learn on their own from the train set. For the purpose of classification, a linear kernel is utilized. It tries to maintain the largest possible gap between the two classes

$$(X) = wT\emptyset(X) + b \qquad (2)$$

Maximum Entropy Classifier. For tweet classification using the maximum entropy classifier, word count is used as an important feature. If a word occurs more often in a particular class, the weight for that particular word-class pair would be higher when compared to the word paired with other classes. Effective entropy of the model is maximized by the classifier by predicting conditional class distribution labels [11]. The relationships between part of speech tag, negation and expressing keywords are used efficiently for the purpose of classification in the feature vector.

4 Results and Discussion

We were able to predict the sentiment of tweets using different methods such as dictionary-based method and also using machine learning algorithms such as Naïve Bayes', SVM, gradient boost, maximum entropy classification method and able to compare the working of each of these models using performance metrics like accuracy, precision, recall and F-measure.

Pie chart represents classification of 100 tweets on OnePlus6T phone using dictionary-based method into positive, negative and neutral as shown in Fig. 5.

The performance of all the classifiers was compared using accuracy, precision, recall and F-measure. Maximum entropy outperformed the other classifiers with regards to accuracy and precision with 92.1% and 98.7%, respectively, whereas Naïve Bayes had a better recall and F-measure of 87% and 92%, respectively. All the algorithms have a similar high precision except the gradient boost which underperformed. This outlines the quality of the feature vector irrespective of the algorithm and change in training data showing less variance. As we increased the size of our training dataset, we observed an increase in the overall accuracy.

The performance of all of these classifiers has been represented effectively in the form of a bar chart in Fig. 6.

Fig. 5 Classification of
tweets using pie chart

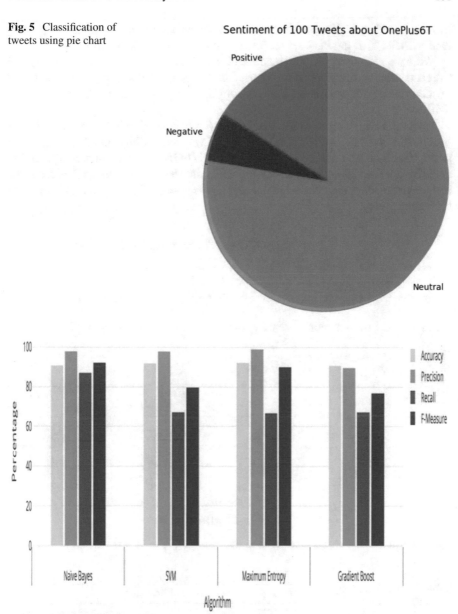

Fig. 6 Bar plot representing algorithm performance

We have used the maximum entropy classifier which was found to be the classifier
with the best accuracy and precision to classify the tweets on the top four mobile
phones of 2019 and have got an accuracy of above 90% for all four mobile phones.

In fact, OnePlus 6T has been titled as the best flagship phone of 2019 till present date in India. This has been represented with the help of a bar chart in Fig. 7.

We have also used the maximum entropy classifier once again to classify the tweets consisting of the top six most used hashtags of 2019 to analyse the sentiment of these tweets. The hashtags that we considered and their meanings are given in Table 4.

As it can be seen clearly from the bar graph representation in Fig. 8 TBT, MOTI-VATIONMONDAY, OOTD, TRAVELTUESDAY have a high number of positive tweet classification results, whereas TBH and VEGAN have comparatively higher negative and neutral tweet classification which is what is expected as they are more serious issues. The accuracy of all the hashtags is above 87% with OOTD having the highest accuracy.

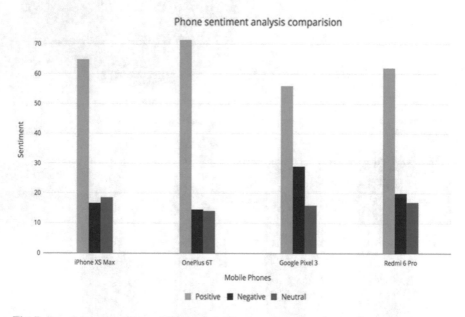

Fig. 7 Bar plot representing mobile phone sentiment analysis

Table 4 Popular hashtags and their meaning

Hashtag	Meaning
VEGAN	Abstaining from food containing animal products
TBT	Throwback thursday
MOTIVATIONMONDAY	Motivating
TBH	To be honest
OOTD	Outfit of the day
TRAVELTUESDAY	About travelling

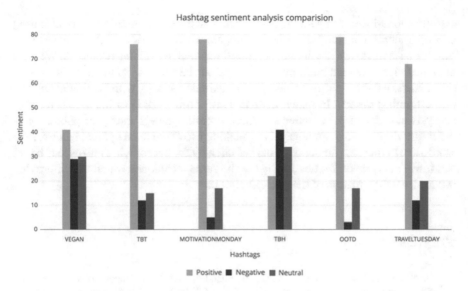

Fig. 8 Bar plot representing hashtag sentiment analysis

5 Challenges

- **Social media trends**: The language (slang) and emoticons used keep changing and updating at a very fast rate in social media such as Twitter as a result of which we must keep ourselves up to date of these new trends and train our models accordingly.
- **Context**: It is hard to accurately analyse the context or polarity of emotion in a particular context due to the presence of sarcasm in a few cases which misleads the system into thinking the opposite of what the opinion was originally. This leads to a false positive case. For example, "Wow, I can't believe I actually bought this product!".
- **Credibility**: The credibility of a particular tweet is hard to determine as it could be a spam of advertisements for promoting something or from someone who wants to spread false news which leads the system to predict the wrong sentiment.

6 Conclusion and Future Work

There are several ways to determine the sentiment of a particular text using NLP techniques. However, in this paper, we mainly focus on the comparison of various sentiment analysers using dictionary-based method and different machine learning algorithms to determine the algorithm which predicts with the highest accuracy rate for correctly classifying tweets of any domain. We have used several datasets of different domains to compare the performance of each technique. We have also used several evaluation methods in order to compare which algorithm is the best performing. We have made use of minimal yet robust features for training the model. We have also made use of human intuition which suggests the accuracy of the classification of our models. All the algorithms have performed satisfactorily with maximum entropy outperforming the rest in many aspects. This entire work is to improve sentiment analysis in order to get a better and more accurate understanding of public sentiment in Twitter through our analysis by making use of the most optimal features. We would like to increase the scope of our research by further foraying into aspect-based sentiment analysis of Twitter data using the same techniques which would provide aspect-level information of the public opinion.

References

1. Dr. Samam K.P., S.N. Vinithra, and Dr. M. Anand Kumar. 2015. Analysis of sentiment classification for hindi movie reviews: a comparison of different classifiers. *International Journal of Applied Engineering Research, 10* (17) (Research India Publications).
2. Dhanya, N.M., and U.C. Harish. 2018. *Sentiment analysis of twitter data on demonitization techniques, computer vision and biomechanics*. Netherlands: Springer.
3. Asghar, Muhammad Zubair, Aurangzeb Khan, Shakeel Ahmad, and Fazal Masud Kundi1. (2014). A review of feature extraction in sentiment analysis. *Journal of Basic and Applied Scientific Research*.
4. Avanco, Lucas Vinicius, and Maria das Gras Volpe Nunes. 2014. Lexicon based sentiment analysis for review of products in Brazilian-Portugese. In *Brazilian Conference on Intelligent Systems*, Oct 2014.
5. Keith Stuart, Ana Botella, and Imma Ferr. 2016. A corpus driven approach to analysis of patient naratives. In *8th International Conference on Corpus Linguistics*, vol. 1, 381–395, CILC 2016.
6. Hsu, Raymond, Bozhi See, and Alan Wu. 2010. Machine learning for sentiment analysis on the experience project. http://cs229.stanford.edu/proj2010/HsuSeeWuMachineLearningForSentimentAnalysis.pdf.
7. Ali Hasan, Sana Moin, Ahmad Karim and Shahaboddin Shamshirband, 2018. Machine learning based sentiment analysis for twitter accounts. *Mathematical and Computational Applications*.
8. Singh, Rashija, and Vikas Goel. 2019. Various machine learning algorithms for twitter sentiment analysis. *Information and Communication Technology for Competitive Strategies*, 763–772.
9. Wang, Yili, KyungTae Kim, ByungJun Lee, and Hee Yong Youn. 2018. Word clustering based on POS feature for efficient twitter sentiment analysis. *Human-Centric Computing and Information Sciences*, 8 (1): 17.
10. Ganesh, B.R., Deepa Gupta, and T. Saikala. 2017. Grammar error detection tool for medical transcription using stop words—POS tags ngram based model. In *International Conference on Computational Intelligence and Informatics (ICCI), Hyderabad*, Sept 2017.
11. https://web.stanford.edu/class/cs124/lec/Maximum_Entropy_Classifiers.pdf.

Multi-objective Faculty Course Timeslot Assignment Problem with Result- and Feedback-Based Preferences

Sunil B. Bhoi and Jayesh M. Dhodiya

Abstract Most of the institution faces the timetabling problem in academic, sports, health,transportation, etc. In particular, academic institution faces the three types of problems as assignment of courses, examination timetabling and school timetabling. To evaluate valid and efficient course timetable is not easy task. In this paper, mathematical model of multi-objective faculty course timeslot assignment problem (MOFCTSAP) is formulated in two phases and its solution is achieved based on average of preferences given by the faculty for courses per hour taught, average of preferences of all faculties, administrator preferences as well as feedback and result based preferences. Fuzzy programming technique is applied to solve multi-objective faculty course timeslot assignment problem (MOFCTSAP) and found results using LINGO 18.0 software.

Keywords Faculty course time slot timetabling · Mixed-integer programming · Fuzzy programming technique

1 Introduction

Every institution such as academic, sports, transportation, and health come across with the timetabling problem over all the world. Timetabling is the allocation of events like course, exam, duties and resources like teacher, worker, doctor over space as classroom, operation room, satisfying some kinds of several constraints. Constraints define solution space and there are two types of constraints, hard constraints which must be satisfied by solution and soft constraints which are to be satisfied as far as possible, i.e., these constraints can be violated.Feasibility of the problem is determined by hard constraints and quality of timetable is determined by soft constraints [1–3]. Many peoples are working on course timetabling where faculty preferences, administrator preferences are taken into consideration. In 1987, Schnierjans and Kim [7] have presented course faculty model and solved it by goal

S. B. Bhoi (✉) · J. M. Dhodiya
S.V. National Institute of Technology, Surat, India

© Springer Nature Singapore Pte Ltd. 2020
Y.-C. Hu et al. (eds.), *Ambient Communications and Computer Systems*, Advances in Intelligent Systems and Computing 1097,
https://doi.org/10.1007/978-981-15-1518-7_9

programming with preferences. This model was extended by Badri [8] in which he has assigned courses to faculty and course faculty taken together to time slot based on preferences for courses as well as time slot in two phases where in first phase courses are assigned to faculties and in second phase course faculty taken together to time slot. Badri et al. [9] have combined two phases in single phase where courses are assigned to faculty and course faculty taken together to time slot with preferences. Werra [3] introduced basic class-teacher model and mentioned how they could be inserted in a general program which produces usable timetables using edge coloring of graph. Astratian and de Werra [4] considered a generalized class-teacher model that extends the basic class-teacher model for timetabling which correspond to some situations frequently occur in the basic training program of universities and schools. Ozdemir and Gazimov [5] proposed three-step method to solve a non-convex multi-objective faculty course assignment problem which take into considerations participants average preferences and presented an effective way to solve the model. In [6], Daskalaki and Birbas presented a two-stage relaxation procedure that solves efficiently the integer programming formulation of a university timetabling problem. The class faculty assignment problem investigated by Yakoob and Sherali [10, 11] resembles many class scheduling problem. In 2013, [11],Yakoob and sherali formulated an integer programming (mixed-integer programming)faculty assignment model and column generation method is developed to solve the linear relaxation model. Bakir and Aksop [12] formulated a 0–1 integer programming model for the problem of course scheduling for the department of Statistics at Gazi University,Turkey, which assigns courses to periods and classroom. In 2017, Ongy [13] also presented faculty course time slot problem in single phase, where courses are assigned as well as faculty course taken together to time slot, in which he has assigned preparations also for faculties.It has been observed from literature that, in solving faculty course assignment problem preferences are being considered but the feedback of the faculty and result of the particular subject taught by faculty member not considered.

1.1 Problem Formulation of Multi-objective Faculty Course Assignment Problem (MOFCAP) Using Faculty Feedback and Result

Parameters: $I = \{1, 2, \ldots, m\}$ courses;
$\{I = \cup I_j , I_j\}$ is the set of courses that faculties j can give; $J = \{1, 2, \ldots, k\}$ faculties; $J_0 = \{1, 2, \ldots, k\}$ tenured faculties; $J_n = \{k + 1, k + 2, \ldots, n\}$ recent faculties; $J = J_0 \cup J_n$ and $k < n$
h_i =total number of lecture hours for the ith course in week;
l_j and $u_j =$ lower and upper bounds for the jth instructor's weekly load,respectively;

t_{ij} = preference level of the ith course by the jth instructor ($t_{ij} \geq 1$, 1indicates the most desired course);

a_{ij} = administrative preference level for the assignment of the course to the jth faculty.

Decision Variable: Decision variable x_{ij} represents assignment of ith course to jth faculty and is defined as .

$$x_{ij} = \begin{cases} 1, & \text{if course i is assigned to faculty j,} \\ 0, & \text{otherwise} \end{cases} \qquad (1)$$

Constraints:

1. Every course must be assigned to only one faculty:

$$\sum_{j=1}^{n} x_{ij} = 1, i = 1, 2, \ldots, m. \qquad (2)$$

2. The weekly load of every faculty must be between his lower and upper limits.

$$l_j \leq \sum_{i=1}^{m} x_{ij} \leq u_j, j = 1, 2, \ldots, n. \qquad (3)$$

Objectives: We generally develop two groups of objectives for the faculties $L_i(x)$, $i = 1, 2, \ldots, l$, and for the administrators $A_i(x)$, $i = 1, 2, \ldots, p$.

1. Minimize the average preference level L_j per hour taught.

$$L_j(x) = \frac{\sum_{i=1}^{m} x_{ij} h_i t_{ij}}{\sum_{i=1}^{m} x_{ij} h_i}, j = 1, 2, \ldots, n. \text{ where } x = x_{ij}. \qquad (4)$$

2. Minimize the average preferences of all faculties.

$$A_1(x) = \frac{\sum_{i=1}^{m} x_{ij} h_i t_{ij}}{\sum_{i=1}^{m} x_{ij} h_i}, \text{ where } x = x_{ij}. \qquad (5)$$

3. Minimize the administration's total preference level:

$$A_2(x) = \sum_{i=1}^{m} \sum_{j=1}^{n} a_{ij} x_{ij}. \qquad (6)$$

4. Minimize the total deviation from the upper load limits of the faculties:

$$A_3(x) = \sum_{j \in J} \left(u_j - \sum_{i=1}^{m} x_{ij} h_i \right). \qquad (7)$$

5. Minimize the feedback of faculty given by students:

$$S_1(x) = \sum_{i=1}^{m} \sum_{j=1}^{n} c_{ij} x_{ij}. \tag{8}$$

6. Preferences based on result of the subject taught:

$$S_2(x) = \sum_{i=1}^{m} \sum_{j=1}^{n} d_{ij} x_{ij}. \tag{9}$$

Thus, the fuzzy multi-objective mathematical model of the the faculty course assignment problem (FMOFCAP) can be formulated as follows:

1.2 Model-1

$$Min[L_1(x), L_2(x), \ldots, L_{11}(x), A_1(x), A_2(x), S_1, S_2]. \tag{10}$$

subject to the constraints:

$$\sum_{j=1}^{n} x_{ij}, i = 1, 2, \ldots, m. \tag{11}$$

$$l_j \leq \sum_{i=1}^{m} x_{ij} \leq u_j, j = 1, 2, \ldots, n. \tag{12}$$

Fuzzy programming technique-based solution approach to solve Mathematical Model for Multi-Objective Faculty Course Assignment Problem(MOFCAP) using Faculty Feedback and result: For finding the solution of the Model-1 by fuzzy programming technique first this model is solved for single objective function and for each objective function find out the positive ideal solution (PIS) and negative ideal solution (NIS) of the model. Now, by positive ideal solution (PIS) and negative ideal solution (NIS) define a membership function $\mu(Z_k)$ for the kth objective function. Here, different membership functions are utilized to find an efficient solution of this multi-objective resource allocation problem and by using this membership function the model is converted into the following model:

1.3 Model-2

When we utilize fuzzy linear membership function

$$\mu\left(Z_k\right) = \begin{cases} 1, & \text{if } f_k\left(x\right) \leqslant l_k, \\ \frac{u_k - f_k(x)}{u_k - l_k}, & \text{if } l_k < f_k\left(x\right) < u_k, \\ 0, & \text{if } f_k\left(x\right) \geqslant u_k, \end{cases} \tag{13}$$

k=1,2, …16. then model-2 structure is as follows:
Maximum λ,
Subject to the constraints:

$$\lambda \leqslant \mu\left(Z_k\right), k = 1, 2, \ldots, 16. \tag{14}$$

Equations (4)–(12). $\lambda \geq 0$ and $\lambda = \min\mu\left(Z_k\right)$

1.4 Model-3

When we utilize exponential membership function

$$\mu\left(Z_k\right) = \begin{cases} 1, & \text{if } f_k\left(x\right) \leqslant l_k, \\ \frac{e^{-S\Psi_k(x)} - e^{-S}}{1 - e^{-S}}, & \text{if } l_k < f_k\left(x\right) < u_k, \\ 0, & \text{if } f_k\left(x\right) \geqslant u_k, \end{cases} \tag{15}$$

where $\Psi_k\left(x\right) \leqslant \frac{f_k(x) - l_k}{u_k - l_k}, k = 1, 2, \ldots, n$ and S is a non-zero parameter, prescribed by the decision maker, then model-3 structure is as follows:
Maximum λ,
Subject to the constraints:

$$\left(e^{-S\Psi_k(x)} - e^{-S}\right) \geqslant \lambda\left(1 - e^{-S}\right) \text{ where, } \Psi_k\left(x\right) \leqslant \frac{f_k\left(x\right) - l_k}{u_k - l_k}, k = 1, 2, \ldots, n$$

Equation (4)–(12). $\lambda \geq 0$ and $\lambda = \min\mu\left(Z_k\right)$.

1.5 Problem Formulation of Multi-objective Faculty Courses Time Slot Assignment Problem (FMOFCTAP) for Theory Courses

h_u = number of hours teaching load. s_j = total load of course to the faculty member at certain time slot. r_l = number of classrooms available for a certain time slot.

In this second stage of the model, the decision variable X_{ijl} represents the faculty member jth is to teach the ith course during the lth timeslot is defined as follows:

$$x_{ijl} = \begin{cases} 1, & \text{if the } j\text{th faculty has to teach } i\text{th course for given} l\text{th timeslot,} \\ 0, & \text{otherwise.} \end{cases} \tag{16}$$

Constraints: 1. The total load of ith course to the jth faculty member at lth timeslot is s_j.

$$\sum_{i=1}^{n} \sum_{l=1}^{k} x_{ijl} = s_j, j = 1, 2, 3, \ldots, m. \tag{17}$$

2. Number of classrooms available for the assigning of ith course to the jth faculty member at lth timeslot.

$$\sum_{i=1}^{n} \sum_{j=1}^{m} x_{ijl} = r_l, l = 1, 2, 3, \ldots, k. \tag{18}$$

3. Number of hours to be assigned for each course to the faculty member for certain preferred timeslots.

$$\sum_{i=1}^{n} \sum_{j=1}^{m} \sum_{l=1}^{k} x_{ijl} = h_u, u = 1, 2, 3, \ldots, g. \tag{19}$$

4. Select one preference from the given three timeslots selected for each of the faculty course taken together.

$$\sum_{l=1}^{k} x_{ijl} = 1, i = 1, 2, \ldots, n \text{ and } j = 1, 2, \ldots, m. \tag{20}$$

1.6 Model-5

When we utilize linear membership function given in (13), then model-5 structure is as follows: Maximum λ,

Subject to the constraints:

$$\lambda \leqslant \mu\left(Z_k\right), k = 1, 2, 3. \tag{21}$$

and Eq. (16)–(20).

1.7 Model-6

When we utilize exponential membership function given in (15), then model-5 structure is as follows:

Maximum λ,

Subject to the constraints:

$$\left(e^{-S\Psi_k(x)} - e^{-S}\right) \geqslant \lambda\left(1 - e^{-S}\right) \text{ where, } \Psi_k(x) \leqslant \frac{f_k(x) - l_k}{u_k - l_k}, k = 1, 2, \ldots, n$$

Equation (16)–(20).$\lambda \geq 0$ and $\lambda = \min \mu\left(Z_k\right)$.

2 Numerical Example

Here we have considered problem faced by XYZ institute,where 34 courses are assigned to 11 faculty members.In phase-I, we will assigned courses to faculties by considering preferences of faculty, administrator and students. Students' preferences are by means of result of particular subject and feedback.In phase-II, based on output of phase-I, preferences for timeslots were collected and prepared model, which will be solved by LINGO 18.0.

2.1 Phase-I

Faculties: Total faculties available are 11. Out of which 2 Professor, 3 Associate professor and 6 Assistant Professor. Hypothetically, professor has 10 hours, associate professor has 12 h and assistant professor 16 h workload.

Teaching Scheme: Every subject has its own teaching methodology. Theory, lab and tutorial are ways in it. Lab and tutorial will be conducted by dividing whole into suitable batches. Duration for practical will be 2 h. For the sake of simplicity, we have labeled all the courses and faculties. Treating practical and tutorial as courses,now total courses to be allocated 34 (Tables 1, 2 and 3).

Table 1 Weekly load of faculties

I	1–16	17–18	19–34
h_i	3	4	6

Table 2 Maximum and minimum load of faculties

Faculty	1	2	3	4	5	6	7	8	9	10	11
Maximum load	5	5	5	5	5	8	7	8	8	8	8
Minimum load	3	3	2	2	2	3	2	3	3	3	3

Table 3 Subject code

Course	Code	Course	Code	Course	Course	Code
101L	01	302L	10	101P	302P	28
102L	02	303L	11	102P	303P	29
103L	03	304L	12	103P	305P	30
104L	04	401L	13	105P	401P	31
201L	05	402L	14	201P	402P	32
202L	06	403L	15	202P	403P	33
203L	07	404L	16	203P	405P	34
204L	08	101T	17	205P		
301L	09	201T	18	301P		

Faculty Preferences: Table 4 shows the preferences given by faculty for courses. P1, P2, P3 and P4 denotes first,second,third and fourth preference. 1 is most desired one.

Administrator Preferences: Table 5 shows the preferences given by administrator for faculty course. 1, 2, 3, 4 are in use where 1 stands for most desired one.

Tables 6 and 7 show the preferences based on result of previous year. Here results are considered as $1 - 0.9 = 0.1$ for minimization of result.

2.2 Solution of Phase-I

When we solve this model of Phase-I by using LINGO18.0 software, then the assignment is given in table, and objective values are given in the table.

In Table 1, faculty courses assigned are shown which is obtained by applying linear membership function. In Table, faculty course assigning are shown which is obtained by applying linear membership function.

Table 4 Faculty preferences for courses

Faculty	Faculty code	P1	P2	P3	P4
Prof1	F1	10, 17, 20	12, 23, 26, 29	9, 15, 21	13, 16, 28, 33
Prof2	F2	24, 27	10, 12, 14, 18, 32	11, 22, 30	–
AssoProf1	F3	16	9, 11, 15, 21, 25, 28	13, 31	9, 26, 34
AssoProf2	F4	12, 15	20, 22, 33	20, 22, 33	10, 17, 29, 14, 27
AssoProf3	F5	9, 14	13, 23, 30	16, 19, 32	11, 24, 28
AssistProf1	F6	7, 26, 34	1, 3, 5, 8, 21, 29	2	18, 31
AssistProf2	F7	1, 6, 22, 33	17, 27	3, 5, 30	24
AssistProf3	F8	1	4, 6, 8, 19, 31	7, 23, 29	21, 25
AssistProf4	F9	5	3, 24	18, 20, 28, 32, 34	1
AssistProf5	F10	3	6, 25, 30, 33	2, 5, 20, 22, 31	17, 27
AssistProf6	F11	8, 21, 29, 32	4, 7, 18	1, 5, 26	2, 23

To solve the model, where we have used linear membership function. Thus, obtained objective values are noted in the table.

When we apply exponential membership function, then the results are as follows:

For different values of s, we get different assignments of courses to faculties. It means that it provides different possibilities. Among all these, we need to choose one which will be suitable.

2.3 Phase-II

The output of first stage as noted in Table 1 is used for faculty course timeslots preferences for each course in Phase-II.

Table 5 Administrator preferences for faculties

Sub code ↓	1	2	4	5	6	7	8	9	10	11
01	–	–	–	–	3	4	–	1	1000	–
02	–	–	–	–	1	–	3	–	4	2
03	–	–	–	–	1	3	–	2	1000	–
04	–	–	–	–	–	–	1	4	–	2
05	–	–	–	–	1	3	–	2	1000	–
06	–	–	–	–	–	3	4	–	1	1000
07	–	–	–	–	2	–	3	–	–	1
08	–	–	–	–	2	–	3	–	–	1
09	1000	–	–	4	–	–	–	–	–	–
10	3	1000	4	–	–	–	–	–	–	–
11	–	1	–	1	–	–	–	–	–	–
12	3	1000	4	–	–	–	–	–	–	–
13	1000	–	–	4	–	–	–	–	–	–
14	–	3	4	1	–	–	–	–	–	–
15	4	–	3	–	–	–	–	–	–	–
16	1000	–	–	4	–	–	–	–	–	–
17	4	–	1	–	–	3	–	–	1000	–
18	–	4	–	–	3	–	–	2	–	1
19	–	–	–	1	–	–	4	–	–	–
20	4	–	3	–	–	–	–	3	4	–
21	3	–	–	–	1000	–	3	–	–	2
22	–	3	1	–	–	1000	–	–	2	–
23	3	–	–	1000	–	–	2	–	–	3
24	–	3	–	1	–	3	–	4	–	–
25	–	–	–	–	–	–	3	–	1	–
26	4	–	–	–	1	–	–	–	–	1000
27	–	3	1	–	–	1000	–	–	2	–
28	–	4	3	–	–	3	–	–	1000	–
29	4	–	1000	–	3	–	2	–	–	1
30	–	3	–	1	–	3	–	–	4	–
31	–	–	–	–	3	–	3	–	1	–
32	–	3	–	1	–	–	–	3	–	4
33	4	–	1	–	–	3	–	–	1000	–
34	–	–	–	–	1	–	–	1000	–	–

Table 6 Result-based preferences for faculties

Sub code	F1	F2	F3	F4	F5	F6	F7	F8	F9	F10	F11
01						4	5		2		3
02						2		5		2	3
03						1	2		1	3	
04								2	4		0
05						2	4		2	1	
06							3	0		2	1
07						2		1			3
08						2		1		1	
09	2		1		4						
10	1	3		6							
11		3	3		3						
12	2	4		2							
13	3		2		2						
14		1		2	3						
15	1		4	7							
16	2		2		1						
17	2			4			3			4	
18		2				1			2		1
19			2		3			1			
20	7			1					2	2	
21	1		2			1		1			2
22		2		1			2			1	
23	2				1			2			1
24		2			1		2		1		
25			1					2		1	
26	3		2			0					2
27		1		2			3			3	
28	1		2		1				2		
29	1			2		2		1			1
30		2			1		4			3	
31			4			1		2		3	
32		2			1				2		3
33	1			2			1			2	
34			1			2			3		

Table 7 Feedback-based preferences for faculties

Sub code	F1	F2	F3	F4	F5	F6	F7	F8	F9	F10	F11
01						7	2		5		1
02						1		1		3	2
03						1	3		4	2	
04								4	3		1
05						1	3		2	1	
06							2	1		2	3
07						2		2			4
08						1		2		4	
09	3		1		1						
10	1	2		1							
11		2	1		2						
12	1	2		3							
13	1		2		3						
14		1		2	1						
15	1		1	2							
16	1		2		1						
17	2			3			3			4	
18		2				1			2		1
19			2		3			2			
20	1			2					1	3	
21	1		2			3		4			3
22		1		2			1			3	
23	3				2			3			4
24		1			2		3		1		
25		2						3		1	
26	1		2			1					3
27		2		1			3			5	
28	4		2		3				1		
29	1			3		3		4			7
30		1			3		2			3	
31		1				4		3		2	
32		1			3				4		1
33	1			2			1			2	
34			4			2			3		

Sr. No.	Faculty	Courses assigned using LMF
1	Prof1	302L, 403L, 101T, 303P
2	Prof2	201T, 301P, 402P
3	Asso1	301L, 303L, 401L, 404L
4	Asso2	304L, 102P
5	Asso3	402L, 201P
6	Asst1	203L, 205P, 405P
7	Asst2	101L, 105P
8	Asst3	202L, 101P, 401P
9	Asst4	201L, 202P, 302P
10	Asst5	102L, 103L, 305P, 403P
11	Asst6	104L, 103P, 204L

Objective function	F1	F2	F3	F4	F5	F6	F7	F8	F9	F10	F11	F12	F13	F14	F15	F15	λ
Objective values	1.75	1.62	2	1.667	1.667	1	1	2	2.2	1.667	1.333	1.671	5080	9	59	62	0.7471264

	S = 0.00005	S = 0.00001	S = 0.001	S = 0.0005	S = 0.0001
F1	1.68	1.315789	1.5454	1.48	1.315789
F2	1.6	1.538462	1.6	1.5384	1.538462
F3	1.83	1.833	2	2	2
F4	1	1	1	1	1.5
F5	1.667	1.75	1.5	1.667	1.667
F6	1	1	1.25	1.5	1
F7	1	1	1	1	1
F8	1.8	1.83	1.833	2	1.8
F9	2	1.75	2	2.1667	2.1667
F10	1.5	1.667	1.67	1.667	1.667
F11	1.4	1.333	1.4	1.333	1.333
F12	1.55	1.510067	1.5406	1.651007	1.9060
F13	5078	3088	4.83	4085	3084
F14	9	9	9	9	9
F15	81	76	85	76	78
F16	61	61	60	60	61
λ	0.77777	0.823091	0.777691	0.77773446	0.777769

2.4 Solution of Phase-II

2.5 Discussion and Conclusion

In this paper, we have solved two-phase multi-objective faculty course timeslot assignment problem with result and feedback. These two phases are solved by using fuzzy programming technique. In first phase, courses are assigned to faculty members using linear membership function to allocate 34 courses to 11 faculty. The linear membership function assigned courses to all faculty members as noted and the optimum values are obtained for objective function as noted. In a second phase, assigned

Courses assigned using exponential function for different values of s					
Faculty	S = 0.0001	S = 0.0005	S = 0.005	S=0.00001	S = 0.000005
Prof1	302L, 101T, 102P, 303P	302L, 101T, 102P, 205P, 303P	101T, 102P, 205P, 303P	302L, 102P, 303P	302L, 401L, 101T, 102P, 303P
Prof2	402L, 201T, 301P	402L, 201T, 301P	302L, 301P, 402P	402L, 201T, 301P	402L, 301P, 402P
Asso1	303L, 401L, 404L, 103P	303L, 401L, 404L, 103P	303L, 401L, 404L, 302P	303L, 404L, 103P, 302P	303L, 404L, 203P, 302P
Asso2	104L, 403L, 105P	304L, 403L	304L, 403L	304L, 403L	304L, 403L
Asso3	301L, 201P	301L, 201P	301L, 402L, 202P	301L, 401L, 103P	301L, 201P
Asst1	203L, 205P, 405P	102L, 203L, 405P	103L, 203L	203L, 205P, 405P	203L, 205P, 405P
Asst2	101L, 202L	101L, 105P	101L, 105P	101L, 105P	101L, 105P
Asst3	102L, 101P, 401P	202L, 101P, 401P	102L, 202L, 101P, 401P	102L, 202L, 101P, 401P	102L, 101P, 401P
Asst4	103L, 201L, 202P, 302P	103L, 201L, 202P, 302P	104L, 201L, 202P	103L, 201L, 202P	104L, 201L, 202P, 401P
Asst5	203P, 305P, 403P	203P, 305P, 403P	203P, 305P, 403P	203P, 305P, 403P	103L, 202L, 305P, 403P
Asst6	104L, 103P, 204L	104L, 402P,204L	201T, 103P	104L, 204L, 402P	201T, 103P, 204L

	S = 0.00001	S = 0.00002	S = 0.00003	S = 0.00005	S = 0.00009
F1	5	5	3	6	6
F2	7	8	9	8	8
F3	14	13	15	14	14
F4	14	16	15	15	12
F5	7	14	17	15	15
F6	29	16	17	14	15
F7	17	17	20	18	18
F8	11	7	9	9	11
F9	11	9	12	6	11
F10	8	9	8	9	8
F11	15	15	12	14	15
F12	14	16	17	15	14
F13	36	36	40	38	36
F14	51	48	60	51	51
F15	68	64	36	60	68
λ	0.6047785	0.615382	0.521735	0.599940	0.5999892

courses are given preferred timeslot by the faculty member preferences as noted in the table. All the faculty members are assigned courses and timeslot as per the preferences given by him/her. Also, the exponential membership function is used to solve three phases at different S values. The assigning of courses to faculty in first stage for different values of S is as, 11 faculty for S = 0.00001, S = 0.00005, S = 0.001 . That is, as S varies the assigning of the courses also changes simultaneously and objective function value obtained are noted in the table . The obtained objective value for the exponential membership function satisfied fuzzy preferences, which is used for better scheduling of courses. In a second phase, assigned courses are allocated with timeslots for different values S.

References

1. Arabinda Tripathy. 1984. School timetabling—a case in large binary integer linear programming. *Management Science, 30*(12)
2. Burke, Edmund Kieran, and Sanja Petrovic. 2002. Recent research in automated timetabling. *European Journal of Operational Research* 140: 266–280.
3. de Werra, D. 1985. An introduction to timetabling. *European Journal of Operational Research* 19: 151–162.
4. Asratian, A.S., and D. de Werra. 2002. A generalized class-teacher model for some timetabling problems. *European Journal of Operational Research* 143: 531–542.
5. Ozdemir, Mujgan S., and Rafail N. Gasimov. 2004. The analytic hierarchy process and multiobjective 0–1 faculty course assignment. *European Journal of Operational Research* 157: 398–408.
6. Daskalaki, S., and T. Birbas. 2005. Efficient solutions for a university timetabling problem through integer programming. *European Journal of Operational Research* 160: 106–120.
7. Schnierjans, Marc J., and Gyu Chan Kim. 1987. A Goal Programming model to optimize departmental preferences in course assignments. *Computers and Operations Research* 2: 87–96.
8. Badri, Massod A. 1996. A two stage multiobjective scheduling model for [faculty-course-time] assignments. *European Journal of Operational Research* 94: 16–28.
9. Badri, Massod A., Donald L. Davis, Donna F. Davis, and John Hollingsworth. 1998. A multiobjective course scheduling model: combining faulty preferences for courses and times. *Computers and Operations Research* 25 (4): 303–316.
10. Al-Yakoob, Salem M., and Hanif D. Sherali. 2006. Mathemaical programming models and algorithms for a class-faculty assignment problem. *European Journal of Operational Research* 173: 488–507.
11. Al-Yakoob, Salem M., and Hanif D. Sherali. A column generation mathematical programming approach for a class-faculty assignment problem with preferences. *Computational Management Science*. https://doi.org/10.1007/s10287013-0163-9.
12. Bakir, M.Akıf, and Cihan Aksop. 2008. A 0–1 integer programming approach to a university timetabling problem. *Hacettepe Journal of Mathematics and Statistics* 37 (1): 41–55.
13. Ongy, Elvira E. 2017. Optimizing student learning: a faculty-course assignment problem using linear programming. *Journal of Science and Technology* 5: 1–14.
14. Ismayilova, Nergiz A., Mujgan sagir, Rafail N. Gasimov, 2007. A multiobjective faculty-course-time slot assignment problem with preferences. *Mathematical and Computer Modeling* 46: 1017–1029.

Simulation of Learning Logical Functions Using Single-Layer Perceptron

Mohd Vasim Ahamad, Rashid Ali, Falak Naz and Sabih Fatima

Abstract As a simplest neural network, the perceptron computes a linear combination of real-valued labeled training samples and predicts the classes for unclassified testing samples. If two different sets of samples can be separated by a straight line, they are called linearly separable. The perceptron can be considered as the binary classifier of unclassified samples based on the supervised machine learning approach. The learning algorithm for the perceptron takes a static value for the learning rate as an input, which can affect the efficiency of the learning process. This work attempts to implement logical operations OR, AND, NOR, and NAND using a single-layer perceptron algorithm. A modified perceptron algorithm is proposed which finds the most optimal value of learning for the learning process. The learning algorithm is provided with a range of learning rate values, and it picks the most suitable learning rate by calculating the number of iterations taken by each of these learning rates and choosing the one which takes the minimum number of the epoch.

Keywords Artificial neural network · Perceptron · Linear classifier · Learning rate · Artificial neuron

1 Introduction

The artificial neural network [1] is a general abstraction of biological nervous system that consists of millions of signal processing elements connected together to form a complex communication network. These signal processing elements are called as (artificial) neurons. The neurons in artificial neural network tend to have fewer connections than biological neurons, and each of them receives a number of input signals coming from other connected neurons. The artificial neural network tends to learn and solve linear, nonlinear, and other complex cognitive and pattern recognition problems. Classification, pattern recognition, forecasting, image processing,

M. V. Ahamad (✉) · R. Ali · F. Naz
Department of Computer Engineering, ZHCET, Aligarh Muslim University, Aligarh, India

S. Fatima
Department of Electrical Engineering, ZHCET, Aligarh Muslim University, Aligarh, India

© Springer Nature Singapore Pte Ltd. 2020   121
Y.-C. Hu et al. (eds.), *Ambient Communications and Computer Systems*, Advances in
Intelligent Systems and Computing 1097,
https://doi.org/10.1007/978-981-15-1518-7_10

and character recognition are some of the fields where artificial neural networks have significant applications. The artificial neural network with single neuron as a processing element is called as the perceptron. The perceptron is considered as the simplest of all artificial neural networks. It is basically used for binary classification of input data. The perceptron, generally, tends to solve the linearly separable problems. The perceptron classifies the input data into two classes based on its inputs, weights, and outputs of the activation function. For this reason, the perceptron is the better method to learn and solve the logical functions such as AND, OR, NOR, and NAND, because they provide output in the form of 0 or 1. The basic perceptron algorithm is based on the static learning rate and may converge after large number of iterations. This work modifies the perceptron algorithm to take dynamic learning rate to implement the logical functions.

2 Related Work

In [2], authors have discussed the structure of single-layer and multilayer perceptrons. They modified the structure of single-layer perceptron to solve the XOR problem which is linearly non-separable problem. A multilayer perceptron is a type of feed-forward neural network having at least three layers—the input layer, the hidden layer, and the output layer [1, 3]. Authors in [3] compared the performances of multilayer perceptron, having different learning algorithms such as back-propagation and delta rule, with the simple perceptron. In [4], authors have used information entropy function-based perceptron to emulate various logical functions, namely NOT, AND, OR, and XOR. Evidences of applications of perceptron and the neural network are found in the classification of the healthcare data [5]. Authors in [5] used multilayer perceptron with support vector machine and neural network to classify the heart disease severity into five classes. Multilayer perceptron can also be used to predict diabetes [6]. In [6], authors have used several learning algorithms with multilayer perceptron for diabetes prediction. Authors in [7] report the study of performance of multilayer perceptron for classifying the qualities of significant compounds in agarwood oil. They have used the scaled conjugate gradient, Levenberg–Marquardt, and resilient back-propagation learning algorithms to implement and enhance the performance of the multilayer perceptron. In [8], authors have reported an empirical analysis of the factors affecting the crime news perceptions using multilayer perceptron model. Authors in [9] generated an intrusion detection and classification model using multilayer perceptron and compared their results with other well-known classification algorithms such as Naïve Bayes and decision tree. The study shows that there are many versions of the basic perceptron algorithm. The motivation behind this work is to use the binary classification capability of the perceptron to learn the simple logical functions such as AND, OR, NAND, and NOR with the help of dynamic learning rate to converge it fast.

3 Basics of Artificial Neural Network

A neural network can be described as a network of biological neurons which are functionally connected to the central nervous system of human [1]. The basic component of a biological neural network is called a neuron. A neuron is a special biological cell whose responsibility is to process the information. As mentioned in Fig. 1 [10], a neuron mainly consists of three sub-components; dendrites, cell body (soma), and axon. Dendrites are the tree-like structure that receives the signal from other connecting neurons, where each dendrite branch is connected to one neuron. Axon is a thin, cylinder-like structure which transmits the signal to other neurons. Dendrite branches are connected to others through synapses. The cell body has a nucleus that contains and processes information.

Artificial neural network (ANN) is a machine learning approach that models human brain [2]. An ANN is composed of artificial neurons that are connected together to form a network. Neuron in ANNs tends to have fewer connections than biological neurons. Each neuron in ANN receives a number of inputs from other neurons. An activation function is applied to these inputs to obtain the output value of the neuron. The neurons are trained by using the knowledge about the domain in the form of training examples.

The neuron model in Fig. 2 receives m input signals from other neurons with m synaptic weights. It, then, calculates the weighted sum of the inputs and weights from all m inputs. The weighted sum is transferred to the activation function. The output of the activation function decides whether neuron is activated or not. The bias b has the effect of applying a transformation to the weighted sum u.

$$u = \sum_{j=1}^{m} w_j * x_j \tag{1}$$

$$v = u + b \tag{2}$$

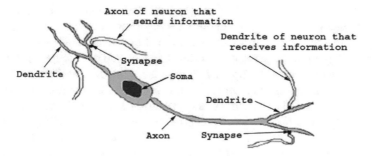

Fig. 1 Biological neuron cell [10]

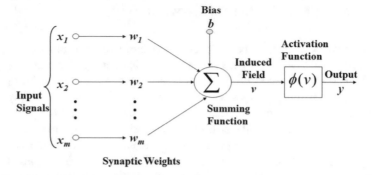

Fig. 2 Artificial neuron model for single neuron (perceptron)

The bias b is an external parameter of the neuron. It can be modeled by adding an extra input. The symbol v is called induced field of the neuron. The combined equation can be written as,

$$v = \sum_{j=0}^{m} w_j * x_j \tag{3}$$

where $w_0 = b$ and $x_0 = 1$.

4   The Perceptron

The most fundamental single-layer feed-forward artificial neural network known so far is the perceptron [3]. A single-layer feed-forward network is defined as a network where connections between various artificial neurons do not form a directed cycle. In this kind of neural network, information moves only in the forward direction, that is, from input nodes to output nodes, through a series of weights. A single-layer feed-forward network contains input and output layers only; there are no hidden layers. The algorithm for the perceptron was invented by Frank Rosenblatt in 1957 [11]. The schematic diagram of the perceptron is shown in Fig. 2. The perceptron follows the supervised machine learning approach to train itself for the classification task. It is considered as a binary classifier for the linearly separable set of inputs [11]. It groups the set of inputs into only two classes, and all examples in the training dataset should belong to only these two classes. Perceptron is a used for classifying the linearly separable set of data only, i.e., a set of data that can be separated by a line in 2-D plane [11]. The perceptron consists of the four components, namely the input layer or input, input weights and bias, weighted sum, and the activation function. Inputs from other neurons are connected with the perceptron and denoted by the input vector $[X_1, X_2, X_3, \ldots, X_{n-1}, X_n]$, where n is the total number of inputs. Input weights are the values that are calculated over the time of training the perceptron. As initial weight values, the perceptron algorithm starts with considering weight values ranging from -1 to 1 and these values get updated for each training error. We represent the weights for perceptron by the weight vector $[W_1, W_2, W_3, \ldots, W_{n-1}, W_n]$. The weighted sum is the addition of the values that are calculated after the multiplication of each weight in weight vector $[W_n]$ associated with the each input vector $[X_n]$. The weighted sum can be calculated by using the formula given in Eq. 4. The activation function is a function used to transform the activation level of neuron (weighted sum of inputs) to an output signal. It controls whether a neuron is 'alive' or 'inactive.' The activation function, mostly used with single-layer perceptron, is STEP function. In STEP function, refer to Eq. 5, the output can be at one of two levels, subject to the condition whether the weighted sum is greater than or less than some threshold value.

$$v = \sum_{j=1}^{n} \left([w_j] T * [x_j] \right) + b \qquad (4)$$

$$\phi(v) = \begin{cases} 0 \text{ if } v < 0 \\ 1 \text{ if } v > 0 \end{cases} \qquad (5)$$

4.1 The Learning Algorithm for Perceptron

As discussed earlier, the perceptron is a supervised learning-based binary classifier of linearly separable input examples. To make it learn, the input set of example is multiplied by the initial weights and the weighted sum is calculated. The activation function is applied to the weighted sum. In this case, STEP function, as an activation function, checks if the output of the weighting function is greater than zero and decision is made if the neuron is activated or not. The learning algorithm for the perceptron is as follows:

1. Initially, assign random weights $(w_1, w_2, \ldots, w_n)$ in the range of $(-0.5$ to $0.5)$.
2. Input a learning dataset sample $(x_1, x_2, \ldots, x_n)$ and its desired output.
3. For each example i in the training dataset sample, perform following steps.

$$y = \text{STEP}\left(\sum_{i=1}^{n} w_i * x_i + b \right)$$
$$e = (d - y)^2$$

4. If there is an error, update weights using the formula $w_i = w_i + \eta * x_i * e$.
5. where η is the learning rate, e is the calculated error, and x_i is the ith example input from training sample.

   a. If output is correct, then 'η' is defined as 0

b. If output is too low, set some positive value
c. If output is too high, set some negative value.

6. Once the modification to weights has taken place, the next piece of training data is presented to the algorithm.
7. Once all examples in the training dataset have been presented, the same process iterates until all the weights are correct and all errors are zero.

5 Proposed Algorithm

The proposed algorithm attempts to implement logical operations OR, AND, NAND, NOR, and NOT using single-layer perceptron algorithm with the motive of minimizing the execution time and making it more efficient with the help of the dynamic learning rate. It also incorporates graphs and animations to make it more interactive and comprehensive so that it can be used in educational fields to train students. As mentioned earlier, the perceptron algorithm is used for binary classification of example data. It is also evident that the output for any logical operation can be either 1 or 0. So, the perceptron can be easily used classifying the input dataset for any Boolean operation. Since the perceptron algorithm comes under supervised machine learning approach, it is divided into two stages:

- Firstly, it is trained with learning example dataset having random values of weights and bias using the algorithms discussed in the Sect. 4.1.
- In the second stage, it is tested with the help of actual inputs from external environment and the class of the given input is examined.

5.1 Algorithm Implementation

The proposed algorithm attempts to attain a learning rate which takes minimum number of iterations for the learning process. While training the perceptron, if we choose larger value of learning rate, it might overshoot the minima whereas smaller values of learning rate might take long time for convergence. To deal with this issue, dynamic values of learning rate are taken and a total number of epochs needed, to get the perceptron trained, are calculated corresponding to each of the learning rate. Then, the learning rate which gives the minimum number of iterations to train the algorithm is chosen. In this way, we shall be able to find a value which gives minimum execution time for the training process. The steps of the modified single-layer perceptron algorithm with dynamic learning rate are as follows:

1. Initialize the random weights $(w_1, w_2, \ldots, w_n)$ in the range of $(-0.5$ to $0.5)$.
2. Initialize the learning rate $\eta = -0.1$.

3. Input a learning dataset sample $(x_1, x_2, \ldots, x_n)$ and its desired output.
4. For each example i in the training dataset sample, perform following steps.

$$y = \text{STEP}\left(\sum_{i=1}^{n} w_i * x_i + b\right)$$
$$e = (d - y)^2$$

5. If there is an error, update the weights using the formula $w_i = w_i + \eta * x_i * e$
   where η is the learning rate, e is the calculated error, and is the ith example
   input from training sample.
6. Once the modification to weights and η has taken place, the next piece of training
   data is presented to the algorithm.
7. Once all examples in the training dataset have been presented, the same process
   iterates until all the weights are correct and all errors are zero.
8. Once all the examples are learnt, iterate from Step 3 with new learning rate until
   $\eta \geq -1.0$.

   a. If output is correct, then 'η' is defined as 0
   b. If the last output corrected is too low, set $\eta = +0.1 * e$
   c. If the last output corrected is too high, set $\eta = \eta - 0.1 * e$.

5.2 The User Interface

This work attempts to implement the logical operations OR, AND, NOR, and NAND
using dynamic learning rate of single-layer perceptron. For the convenience of users,
the overall architecture is divided into two parts—the user interface layer and the
functional layer. To provide the realistic view of mathematical calculations in the
functional layer, animation is provided at the user interface. This will help in understanding the concept of perceptron learning algorithm easily. The real time calculation
of the logical OR operation on input $(1, 1)$ and initial weights $(0.6, 0.4)$ with bias
value of -0.32 are shown in Fig. 3. Other inputs of the logical OR, AND, NAND,
and NOR can be viewed similarly. The weighted sum is calculated using the formula
shown in Eq. 4, as $(0.6 * 1 + 04 * 10 - 0.32 = 0.68)$. The calculated value of the
summation function is transmitted to the activation function. The activation function
calculates the final output, using the formula shown in Eq. 5, which is 1. Similarly,
other inputs are given to the algorithm through the user interface and final output is
tested. For wrong outputs, it follows the algorithms steps as discussed in the Sect. 5.

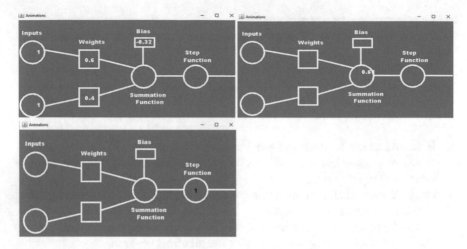

Fig. 3 Animation of the logical OR operation on input (1, 1) and weights (0.6, 0.4)

6 Results and Discussion

It is already discussed in previous sections that the perceptron algorithm is used for binary classification of data. In this work, the modified perceptron learning algorithm is used to classify the input sets of logical operations OR, AND, NOR, and NOR into two classes. The first class is the one where the output is 0 and the other being the one where output is 1. During the training of the perceptron, authors are actually interested in finding the slope and intercept of the line which correctly classifies the two classes of data. To differentiate between the two classes, authors are representing the input set whose output is 1 by red dots and the input set whose output 0 by black dots. Initially, any random weights as well as the bias are selected, so an arbitrary line is drawn. From Figs. 4, 5, 6, and 7, slops and intercepting lines are drawn for the logical operations OR, AND, NOR, and NOR, respectively. The perceptron learning algorithm keeps updating the weights until the two classes of data are linearly separable, as it can be seen in the following figures.

The learning algorithm is provided with a range of learning rate values, and it picks the most suitable learning rate by calculating the number of iterations taken by each of these learning rates and choosing the one which takes the minimum number of epoch. In Fig. 8, the total number of epoch is calculated with respect to learning rates ranging from -0.1 to -1.0, weights $w_1 = 0.4$ and 0.2, and the bias $b = -0.3$. The learning rate in which least number of epoch is taken is considered as the most optimal learning rate.

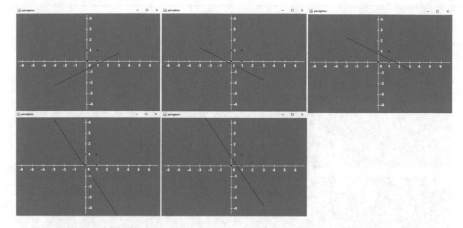

Fig. 4 Learning slop of logical OR

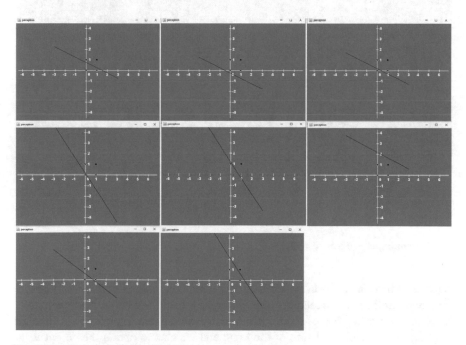

Fig. 5 Learning slop of logical NAND

7 Conclusion and Future Directions

A neural network is defined as a complex network of biological neurons which are functionally connected to the central nervous system of living beings. An artificial neural network is comprised of artificial neurons that are functionally connected

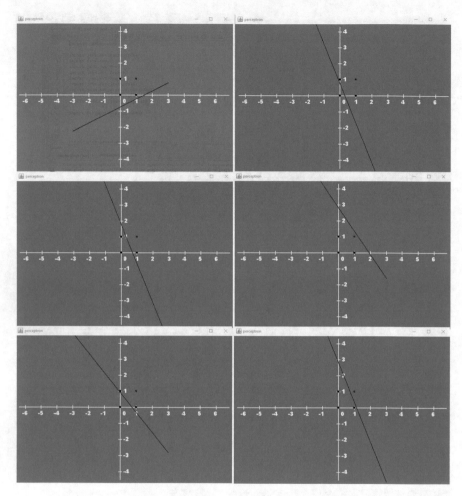

Fig. 6 Learning slop of logical AND

together to form a network. Neurons in an artificial neural network (ANN) tend to have less number of connections than biological neural networks. Each neuron in ANN receives a number of inputs from other neurons, and an activation function is applied to the weighted sum of the input and weights to obtain the output value of the neuron. These neurons are trained by using training example datasets. As a binary classifier, the perceptron groups the set of inputs into only two classes and all examples in the training dataset should belong to only these two classes. The proposed algorithm implements logical operations OR, AND, NAND, and NOR using single-layer perceptron with the motive of minimizing the execution time and making it more efficient with the help of the dynamic learning rate. This work also provides graphs and animations to make it more interactive and comprehensive so that it can be used to provide visuals of calculations and results. It also attempts to attain a

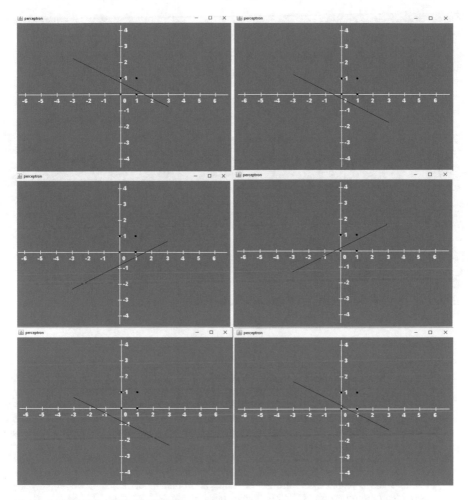

Fig. 7 Learning slop of logical NOR

learning rate which takes minimum number of iterations to train the perceptron to learn the logical operations. To deal with this, dynamic values of learning rate are taken at each iteration and then calculated the total number of epochs needed to get the perceptron trained w.r.t to each learning rate. Then, the learning rate which gives the minimum number of iterations to train the algorithm is chosen. To visualize the linearly separable classes with slope and intercepts during the training of the perceptron, graphical results are also provided for the previously mentioned logical operations. Initially, any random weights as well as the bias are selected, so an arbitrary line is drawn. After that, weight is updated according to the weight updation rule discussed in the proposed algorithm. In future, we need to improve results using multilayer feed-forward neural networks and to implement more logical operation using the proposed method as well as the multilayer perceptron.

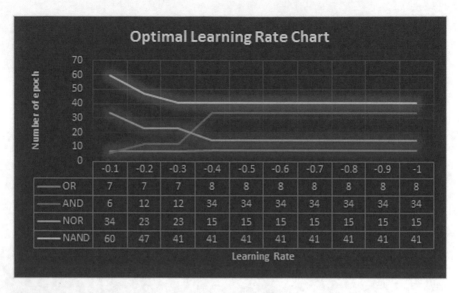

Fig. 8 Total number of epoch w.r.t different learning rates

References

1. Ahamad, Mohd Vasim, Misbah Urrahman Siddiqui, Tariq Ahmed, and Asia Mashkoor. 2017. Clustering and classification algorithms in data mining. *International Journal of Advance Research in Science and Engineering* 6 (7): 1110–1117.
2. Yanling, Zhao, Deng Bimin, and Wang Zhanrong. 2002. Analysis and study of perceptron to solve XOR problem. In *Proceedings of the 2nd International Workshop on Autonomous Decentralized System*. ISBN: 0-7803-7624-2/02.
3. Khalil Alsmadi, Mutasem, Khairuddin Bin Omar, Shahrul Azman Noah, and Ibrahim Almarashdah. 2009. Performance comparison of multi-layer perceptron (back propagation, delta rule and perceptron) algorithms in neural networks. In *IEEE International Advance Computing Conference (IACC 2009)*, Patiala, India. ISBN: 978-1T-4244-1888-6/08.
4. Melnychuk, Stepan, Mykola Kuz, and Serhiy Yakovyn. 2018. Emulation of logical functions NOT, AND, OR, and XOR with a perceptron implemented using an information entropy function. In *14th International Conference on Advanced Trends in Radioelectronics, Telecommunications and Computer Engineering (TCSET-18)*, Lviv-Slavske, Ukraine. ISBN: 978-1-5386-2556-9/18.
5. Naraei, Parisa, Abdolreza Abhari, and Alireza Sadeghian. 2016. Application of multilayer perceptron neural networks and support vector machines in classification of healthcare data. In *2016 Future Technologies Conference (FTC)*, 6–7 Dec 2016, San Francisco, USA. ISBN: 978-1-5090-4171-8/16.
6. Saji, Sumi Alice, and K. Balachandran 2015. Performance analysis of training algorithms of multilayer perceptrons in diabetes prediction. In *2015 International Conference on Advances in Computer Engineering and Applications (ICACEA) IMS Engineering College*, Ghaziabad, India. ISBN: 978-1-4673-6911-4/15.
7. Zubir, N.S.A., M.A. Abas, Nurlaila Ismail, Nor Azah M. Ali, M.H.F. Rahiman, N.K. Mun, N.T. Saiful, and M.N. Taib. 2017. Analysis of algorithms variation in multilayer perceptron neural network for agarwood oil qualities classification. In *IEEE 8th Control and System Graduate Research Colloquium*, Aug 2017, Shah Alam, Malaysia. ISBN: 978-1-5386-0380-2/17.

8. Ravichandran, K., and S. Arulchelvan. 2017. The model of multilayer perceptron analysed the crime news awareness in India. In *International Conference on Advanced Computing and Communication Systems*. ISBN: 978-1-5090-4559-4/17.
9. Mubarek, Aji Mubalaike, Esref Adali. 2017. Multilayer perceptron neural network technicque for fraud detection. In *2nd International Conference on Computer Science and Engineering* 383–387. ISBN: 978-1-5386-0930-9/17.
10. Hasan, Ali T., Hayder M.A.A. Al-Assadi, and Ahmad Azlan Mat Isa. 2011. Neural networks' based inverse kinematics solution for serial robot manipulators passing through singularities. *Artificial Neural Networks—Industrial and Control Engineering Applications*, Kenji Suzuki, IntechOpen. https://doi.org/10.5772/14977.
11. Merikle, Philip M., D. Smilek, and John D. Eastwood. 2001. Perception without awareness: perspectives from cognitive psychology. *Science Direct* 79 (1–2): 115–134.

Applicability of Financial System Using Deep Learning Techniques

Neeraj Kumar, Ritu Chauhan and Gaurav Dubey

Abstract To predict the future stock price is not a slipshod task due to the unpredictable behavior of stock market. The financial market always tends to be a great challenge for experts to predict the future stock prices. Hence, market experts use varied differential techniques such as fundamental or technical to determine the financial forecasting. In past, there exist various regression models, such as ARIMA, ARCH, and GARCH which are used for technical analysis, but they tend to be less robust for non-stationary data. Moreover, varied machine learning algorithms are applied in the past which include SVM and random forest for financial forecasting. However, results retrieved from these algorithms are still questionable with time series data. In the current study of approach, we have utilized deep learning models which have a tendency to be more focused on right features by themselves; it requires very little intervention of developer. Hence, deep learning methods can increase the ability of technical analysis for future stock prediction. In this paper, varied comparison is drawn among the two deep learning approaches such as LSTM and CNN which are applied to predict the future stock prices of Infosys company. Empirically, it has been proved that the LSTM outperforms the CNN.

Keywords Long short-term memory · Convolutional neural network · Deep learning · Neural network · Recurrent neural network · Financial data

1 Introduction

Stock market prediction is a trusted area of research in the field of data analytics where the challenge is to variable patterns for future knowledge discovery. In past, several studies have suggested that stock market fluctuation is very random in nature, so retrieval of patterns is not easy to predict for any financial instrument. Moreover, there are various internal and external factors which affect the financial assets of a

N. Kumar · R. Chauhan
Amity University Uttar Pradesh, Noida, India
e-mail: nkumar8@amity.edu

G. Dubey (✉)
ABES Engineering College, Ghaziabad, India

© Springer Nature Singapore Pte Ltd. 2020
Y.-C. Hu et al. (eds.), *Ambient Communications and Computer Systems*, Advances in Intelligent Systems and Computing 1097,
https://doi.org/10.1007/978-981-15-1518-7_11

company. Some internal factors may include plans and policies, human resource, internal technologies, organizational structure, and physical assets. As well as external factors can be customers, economics, marketing and media, political environment, competitors, etc. Inclusion of such forecasting features in the stock price of any company is ad jointly emanates beneath the fundamental analysis. Hence, fundamental analysis of any data determines varied method of forecasting for the future price movements in correspondence with financial instrument based on these internal and external factors. Likewise, technical analysis tends to be a similar method which can be exploited for the future price movement of any stock market. Technical analysis depicts the price movement and predicts the future price on the basis of past data available. Unlike fundamental analysis, technical analysis does not include any other factor for future forecasting except the past data of price movement.

There are various predictive models which are existing from past decades and utilizing variable technology such as ARIMA (Auto Regressive Integrated Moving Average), ARCH (Autoregressive Conditional Heteroscedasticity), and GARCH (Generalized Autoregressive Conditional Heteroscedasticity). The proportion of effort has been done over these models in past two decades for future forecasting. However, ARIMA has proved itself one of the best-known models for time series analysis. There are some other machine learning techniques also like support vector machine [1], neuro-fuzzy [2], random forest [3]-based system, but due of nonlinear behavior of stock market data, these methods are rarely utilized due to applicability in depicting the predicting results.

Recently, a wide range of research is focused on utilizing deep learning approach for image analysis, time series analysis, and natural language processing. The concept of deep learning method is based on the multi-layer neural network, the multi-layer arrangement one is input layer, and other is output layer, and between these two layers, there are various hidden layers, and the number of hidden layers is depending upon the nature of the problem. Any neural network where the number of hidden layers is more than two are known as deep neural network (deep learning). Output of one hidden layer is given as input to other hidden layer for more refinement and accuracy number of hidden layers can be increased. The previous studies show the capabilities of deep learning algorithms to handle the nonlinear stock market data in a much efficient way which cannot be done with the help of traditional financial forecasting methods.

This paper discusses a long short-term memory (LSTM) model and convolutional neural network (CNN) model. LSTM is a type of recurrent neural network, and it is one of the well-known techniques for sequence prediction. CNN is also a deep learning method, and it is generally used for image classification. CNN proves itself one of the best methods of feature extraction from image dataset. In this paper, we check the predictive capabilities of these two models with time series stock market data and compare the results. The rest of the paper is discussed as background literature work is laid in Sect. 2, prediction related to stock market is discussed in Sects. 3 and 4 has overall methodology, and Sect. 5 has predicted results, and in last section, conclusion is discussed.

2 Related Work

Forecasting in financial research is very much anticipated from past decades, so lot of work has been proposed by the researchers in this area. A learning-based study was designed to enhance the performance of reinforcement-learning-based system, which incorporates multiple Q-learning agents to effectively divide and conquer the stock trading problem [4]. A similar study was proposed by a stock movement prediction approach based on a newly defined instantaneous frequency and signal decomposition; they used Hang Seng Index of Honk Kong stock exchange [5]. ARIMA model of time series forecasting for prediction of stock market movement is a univariate model [6]; one of the shortcoming of this model is that it cannot work with non-stationary data. Some variants of neural networks like TDNN, PNN, and RNN were compared and evaluated them against a conventional method of prediction and also, a predictability analysis of the stock data is presented and related to the neural network results [7]. A framework is proposed to predict the directions of stock prices by using financial news and sentiment analysis; this paper includes the proposal of a novel two-stream gated recurrent unit network [8]. An intelligent time series prediction system was proposed that uses sliding window metaheuristic optimization for the purpose of predicting the stock prices [9]. Few artificial intelligence algorithms (BPNN, extreme learning machines, and radial basis function neural network) were compared to check their predictive performance with high-frequency stock data, and they conclude that the deep learning methods to predict stock index futures outperform BPNN [10]. The performance of is also verified by the researchers. Support vector machine (SVM) is also verified in financial forecasting, and after implementing the SVM, they compare the results with multi-layer BPNN and find that SVM outperforms the BPNN [11]. By implementing LSTM in the experiments and comparing the results with RTRL, BPTT, Elman Nets, and Neural Sequence Chunking, LSTM leads to many more successful run and found that LSTM can solve complex long time lag task [12]. A tensor-based information framework is also used to predict stock movement; for this purpose, they have used a global dimensionality-reduction algorithm to capture the links among various information sources in a tensor [13]. In another study, they utilize a weighting scheme to combine qualitative and quantitative features of financial reports together to predict the short-term stock price movement; the empirical results show that the proposed method outperforms the SVM, Naïve Bayes, and PFHC [14]. Another research is related to the sentiment analysis of financial news articles; they used different textual representation such as bag of words, noun phrases, and named entities, and SVM is used as a classification tool [15]. This research deals with the application of SVM with financial forecasting; they have implemented the SVM over financial stock data and compared the result with BP and RBF neural network and found that SVM outperforms BPNN and RBF. A new method was introduced to simplify financial time series data by sequence reconstruction; they also utilized convolutional neural network to capture spatial structure of time series. Author claimed 4–7% of improved accuracy compared with traditional signal process methods and frequency trading patterns modeling approach [16].

3   Stock Market Prediction

Stock market plays a very important role as an economic indicator of a country; hence, predicted results can overall change the global exchange. We can also say that economic condition of a country can be easily identified by the stock prices of some major companies of that country. If the stock price of these companies goes up, it means the economic condition of that country is good or vice versa. A share or equity market is a place known for selling, issuing, and buying of stocks. Issuing shares means a company wants to sell some portion of its market value to others; generally, company do it to raise money to expand the business. Customers can also buy the new stocks and sell the existing stock with him in the stock market; India is one of the major economies in the world.

In India, there are two stock exchanges—Bombay Stock Exchange (BSE) and National Stock Exchange (NSE) [6]. BSE is oldest stock exchange, and it started its operation from 1875, whereas the NSE comes into existence much later than BSE, in 1994 NSE [6] started its services in India. All the major companies of India are listed in these to stock exchanges. Around 5000 companies are listed under BSE and 2000 in NSE. As NSE is new, the numbers of companies listed in it are less as compared to BSE. There are two major indexes in the Indian stock market—Nifty and Sensex. These indexes are the indicator of the stock market movement. Sensex is the indexing used by the BSE, and Sensex index is calculated on the basis of top thirty companies listed under BSE. Nifty is the indexing of NSE and Nifty index is calculated on the basis of top fifty listed in NSE. Up and down movements of Nifty and Sensex indicate that the most of the shares in NSE and BSE are going up or down, respectively. Prediction over stock market means we can predict the future price of a stock in any stock exchange.

4   Deep Neural Network

The simplest deep neural network is combination of back-propagation with gradient descent with more than two hidden layers. Fundamentally, deep neural network consists of one input layer and one output layer, and between these two layers, there maybe multiple or more than two hidden layers. The output layer is the deciding layer; results are taken from the output layer which is computed by the all hidden layers; in some cases, a hidden layer can also work as an output layer.

Figure 1 shows a typical deep neural network with one input layer which consists of three neurons and three hidden layers with four neurons each and one output layer with two neurons. Every neuron of each hidden layer is connected with every neuron of next hidden layer. Output of one hidden layer is given as input to next hidden layer; as the number of hidden layers increased, we move toward the more accurate results.

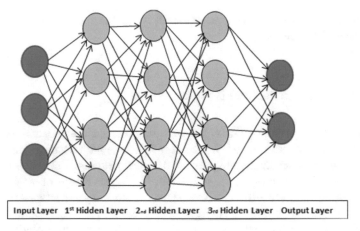

Input Layer   1ˢᵗ Hidden Layer   2ₙₐ Hidden Layer   3ᵣₐ Hidden Layer   Output Layer

Fig. 1 A typical deep neural network

5   LSTM (Long Short-Term Memory)

LSTM is a type of recurrent neural network (RNN), they are the state-of-the-art algorithm for sequential data [12], sequential data means that the data generate sequences, such as music generation, text sequence, etc. RNN is generally used in Apple Siri, Google Assistant, and Amazon Alexa. The structure of recurrent neural network looks like this as shown in Fig. 2.

RNN generally works with respect to time stamp; it works on the policy that for each and every input at a particular time there is an output, as given in Fig. 2 x_{t-1}, x_t and x_{t+1} are input at different timestamps and y_{t-1}, y_t and y_{t+1} are there output for the given inputs, respectively. The output of one timestamp is reefed with the input

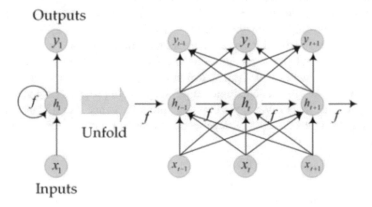

Fig. 2 Structure of recurrent neural network

of next timestamp. That is the reason this RNN is capable to solve complex problems like sequence of datasets.

6 Convolutional Neural Network (CNN)

A CNN [2] is most popularly known for analyzing images. But they can also be used for other data analytics problems. We can see CNN as an artificial neural network that has some type of specialization to pick out or detect patterns. This pattern detection makes CNN better for image analysis. A CNN is a special type of artificial neural network which makes it different from standard multi-layer perceptron. In CNN, there are some hidden layers, and these hidden layers are called convolutional layers; along with the convolutional layers, the CNN may have non-convolutional layers, but the base of CNN is convolutional layers. What these convolutional layers do? It receives input trends form the input someway, and output trend forms to the next layer. For convolutional layer, this trend formulation is a convolutional operation. Earlier we said convolutional layers are used to identify patterns, but these are the filters with each convolutional layer to identify the patterns, and the number of filters in each convolutional layer should be decided in advance. If we talk about the pattern in an image, there are various types of patterns such as shapes, edges, textures, and corners. So there are different types of filters for different patterns. Some may detect edges, some may detect corners, and so on. When we go deeper in the convolutional layers, more sophisticated filters are used to identify objects or parts of objects such as human being in an image and eyes or face of that human.

CNN can be applied on time series forecasting also; in this paper, we are using CNN for stock market data which is also time series data. As we know CNN works well with images, and images are matrix of pixels in the same way we have to represent the time series data in a matrix form. Let us see how CNN converts that time series data into a matrix form. CNN divides the complete input sequence in the form of inputs and outputs; suppose we have an input sequence $x_1, x_2, x_3, x_4, \ldots, x_n$; then, CNN firstly divides this input sequence into multiple inputs and outputs. Let us say x_1 is input and then x_2 is output for x_1, if x_2 is input, then x_3 is output for x_2, and so on. We can increase the split size of input sequence also where we can choose more than one input items and single output (e.g., sequence x_1, x_2, x_3 is input, and x_4 is output). So, CNN divides the complete input sequence like this into a matrix of inputs and outputs. These matrices work as trained examples for preparation of training data.

7 Method and Result

The analytical results retrieved were focused on Infosys stock price market data for last five years from May 2014 to April 2019. The deep learning techniques are applied utilizing univariate LSTM model and CNN model for future predictable forecasting of financial data. The error percentage of each model is compared to determine the applicability of each model in correspondence with financial data.

7.1 Implementation of LSTM

We are using univariate LSTM model; it is a type of RNN. LSTM works on the technique where the past observations are divided as input and output observations. These observations are divided to create the examples for training dataset. We can divide the complete dataset into input/output sequence; as the size of dataset is large, we use sixty time steps as input and one time step as output. We have used a split sequence function to split the data into input and output sequences. There are four hidden layers, one input and one output layer. Each hidden layer contains fifty neurons. Model is fit using Adam optimizer. Dataset is in the CSV format, and the complete dataset is divided into two parts—training and test dataset. The size of training dataset is 67% of the total data, and rest of the 33% data is used for testing purpose. The model is trained over the 67% data, and rest of the 33% testing data is to validate the model. We have implemented the LSTM and predict the future values of stock price on the known dataset; for this, a sample data is taken for the month of April 2019 from April 01, 2019 to April 30, 2019 and graphically represent the predicted values and the actual values of stock price for sample data. We have also represented the matrix of predicted values and actual values and the percentage of deviation of predicted values from actual values.

In Fig. 3, we have shown the graph of Infosys stock price for last five years from May 2014 to April 2019. On x-axis, there are number of observation and on y-axis

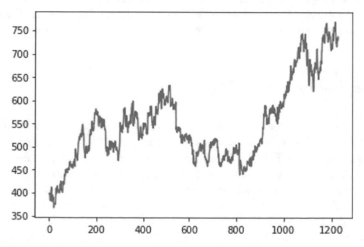

Fig. 3 Price movement of Infosys technologies for last five years

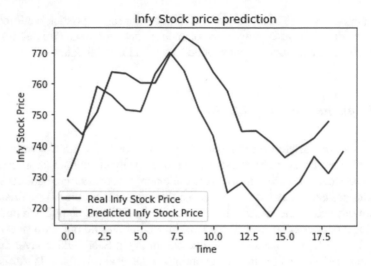

Fig. 4 Comparison graph between actual and predicted prices for LSTM

the opening price; for sake of easiness, we are using number instead of years and months on *x*-axis. As we can see in the graph, there is no clear-cut trend; initially, the graph shows upward movement, then down, then again up, extremely down in the middle, and then up till the end.

Figure 4 shows the graph of sample data of April 2019 used for prediction. It shows the difference between actual stock price and the stock price predicted by the LSTM model. The red line in graph is for actual price, and blue line is for predicted price. We can see the small difference between actual and predicted price.

Same thing we have represented in the tabular form also in Table 1. Table shows the difference and error percentage between actual and predicted stock. If we see the error percentage, it is very marginal. The mean value of error percentage is only 1.65% which clearly shows the accuracy of prediction in the LSTM model.

7.2 Implementation of CNN

Generally, CNNs are used with image data, but it can also be utilized with univariate time series data. The model which is designed using the CNN takes the past observations sequence and predicts the future values in sequence. Just like we do in LSTM, CNN also divides past observations in input and output observations. Again, the same we divide the data into input/output observations to create examples for training data. These multiple input/output sequences are called samples. We use the split sequence function to divide data in input and output form. In this, one-dimensional CNN model has a hidden convolution layer that works for one-dimensional sequence; there may be a second convolutional layer which can add if required, but in this case, we are

Table 1 Actual and predicted stock price, difference, and error percentage

Predicted price	Actual price	Difference	Error %
748.18	730	18.18	2.490410959
743.45	742	1.45	0.19541779
750.68	759	8.32	1.096179183
763.67	756	7.67	1.014550265
763.15	751.400024	11.749976	1.563744427
760.14	750.900024	9.239976	1.230520136
760.12	763.049988	2.929988	0.383983755
769.08	770	0.92	0.119480519
775.2	764	11.2	1.465968586
771.96	751.75	20.21	2.688393748
763.77	743.099976	20.670024	2.781593953
757.49	724.700012	32.789988	4.524629151
744.51	727.900024	16.609976	2.281903483
744.65	722.5	22.15	3.065743945
741.1	717.049988	24.050012	3.354021672
736.11	723.900024	12.209976	1.686693686
739.34	728.200012	11.139988	1.529797833
742.37	736.450012	5.919988	0.803854695
747.71	731	16.71	2.285909713

using only one convolutional layer. After convolutional layer, there is pooling layer that is used to refine the output of convolutional layer; after pooling layer, there is a flatten layer, which is used in reduction of features to a single one-dimensional vector. After the flatten layer, we add fully connected dense layer; this layer is used to translate the features extracted by the convolutional layer. The model expects the same number of time steps in each sample and same number of features. As we are working with one-dimensional model, the number of features is one in each sample. In this paper, we design a convolution layer with 64 filters and kernel size 1. An output layer is used which predicts the 20 values. The model is fit using Adam optimizer. In this research, we used multi-step CNN; it can take multiple time steps as input and predict multiple future values. We have used 20 values as input and 20 values as output. The model used 67% data for training set and rest 33% to validate the model.

In Fig. 5, we have shown the comparison graph of one month (May 2019) sample data we have taken for prediction. It shows the difference between actual price and the price predicted by CNN. Table 2 shows the actual and predicted price and difference between them and also the error percentage in each price prediction. The total error percentage for CNN is 2.82%

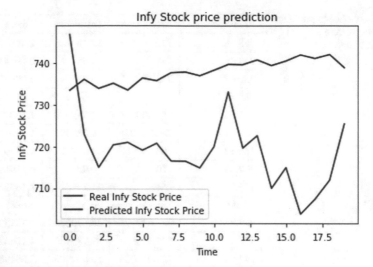

Fig. 5 Comparison graph between actual and predicted prices for CNN

Table 2 Actual and predicted stock price, difference, and error percentage

Predicted price	Actual price	Difference	Error %
733.51	747	13.49	1.8058902
736.12	722.900024	13.219976	1.828742
733.97	715	18.97	2.6531469
735.18	720.400024	14.779976	2.0516346
733.55	721	12.55	1.740638
736.46	719.150024	17.309976	2.4070049
735.79	720.799988	14.990012	2.0796354
737.74	716.549988	21.190012	2.9572273
737.9	716.5	21.4	2.9867411
737.02	714.900024	22.119976	3.0941356
738.36	720	18.36	2.55
739.77	733.150024	6.619976	0.9029497
739.7	719.700012	19.999988	2.778934
740.85	722.650024	18.199976	2.5185049
739.49	710	29.49	4.1535211
740.59	715	25.59	3.579021
742.08	703.799988	38.280012	5.439047
741.21	707.400024	33.809976	4.7794706
742.18	712	30.18	4.238764
738.99	725.5	13.49	1.8594073

8 Conclusion

Predicting the stock market price is complicated task due to high frequency of data and various external factors which affect the stock prices. For this paper, we have taken 5 years of data from May 2014 to April 2019 for training purpose and designed two different models using RNN and CNN. On the basis of designed model, we predict the future stock price; for RNN, we predict the stock price for April 2019, and for CNN, we do it for May 2019. As the results shown in the previous section, the error percentage of RNN is about 1.65%, and on the other hand, error percentage for CNN is 2.82%, which is quite high in comparison with RNN. If you see the graph in Fig. 5, all predictions are in a particular range irrespective of price movement, but if we see the RNN, it does not follow any particular pattern and the predicted value graph is almost parallel to actual value graph. On the basis of the results' extract, we can conclude this paper that RNN works better then CNN for time series stock market data.

References

1. Huang, Wei, Yoshiteru Nakamori, and Shou-Yang Wang. 2005. Forecasting stock market movement direction with support vector machine. *Computers and Operations Research* 32 (10): 2513–2522.
2. Lin, Sangdi, and George C. Runger. 2018. GCRNN: group-constrained convolutional recurrent neural network. *IEEE Transactions on Neural Networks and Learning Systems* 29 (10): 4709–4718.
3. Lauretto, Marcelo S., B.C. Silva, and P.M. Andrade. 2013. Evaluation of a supervised learning approach for stock market operations. arXiv: 1301.4944 [Stat.ML].
4. Lee, Jae Won, Jonghun Park, O. Jangmin, Jongwoo Lee, and Euyseok Hong. 2007. A multiagent approach to Q-learning for daily stock trading. *IEEE Transactions on Systems, MAN and Cybernetics—Part A: Systems and Humans* 37 (6): 864–877.
5. Zhang, Liming, Na Liu, and Pengyi Yu. 2012. A novel instantaneous frequency algorithm and its application in stock index movement prediction. *IEEE Journal of Selected Topics in Signal Processing* 6 (4): 311–318.
6. Idrees, Sheikh Mohammad, M. Afshar Alam, and Parul Agarwal. 2019. A prediction approach for stock market volatility based on time series data. *IEEE Access* 7: 17287–17298.
7. Saad, Emad W., Danil V. Prokhorov, and Donald C. Wunsch. 1998. Comparative study of stock trend prediction using time delay, recurrent and probabilistic neural networks. *IEEE Transactions on Neural Networks* 9 (6): 1456–1470.
8. Minh, Dang Lien, Abolghasem Sadeghi-Niaraki, Huynh Duc Huy, Kyungbok Min, and Hyeonjoon Moon. 2018. Deep learning approach for short-term stock trends prediction based on two-stream gated recurrent unit network. *IEEE Access* 6: 55392–55404.
9. Chou, Jui-Sheng, and Thi-Kha Nguyen. 2018. Forward forecast of stock price using sliding-window metaheuristic-optimized machine-learning regression. *IEEE Transactions on Industrial Informatics* 14 (7): 3132–3142.
10. Chen, Lin, Zhilin Qiao, Minggang Wang, Chao Wang, Ruijin Du, and Harry Eugene Stanley. 2018. Which artificial intelligence algorithm better predicts the chinese stock market? *IEEE Access* 6: 48625–48633.
11. Tay, Francis E.H., and Lijuan Cao. 2001. Application of support vector machines in financial time series forecasting. *Omega* 29: 309–317.

12. Hochreiter, Sepp, and Jurgen Schmidhuber. 1997. Long short term memory. *Neural Computation* 9 (8): 1735–1780.
13. Li, Qing, Yuanzhu Chen, Ping Li, and Hsinchun Chen. 2016. A tensor-based information framework for predicting the stock market. *ACM Transactions on Information Systems* 34 (2): 11.
14. Lin, Ming-Chih, Anthony J.T. Lee, Rung-Tai Kao, and Kuo-Tay Chen. 2011. Stock price movement prediction using representative prototypes of financial reports. *ACM Transactions on Management Information Systems* 2 (3): 19.
15. Schumaker, Robert P., and Hsinchun Chen. 2009. Textual analysis of stock market prediction using breaking financial news: the AZFin text system. *ACM Transactions on Information Systems* 27 (2): 12.
16. Cao, L.J., and Francis E.H. Tay. 2003. Support vector machine with adaptive parameters in financial time series forecasting. *IEEE Transactions on Neural Networks* 14 (6): 1506–1518.

Intelligent Hardware and Software Design

Novel Reversible ALU Architecture Using DSG Gate

Shaveta Thakral and Dipali Bansal

Abstract In the era of giant-scale integration and technology advancement, it will not be economically feasible to cope up as power dissipation is a fundamental threat to the technology. The impending end of Moore's law is motivating to seek for better technology. Reversible logic provides dramatic improvements in energy efficiency and proves to be a promising solution for the future. The most promising application of reversible logic will be for smart and massive quantum computing applications in near future. Implementing reversible logic is an intractable research problem, but many researchers have chosen this path to save energy and decided to blaze into this new trail of technology. The arithmetic and logic unit (ALU) is command center of all computing environment. The most important aspects of ALU like number of operations, ancillary inputs, garbage outputs and quantum cost are mostly discussed to compare whether any reversible logic-based ALU is more efficient over other. This paper presents a novel reversible ALU where these aspects have been examined, and optimization metrics comparison shows the significant improvements over existing designs. The proposed ALU design is coded in Verilog HDL, synthesized and simulated using electronic design automation (EDA) tool—Xilinx ISE design suit 14.2. RCViewer + tool has been used to validate quantum cost of proposed design.

Keywords Power dissipation · ALU · Reversible logic · Optimization · Ancillary

S. Thakral (✉) · D. Bansal
Department of Electronics & Communication (FET), Manav Rachna International
Institute of Research & Studies, Faridabad, India
e-mail: shaveta.fet@mriu.edu.in

D. Bansal
e-mail: dipali.fet@mriu.edu.in

© Springer Nature Singapore Pte Ltd. 2020
Y.-C. Hu et al. (eds.), *Ambient Communications and Computer Systems*, Advances in
Intelligent Systems and Computing 1097,
https://doi.org/10.1007/978-981-15-1518-7_12

149

1 Introduction

The impending end of Moore's law due to miniaturization touching atomic levels is a big cause of switching from conventional logic-based computation to reversible logic-based computation. Landauer's principle [1] states that conventional logic leads to loss of information and loss of information in turn makes system inefficient in terms of power/energy loss. Dr. Charles Bennett [2] in 1973 gave idea about reversible computing to save energy, and nowadays, researchers have declared that this is the only way by which conventional computing can be improved in terms of cost and energy efficiency. Many reversible logic gates have been investigated by popular scientists Feynman, Fredkin and Toffoli. Feynman gate, Toffoli gate and Fredkin gates have significant contribution in history of reversible computing. Reversible logic-based arithmetic and logic unit is demanded in almost all types of computing environment, and many researchers have given their significant contribution in this field. Syamala and Tilak [3] have proposed two ALU architectures, but both circuits have low functionality and high quantum cost. Morrison and Ranganathan [4] have proposed reversible logic-based ALU with quantum cost 35 that can perform nine arithmetic and logical operations using eight gates, and their circuit took two constant input lines and produced six garbage output lines. Singh et al. [5] have improved ALU architecture with less ancillary inputs, garbage outputs and quantum cost. Moallem et al. [6] designed ALU with improvement in functionality as well as reduction in quantum cost. Sen et al. [7] designed ALU with more number of operations, but circuit suffers with very high quantum cost. Other existing ALU designs [8–14] have trade-off between functionality, quantum cost, ancillary inputs and garbage outputs. Indeed the improvement scope in this paper arises from designing a novel reversible ALU architecture with reduced quantum cost. This paper aims in designing a novel reversible ALU architecture using proposed DSG gate. A brief insight into all existing reversible logic-based gates used for proposed architecture is given in Table 1. A brief information about proposed DSG gate is given in Sect. 2. Methodology of proposed ALU architecture is given in Sect. 3. Proposed ALU architecture and its explanation are covered in Sect. 4. Comparison and results are covered in Sect. 5 followed by conclusion in last section.

2 Proposed DSG Gate

Proposed novel reversible DSG gate is 4*4 gate which can function as universal reversible logic-based gate. The input vector of DSG gate is taken as Ai, Bi, Ci, Di, and its output vector is taken as Po, Qo, Ro, So. The DSG gate under operation is shown in Fig. 1, and its optimized quantum implementation is shown in Fig. 2.

Table 1 Existing reversible logic-based gates used in proposed ALU architecture

Reversible logic gate	Expression	Quantum cost	NCV equivalence
Peres/NTG	$P = A$ $Q = A \oplus B$ $R = AB \oplus C$	4	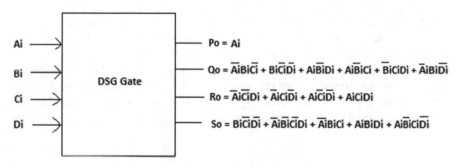
Feynman	$P = A$ $Q = A \oplus B$	1	
RMUX1 $=$	$P = A$ $Q = \overline{A}B + AC$ $R = \overline{A}C + A\overline{B}$	4	

Po = Ai

Qo = $\overline{Ai}Bi\overline{Ci} + Bi\overline{Ci}Di + Ai\overline{Bi}Di + Ai\overline{Bi}Ci + \overline{Bi}CiDi + \overline{Ai}Bi\overline{Di}$

Ro = $\overline{Ai}Ci\overline{Di} + \overline{Ai}\overline{Ci}Di + Ai\overline{Ci}Di + AiCi\overline{Di}$

So = $Bi\overline{Ci}\overline{Di} + \overline{Ai}\overline{Bi}\overline{Ci}Di + \overline{Ai}Bi\overline{Ci} + Ai\overline{Bi}Di + Ai\overline{Bi}\overline{Ci}\overline{Di}$

Fig. 1 Proposed novel reversible DSG gate

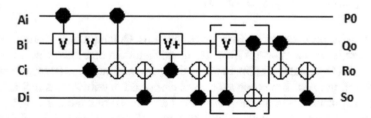

Fig. 2 Optimized quantum implementation of DSG gate

3 Methodology

ALU architecture is proposed based on eight reversible logic gates including two Peres gates, two Feynman gates, one DSG gate and three RMUX1 gates. The quantum cost of proposed DSG gate is nine. The quantum cost of Peres, RMUX1 gate is 4, and Feynman gate is 1. Proposed ALU architecture is shown in Fig. 3. Peres gate 1 is passing 0 or 1 or B or B′ as per desired logic depending upon select lines S3 and S2. Peres gate 2 is passing 0 or 1 or Cin/Bin or Cin′/Bin′ as per desired logic depending upon select lines S1 and S0. Feynman gate 1 is passing B or B′ as per desired logic depending upon select line S4. Feynman gate 2 is passing B XOR S3 or NOT (B XOR S3) as per desired logic depending upon select lines S3 and S5. T2 is passing B XOR S3 signal, and T11 is passing S5. If S3 is put zero, then B′ will be passed, and if S3 is put 1, then B is passed. DSG gate is operated under different combinations of inputs, and it is able to perform twelve logical operations like passing Ai, AiBi′, NOR, XNOR, OR, Ai′Bi, AND, XOR, NAND, Ai + Bi′, Ai′BiCin′ + AB′Cin, A′B′Cin + ABCin′. It is also able to perform four arithmetic operations like Ai plus Bi plus Cin (full adder operation), Ai plus Bi (half-adder operation), Ai minus Bi minus Bin (full subtractor operation) and Ai minus Bi (half-subtractor operation) where Ai and Bi are operand on which various arithmetic and logical operations can be performed, and Cin and Bin are initial carry and initial borrow, respectively. Cout is output carry obtained after addition, and Bout is output borrow obtained after subtraction. Notation '+' is used for logical OR operation, 'plus' is used for arithmetic addition, and 'minus' is used for arithmetic subtraction. RMUX1 gate 1, RMUX1 gate 2 along with RMUX1 gate 3 make 4:1 multiplexer and passing desired function on Func output line depending upon select lines S6 and S5.

4 Proposed ALU Architecture

Proposed ALU architecture performs 20 operations. Quantum cost of proposed ALU architecture is 31. Complete ALU is designed using ten input lines including three input bits A (Operand1), B (Operand 2), Cin (Carry input)/Bin (Borrow input) and seven selection lines. The proposed circuit produces ten output lines including nine garbage output lines and one desired Func line. Circuit took no constant input and produces nine garbage outputs to maintain reversibility. Proposed novel reversible ALU architecture using DSG gate is shown in Fig. 3. Simulation waveform of proposed ALU architecture is shown in Fig. 4. It shows the case that if inputs A and B are equal, then Func output will be equal to one. To verify this condition, S5 and S4 are put to 1, and rest selections lines are zero. The functions performed by proposed ALU architecture are shown in Table 2.

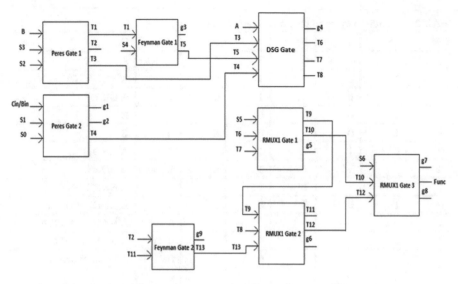

Fig. 3 Proposed ALU architecture

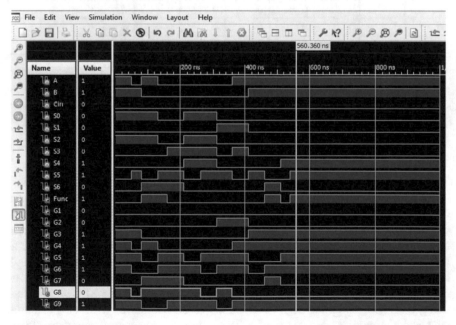

Fig. 4 Simulation waveform

Table 2 Functions performed by proposed ALU architecture

S6	S5	S4	S3	S2	S1	S0	Func
0	0	0	0	1	0	1	A NOR B
0	1	0	0	1	0	1	A XNOR B
1	0	0	0	1	0	1	A OR B
1	1	0	0	0	0	0	B'
1	1	0	1	0	0	0	B
0	0	1	1	1	0	1	A AND B
0	1	1	1	1	0	1	A XOR B
1	0	1	1	1	0	1	A NAND B
0	1	0	0	0	1	0	A plus B plus Cin
1	0	0	0	0	1	0	$A'B'Cin + AB Cin'$
0	0	0	1	0	1	0	$A'BCin' + AB'Cin$
0	1	0	1	0	1	0	A minus B minus Bin
0	1	0	0	0	0	0	A plus B
1	0	0	0	0	0	0	A multiply B
0	0	1	0	0	0	0	AB'
0	1	1	0	0	0	0	A = B
1	0	1	0	0	0	0	A > B
1	0	1	0	0	0	1	A'B
0	0	1	0	0	0	1	A + B'
0	1	1	0	0	0	1	A minus B

5 Comparison and Results

The proposed novel reversible ALU architecture using DSG gate is compared with other designs available in the literature, and some parameters like number of gates used, number of operations performed, quantum cost, ancillary inputs used and produced garbage outputs are analyzed. Design IV is found to be optimum balance in terms of all aspects of design and used to perform 12 arithmetic and logical operations. Design V is the best in terms of high functionality, yet it suffers from too high quantum cost and took many ancillary inputs and produces more garbage outputs as compared to other existing designs. The proposed novel reversible ALU architecture using DSG gate holds the high functionality of 20 operations with least quantum cost, and it is ancillary free circuit also. The optimization aspects comparison of proposed ALU design with other designs available in the literature is given in Table 3. A comparison of optimization aspects is presented in tabular form proves that the proposed ALU design is efficient in terms of all optimization aspects as compared to other existing designs as shown in Fig. 5.

Table 3 Optimization aspects comparison of ALU designs

ALU design	No. of gates	Quantum cost	Number of operations	Garbage outputs	Ancillary inputs
Design I [4]	8	35	9	6	2
Design II(a) [3]	9	41	5	6	1
Design II(b) [3]	7	34	4	9	5
Design III [5]	8	24	8	3	2
Design IV [6]	6	25	12	5	4
Design V [7]	14	135	20	15	10
Proposed ALU architecture	8	31	20	9	0

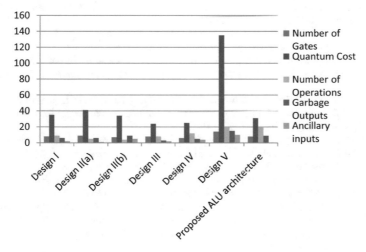

Fig. 5 Optimization aspects comparison chart

6 Conclusion

Proposed ALU architecture performs 20 arithmetic and logical operations. Quantum cost of proposed circuit is 31. Circuit took no ancillary inputs and produces nine garbage outputs to maintain reversibility. Proposed ALU is found to be better than existing designs I to V in terms of number of operations. Existing design VI performs 20 operations, but quantum cost is too high. Proposed ALU architecture is found to be optimum in terms of number of gates, quantum cost, ancillary inputs and garbage outputs as well as functionality. The future scope of this research work is to design ancillary and garbage-free ALU with enhanced functionality and incorporate fault tolerance for smart and massive computing applications.

References

1. Landauer, R. 1961. Irreversibility and heat generation in the computing process. *IBM Journal of Research and Development* 5: 183–191.
2. Bennett, C. 1973. Logical reversibility of computation. *IBM Journal of Research and Development* 17: 525–532.
3. Syamala, Y., and A. Tilak. 2011. Reversible arithmetic logic unit. In: *3rd international conference on electronics computer technology (ICECT)*, 207–211, IEEE.
4. Morrison, M., M. Lewandowski, R. Meana, and N. Ranganathan. 2011. Design of a novel reversible ALU using an enhanced carry lookahead adder. *2011 11th IEEE international conference on nanotechnology*, 1436–1440. Portland, Oregon, USA: IEEE.
5. Singh, R., S. Upadhyay, K. Jagannath, and S. Hariprasad. 2014. Efficient design of arithmetic logic unit using reversible logic gates. *International Journal of Advanced Research in Computer Engineering & Technology (IJARCET)* 3.
6. Moallem, P., M. Ehsanpour, A. Bolhasani, and M. Montazeri. 2014. Optimized reversible arithmetic logic units. *Journal of Electronics (China)* 31: 394–405.
7. Sen, B., M. Dutta, M. Goswami, and B. Sikdar. 2014. Modular design of testable reversible ALU by QCA multiplexer with increase in programmability. *Microelectronics Journal* 45: 1522–1532.
8. Guan, Z., W. Li, W. Ding, Y. Hang, and L. Ni. 2011. An arithmetic logic unit design based on reversible logic gates. In: IEEE Pacific rim conference on communications, computers and signal processing (PacRim), 925–931, IEEE.
9. Gupta, A., U. Malviya, and V. Kapse. 2012. Design of speed, energy and power efficient reversible logic based vedic ALU for digital processors. In *(NUiCONE)*, 1–6, IEEE.
10. Gopal, L., N. Syahira, M. Mahayadin, A. Chowdhury, A. Gopalai, and A. Singh. 2014. Design and synthesis of reversible arithmetic and logic unit (ALU). In *International conference on computer, communications, and control technology (I4CT)*, 289–293, IEEE.
11. Thakral, S., and D. Bansal. 2016. Fault tolerant ALU using parity preserving reversible logic gates. *International Journal of Modern Education and Computer Science* 8: 51–58.
12. Sasamal, T., A. Singh, and A. Mohan. 2016. Efficient design of reversible ALU in quantum-dot cellular automata. *Optik* 127: 6172–6182.
13. Shukla, A., and M. Saxena. 2016. Efficient reversible ALU based on logic gate structure. *International Journal of Computer Applications* 150 (2): 32–36.
14. Thakral, S., D. Bansal, and S. Chakarvarti. 2016. Implementation and analysis of reversible logic based arithmetic logic unit. *TELKOMNIKA (Telecommunication Computing Electronics and Control)* 14: 1292–1298.

An Appraisal on Microstrip Array Antenna for Various Wireless Communications and Their Applications

Gurudev Zalki, Mohammed Bakhar, Ayesha Salma and Umme Hani

Abstract This paper represents a unique study of different microstrip array antenna designs for various wireless communications and their applications. The objective is to achieve better gain, return loss, bandwidth and other parameters of antenna, in order to use them in wireless communication applications. So compared to single microstrip antenna, we prefer to go for the array designs which help to improve the gain, return loss, bandwidth and other parameters that are important for various wireless communications. The design techniques discussed and make antenna array elements to operate at various wireless communication bands such as L, S, C, X and Ka, which include various wireless communication applications. Nowadays, miniaturization is so important for microstrip antenna because of their small volume, low cost and low profile become very attractive. This paper will make readers easy to understand different kinds of array antenna used in wireless communication applications.

Keywords Microstrip antenna · Array antenna · Feed types · Parameters · Wireless communication

G. Zalki (✉)
Department of E&CE, Sharnbasva University, Kalaburagi, India

M. Bakhar
Department of E&CE, GNDCE, Bidar, India

A. Salma · U. Hani
Department of Digital Electronics, Sharnbasva University, Kalaburagi, India

© Springer Nature Singapore Pte Ltd. 2020
Y.-C. Hu et al. (eds.), *Ambient Communications and Computer Systems*, Advances in Intelligent Systems and Computing 1097,
https://doi.org/10.1007/978-981-15-1518-7_13

1 Introduction

Antenna is an interconnection between guided wave and freespace. It consists of elevated conductors which resemble the transmitter or receiver [1]. Antenna exhibits the property of reciprocity, whether it is transmitting or receiving antenna that possesses the same characteristics. Regardless of the nature of the antenna, they possess some basic properties like directive gain, power gain, beamwidth, bandwidth, radiation resistance, polarization and radiation pattern. Antenna application includes fields such as telecommunication, RADAR and biomedical. In order to enhance the gain, array antenna can be used in the system [2]. The array antenna takes the combined or processed signals to obtain the improved performance over a single antenna. In this paper, complete study of microstrip array antenna (MAA) explains the ways to develop the various characteristics of antenna and their application in the field of wireless communication is presented. The array structure is the simple form which is available. Antenna component includes one PCB with radio frequency connector along with the load. Microstrip antenna constructed with low-cost, closed-controlled substrate material. Microstrip antennas flexible in nature, and they can be used for applications required in obtaining pencil beams, fan beams and omnidirectional coverage [3].

2 Structure of Microstrip Array Antenna

2.1 Linear Microstrip Array Antennas

The linear array provides three major feed methods, and corporate feed contains power divider along with transmission lines for all the radiators. The technique is expensive and physically large, and it is wideband in nature. Phased arrays present in corporate feed with phase shifter steer the beam electrically in different directions. Travelling feed contains radiators that are connected to individual transmission line and along the length through line energy decreases. To obtain similar radiation pattern, coupling factors should be made asymmetrical. The remaining energy is captured by the absorptive load. The bandwidth is limited by permitting beam pointing variations. Resonant feed includes radiators that are purposely spaced in multiple of half guide wavelength separated to yield broadside directing beam. It consists of shunt conductance and series resistance as the radiators. Absorptive loads are not used, and radiator coupling factors made symmetrical at the symmetrical power dissemination point known as aperture [3].

2.2 *Microstrip Planar Array Antennas*

Microstrip planar array antennas are used in applications where there is a need to obtain beam in pencil beam direction and elements of arrays can be designed in different ways. Corporate feed is used to operate every single array element individually. They usually consist of two elements such as network of microstrip lines and patch radiators. The feed line lengths should be the same to achieve beam directing in broadside to antenna array for various frequencies. The planar resonant feed can be used in structures like hexagons, squares and equilateral triangles. The length of the line half of wavelength in medium and generated voltage at the feed for each line is repetitive at the joining. From resonant feed, radiation efficiency is relatively high. The performance can be further improved by accomplishing one-fourth wavelength stub with a short-circuited ground by avoiding spurious radiation along with array boundary. It leads to increased radiation efficiency and reduced side lobes [4].

3 State-of-the-Art Reviews

The study below explains about different antenna array design concepts by using different geometries, material, techniques employed that are used for operating at various wireless communication application bands. Each paper describes a unique technique to obtain improved parameters for antenna such as gain, bandwidth, return loss, VSWR and efficiency. The study helps reader to understand how antenna element and its array geometries lead to get better results in comparison with single microstrip patch. communication applications such as smart vehicles, Mission-critical control applications (Fig. 1).

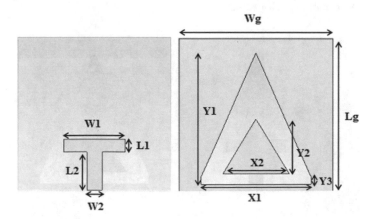

Fig. 1 Proposed dimensions and geometry of 5G antenna for dual band

Yusnita Rahayu and Mohammed Ibnu Hidayat designed antenna array with single, two, four and six elements which resonate at frequency of 28/38 GHz. The gain estimated is 7.47 dBi with −30.7 dB return loss at 28 GHz and 12.1 dBi gain with −34.5 dB return loss at 38 GHz that can be used in 5G wireless communication applications [5] (Fig. 2).

Uzma Uddin, W. Alarm and Md Rashid Ansari suggested the analysis of rectangular microstrip antenna array operating at 5.9 GHz. The return loss obtained is −26.91 dB of side lobe level as −20.3 dB and radiation efficiency as 77c/o with Directivity of 6.826 dB. This designed was analysed for vehicular communication [6] (Fig. 3).

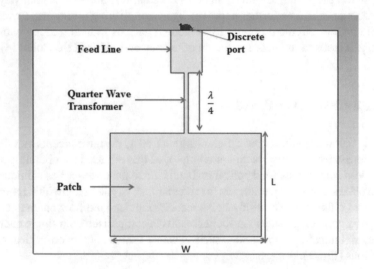

Fig. 2 Single microstrip patch antenna

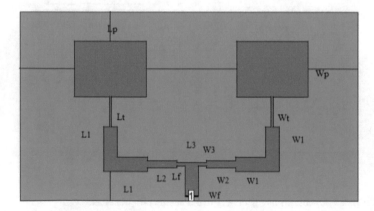

Fig. 3 Dual-microstrip array antenna

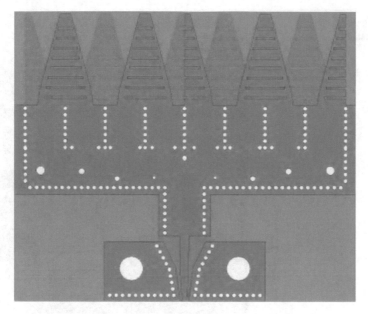

Fig. 4 Tapered slot antenna array with microstrip line SIW transition

Prahlada Rao, Vani R.M and P.V Hunugund analysed the effect of electromagnetic gap (ECG) on dual elements array antenna. The antenna array design is such that the slot with square design END ground plane structure and patch with fractal END generates 8.78 dB of gain with −50.21 db mutual coupling. The resonant frequency is used at 5.53 GHz with mutual coupling and −17. 83 db is used for satellite communications [7] (Fig. 4).

Pengfei Liu, Xiaomi Zhu, Hongjun Tang, Xiang Wang and Wei Hong analysed a tapered slot antenna which is resonating at frequency of 42 GHz. The result estimated is half beamwidth power in H and E planes are 87° and 11° with gain as 0.7 dB that can be used in 5G wireless communication applications [8] (Fig. 5).

W.A.W Muhammad, R. Nagh, M. Fjamlos, P.J. Soh, H. Lago analysed rectangular diamond-shaped microstrip grid array antenna operating at the frequency of 28 GHz. The gain estimated is 11.324 dBi with reflection coefficient of −24.08 dB. The antenna can be used in 5G wireless communication applications [9] (Fig. 6).

S. Latha, P.M. Rubeshanand designed circular polarized microstrip patch antenna array operating at the frequency of 2.4 GHz for 4 element broad side arrays. The gain estimated is 3.2 dBi for single element and 8.0 dBi for array with return loss of −18.1 and −17.3 dB which is suitable for use in WLAN applications [10] (Fig. 7).

G. Mustiko Aji, Md Ammar Wibisono and Munir designed high gain 4 × 4 rectangular patch array antenna with corporate feeding which operates at the frequency of 2.4 GHz. Gain estimated is 15.59 dB with return loss of 19.52 dB, bandwidth of 130 MHz and VSWR of 1.24 for rural area applications [11] (Fig. 8).

Fig. 5 Single patch grid
array antenna

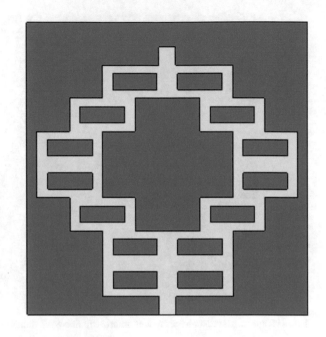

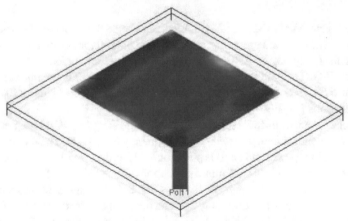

Fig. 6 Diagonal fed microstrip patch antenna layout and its current distribution

Kamil Yauvz Kapusuz, Yakup Sen, MetehenBulut, Iiter Karadede, UgurOguz
analysed 8 × 8 phased antenna array with orthogonal feeding for ground plane
and bottom layer operate at the frequency of 10.7–12.75 GHz. The gain estimated
is 20 dBi with polarization below −10 dB that can be used for mobile satellite
communication [12] (Fig. 9).

Dr. Jagtar Singh, Sunita Rani, Mandeep Singh analysed 8 × 1 rectangular and circular microstrip array antenna with series feed operating at the frequency of 10 GHz.
The gain estimated is 10.19 and 11.650 dBi with return loss of −24.36 and −16.36 dB,

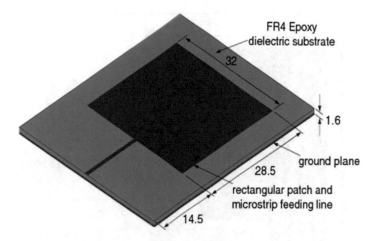

Fig. 7 Single patch antenna element

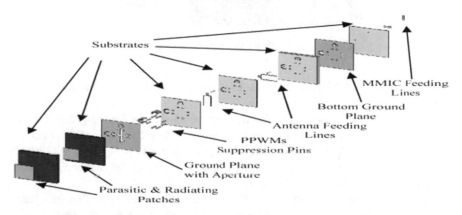

Fig. 8 Expanded view of single radiating antenna element

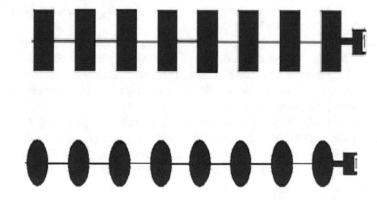

Fig. 9 8 × 1 microstrip array antenna in rectangular and circular patch

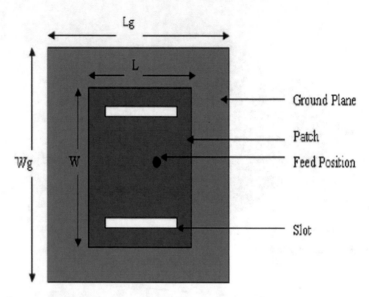

Fig. 10 Geometry of dual-slot microstrip patch antenna

the efficiency of 88.82% and 87.85% for use in speed detection for traffic management [13] (Fig. 10).

Rathindranath Biswas, Asim Kar designed dual-slot microstrip patch antenna array with co-axial feeding having two parallel slots near the edges which operates at the frequency of 2 and 1.982 GHz. The gain estimated is 7.12 dB with return loss of −40.96 dB, bandwidth of 28.7 MHz and HPBW of 76° which is used for microwave and millimetre applications [14] (Fig. 11).

Chandan Kumar Ghosh, Susanta Kumar Parvi designed a 2 × 2 microstrip patch antenna array that has two elements placed at a distance of 6.2 cm with co-axial feeding operates at frequency of 2.45 GHz. The gain estimated is 3.4 dBi for single element and 11.80 dBi for array with return loss of −26.30 dB that can be used for WLAN/MIMO applications [15] (Fig. 12).

A. Asrokin, M. H. Jamaluddin, M. K. A. Rahim, M. R. Ahmad, M. Z. A. Abdul Aziz and T. Married analysed the design of microstrip patch antenna array which resonates at frequency of 5.8 GHz. The return loss estimated is −30.42 dB with 15c/o bandwidth and gain of 16 dB with 9° half power beamwidth for use in point-to-point communication [16] (Fig. 13).

B. Sandhya Reddy, V. Senthil Kumar, V. V. Srinivasan designed 2 × 2 rectangular linearly polarized microstrip patch antenna array with corporate feeding which resonates at frequency of 5.8 GHz. The gain estimated is 9.29 dBi for single element with bandwidth of 340 MHz and 14.3 dBi for 2 × 2 array with bandwidth of 490 MHz for solar power satellite applications [17] (Fig. 14).

Yusnita Rahayu, Ivan Rafli Mustofa suggested 2 × 2 MIMO rectangular patch antenna array with microstrip feed operates at the frequency of 28 GHz. The gain

Fig. 11 Geometry of array with ground plane and its feed point

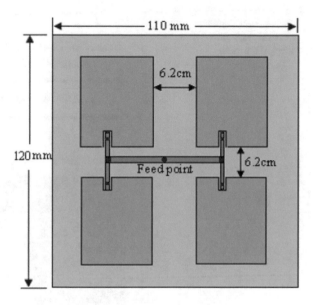

Fig. 12 Single element rectangular patch antenna

estimated is 18.5 dBi with return loss of −44.71 dB having an impedance of 1.95 GHz for application in 5G wireless communication [18].

4 Result and Discussions

Table 1 signifies the different antenna configuration and techniques employed in designing antenna array structures. It represents the various microstrip array antenna

Fig. 13 Rectangular microstrip antenna (i) side view (ii) top view

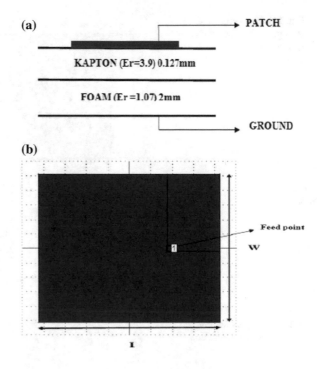

Fig. 14 2 × 2 MIMO rectangular patch antenna array

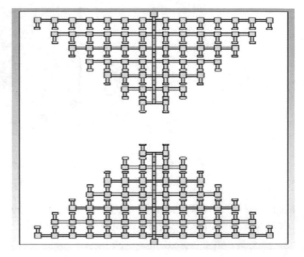

design structures with different dielectric materials operating at various frequency bands that can be used for different wireless communication applications.

The Table 1 represents different antenna configurations and what are the gain, return loss values and in which wireless application the antenna array structures can be used. The study shows techniques for improving major parameters such as

Table 1 Comparative study of different methods

Sl. No.	Antenna configuration	Resonant frequency (GHz)	Return loss (dBi)	Gain (dBi)	Remarks
1.	Dual-band Triangular-shaped slot MAA	28/38	−34.5	12.1	5G wireless communication applications
2.	Rectangular MAA	5.9	−35.99	8.419	Vehicular communication
3.	Two element linear MAAs	5.53	−22.60	8.78	Satellite communications
4.	Tapered slot MAA	42	−43	12.8	5G wireless communication applications
5.	Rectangular diamond-shaped microstrip grid array antenna	28	−24.08	11.32	5G wireless communication applications
6.	Circular polarized MAA	2.4	−17.3	8.0	WLAN applications
7.	4 × 4 rectangular patch antenna array	2.4	−19.52	15.59	Rural area applications
8.	8 × 8 phased antenna array	12.75	–	20	Mobile satellite communication
9.	8 × 1 rectangular and circular MAA	10	−24.36	11.65	Speed detection for traffic management
10.	Dual-slot MAA having two parallel slot near the edges	2	−40.96	7.12	Microwave and millimetre applications
11.	2 × 2 microstrip patch antenna array	2.45	−26.30	11.80	WLAN/MIMO applications
12.	Microstrip patch antenna array	5.8	−30.42	16	Point-to-point communication
13.	2 × 2 rectangular linearly polarized MAA	5.8	−7	14.3	Solar power satellite applications
14.	2 × 2 MIMO rectangular patch antenna array	28	−44.71	18.5	5G Wireless communication

bandwidth, gain, return loss, efficiency, directivity, coupling and polarization. The Table 1 briefs about array structures, their resonating frequency and application area.

5 Conclusion

This paper is attempted to provide the study of designing antenna array structures which overcome the limitations of microstrip antenna. The microstrip antenna array structures are capable of providing improved gain, return loss, bandwidth, beamwidth and efficiency. The antenna array is designed with different feeding techniques, and operating at different resonant frequencies can be used in various wireless communication applications. This paper elaborates the present research in the use of antenna arrays to meet the required need for increased channel capacity by interpreting how an array can be used in various wireless communication systems. The study shows theoretical, experimental, and computer-simulated studies that show the usefulness of such systems.

Acknowledgements I extend my heartfelt thanks to my all dear colleagues who supported throughout my work. I extend my sincere acknowledgement to my parents for supporting throughout the work.

References

1. Review on various types of microstrip antennas for wireless communication. *IOSR Journal of Electronics and Communication Engineering (IOSR-JECE)* 12 (4): 36–42. e-ISSN: 2278-2834, ISSN: 2278-8735. Ver. II (Jul.–Aug. 2017).
2. Kraus, John D., and K. D Prasad. *Antennas and wave propagation.*
3. A practical guide to the design of microstrip antenna arrays. *Microwave.* journal/articles/3144/ February 1, 2001.
4. Gibson, P.J., and R. Hill. A planar microstrip array. In *IEEE colloquium advances in printed antenna design and manufacture digest*, 3/1–3/3, No. 1982/19.
5. Rahayu, Yusnita, and Muhammad Ibnu Hidayat. 2018. Design of 28/38 GHz dual-band triangular-shaped slot microstrip antenna array for 5G applications. In *2018 2nd international conference on telematics and future generation networks (TAFGEN)*. 978-1-5386-1275-0/18/$31.00 ©2018, IEEE.
6. Uddin, Uzma, W. Akaram, and Md Rashid Anasri. 2018. Design and analysis of microstrip patch antenna and antenna array for vehicular communication system. In *Proceedings of the 2nd international conference on inventive communication and computational technologies (ICICCT 2018)* IEEE Xplore Compliant - Part Number: CFP18BAC-ART; ISBN: 978-1-5386-1974-2 978-1-5386-1974-2/18/$31.00 ©2018 IEEE.
7. Prahlada Rao, K., P.V. Hunugund, and R.M. Vani. 2018. Design of enhanced gain two element linear microstrip antenna array. In *2018 twenty fourth national conference on communications (NCC)*. 978-1-5386-1224-8/18/$31.00 ©2018 IEEE.
8. Liu, Pengfei, Xiaowei Zhu, Hongjun Tang, Xiang Wang, and Wei Hong. 2017. Tapered slot antenna array for 5G wireless communication systems. 978-1-5386-1608-6/17/$31.00 (c) 2017 IEEE.

9. Muhamad, W.A.W., R. Ngah, M.F. Jamlos, P.J. Soh, and H. Lago. 2016. Gain enhancement of microstrip grid array antenna for 5G applications. In *2016 URSI Asia-Pacific Radio Science Conference*, August 21–25, 2016/Seoul, Korea 978-1-4673-8801-6/16/$31.00 ©2016 IEEE.

10. Latha, S., and P.M. Rubesh Anand. 2016.Circular polarized microstrip patch array antenna. For wireless LAN applications. In *IEEE WiSPNET 2016 conference*, 978-1-4673-9338-6/16/$31.00_c 2016 IEEE.

11. Aji, Galih Mustiko, Muhammad Ammar Wibisono, and Achmad Munir. 2016. High gain 2.4 GHz patch antenna array for rural area application. In *The 22nd Asia-Pacific conference on communications*, 978-1-5090-0676-2/16/$31.00 ©2016 IEEE.

12. Kapusuz, Kamil Yavuz, Yakup Sen, Metehan Bulut, Ilter Karadede, and Ugur Oguz. 2016. Low-profile scalable phased array antenna at Ku-band for mobile satellite communications, 978-1-5090-1447-7/16/$31.00 ©2016 IEEE.

13. Singh, Jagtar, Er. Sunita Rani, and Er. Mandeep Singh. 2015. Design of microstrip patch antenna array for traffic speed detector application. In *2015 fifth international conference on advanced computing & communication technologies*, 2327-0659/15 $31.00 © 2015 IEEE https://doi.org/10.1109/acct.2015.65.

14. Biswas, Rathindra Nath, and Asim Kar. 2008. A novel PSO-IE3D based design and optimization of a low profile dual slot microstrip patch antenna. In *TENCON 2008, IEEE Region 10 Conference*. https://doi.org/10.1109/tencon.2008.4766781.

15. Ghosh, Chandan Kumar, and Susanta Kumar Parvi. 2009. Design and study of a 2×2 microstrip patch antenna array for WLAN/MIMO application. In *2009 international conference on emerging trends in electronic and photonic devices & systems (ELECTRO-2009)*.

16. Rahim, M.K.A., A. Asrokin, M.H. Jamaluddin, M.R. Ahmad, T. Married, and M.Z.A. Abdul Aziz. 2006. Microstrip patch antenna array at 5.8 GHz for point to point Communication. In *2006 international RF and Microwave conference proceedings*, September 12–14, 2006, Putrajaya, Malaysia.

17. Sandhya Reddy, B., V. Senthil Kumar, and V.V. Srinivasan. 2012. Design and development of a light weight microstrip patch antenna array for solar power satellite application. In *IEEE transaction and proceedings on antenna and propagation*.

18. Rahayu, Yusnita, and Ivan Rafli Mustofa. 2017. Design of 2×2 MIMO microstrip antenna rectangular patch array for 5G wireless communication network. In *2017 progress in electromagnetics research symposium*, Fall (PIERS-FALL), Singapore, 19–22 November.

Workload Aware Dynamic Scheduling Algorithm for Multi-core Systems

Savita Gautam and Abdus Samad

Abstract The key challenge of scheduling in multi-core systems is to map highly irregular processes that require the inspection of thread behavior and efficiency of multi-core systems. The motive is to schedule multiple tasks on multiple cores. In this paper, a novel scheduling technique is proposed that works on execution technique for tree type tasks structures that are mapped on different multi-core systems designed using different multiprocessor systems. In particular, the performance is evaluated by applying the proposed technique to a particular class of multiprocessor system known as hybrid multiprocessor systems that are used as basic building blocks of a multi-core system. The scheduling algorithm is applied by dividing tasks in terms of computation efficiency of these systems. The key novelty of the proposed method is that tasks which are executed partially may be migrated on systems which are under-loaded and having good efficiency. In other words, the scheduling of tasks to cores must be automated to adapt to the changing program behavior and current load on the system. Before migration of remaining tasks, the efficiency of core on such systems is evaluated. A comparative study is carried out by applying other standard scheduling algorithms on the same multi-core systems. Simulation results show that the proposed algorithm gives better performance while executing tasks on various multi-core systems having different computational efficiency. In particular, the load imbalance is improved by 20–30% and execution time is reduced by 35–55% as compared to traditional algorithms. Further, we show that in many cases, our approach is able to deliver better performance by combining it with classical scheduling algorithms.

Keywords Multi-core systems · Scheduling algorithm · Computational capability · Load imbalance · Execution time · Performance parameter · Multiprocessor interconnection networks · Packet size

S. Gautam · A. Samad (✉)
F/O Engineering & Technology, University Women's Polytechnic,
Aligarh Muslim University, Aligarh 202002, India

© Springer Nature Singapore Pte Ltd. 2020   171
Y.-C. Hu et al. (eds.), *Ambient Communications and Computer Systems*, Advances in
Intelligent Systems and Computing 1097,
https://doi.org/10.1007/978-981-15-1518-7_14

1 Introduction

In multiprocessor systems, the conventional applications scheduling mechanisms allocate most of the applications to distribute the load among different nodes evenly. The scheduling plays a vital role when these systems employee multi-core systems. In heterogeneous processing devices which are equipped with multi-core CPUs, the efficiency is further improved by efficient utilization of each multi-core system. The success of such systems depends upon proper utilization of computing resources, lower execution time and high throughput. The computing power of each multi-core system is to be taken into consideration while allocating load on different processing elements. The high speedup execution using multi-core processors is suitable for those applications which require an excessive amount of parallel computations [1, 2]. Therefore, parallel programming of heterogeneous systems has emerged as a research problem in multi-core systems.

Different topologies that provide high performance in super computers, multiprocessor systems and other high-performance machines are reported and used in all those systems which operate parallel applications. The conventional scheduling mechanisms generally adhere to manage distribution of load among CPUs as multiple units. However, the computing power of each node is not taken into consideration. This results in underutilization of nodes during the execution of parallel tasks. Therefore, within a heterogeneous machine, the scheduling of tasks requires balancing of nodes while utilizing the computing resources to reduce overall makespan. For example, in a cloud computing environment, the central scheduler plays a significant role along with the application developer [3]. On the other hand, scheduling techniques that work on single job are generally based on specific job and do not require a low-level scheduling mechanism [4, 5]. Hence, there is a need to design schedulers which are computational-intensive and require offline profitability of computing devices [6].

In the present work, a novel scheduling algorithm that maps parallel applications on heterogeneous machines is proposed. The proposed scheme manages the allocation of load by observing the computing capability of machine as well as the overall distribution of load on the system. The main objective is to use the computing resources efficiently. Applications which require slow execution are scheduled on lesser computing devices, whereas those require fast execution or having heavy load are scheduled on high computing devices. The balancing of tasks is carried out with reduced execution time and evenly distribution overall load on the system.

The paper is organized in five sections as follows. Section 1 is dedicated for introduction. Section 2 gives an overview of the related work in brief. The proposed methodology is detailed in Sect. 3. The experimental results are analyzed by implementing the proposed algorithm and discussed in Sect. 4. The paper is concluded in Sect. 5.

2 Related Work

Different scheduling techniques are available for multi-core machines which are designed to reduce the overall makespan and with improved device utilization. Conventional algorithms may be either static or dynamic. The static algorithms are generally designed for predefined load or fixed load, whereas performance of dynamic algorithms is better for an unpredictable type of load where decisions are taken on fly [7]. The scheduling decision also depends upon the variety of application models. A number of parameters such as machine architectures, computing devices and task structures affect the overall performance of the system.

Grewe et al. (2011) proposed a scheduling algorithm in which task partitions are made depending upon their behaviors. The nature of tasks then used to take scheduling decisions. The overall speedup over a single-core processor demonstrates the significance of program-specific task mapping [8]. Using this concept, a machine learning-based scheduling approach is designed which works on the principle of branch divergence. The author claims that control flow deviation has a significant effect over the way used in program partitions [9]. Another scheduling technique is proposed that works on profiling the job. In this method, the concept of offline profiling is used to obtain reduced execution time data dependencies [10]. Concept of offline profiling is also used in HDSS scheduling algorithm which improves the execution time by partitioning workload between CPU and GPUs [11]. The advantage of mapping multiple command queues and impact of different partitions set algorithms for parallel execution are evaluated and analyzed. HDSS mechanism enables a load balance execution on heterogeneous type of machines. Our approach supports the concept without code change and offline splitting of tasks.

Another approach that schedules the tasks by dividing them into chunk sizes is reported [12]. In this algorithm, the tasks are divided into variable sizes of packets also known as chunks and scheduled among CPU and GPUs dynamically. The chunk sizes are adjusted based on the previous state of execution and computing capabilities of the system. Our approach also used the concept of packets; however, packets are distributed among different processing elements having different computing power. Therefore, scheduling of tasks is carried out depending upon the current workload and available computing resources.

3 Proposed Algorithm

To reduce the execution time, it is essential to ensure minimum load imbalance of tasks while mapping them onto different computing devices. Each computing device has a parallelism potential by incorporating multi-core architecture. The underutilization of computational resource leads small computation which ultimately results inefficient resource utilization. For applications with small data size the execution similar performance as that of traditional approaches. However, for large data size applications, performance could be improved by improving device utilization based on computing power. In addition to that, the load imbalance factor should also be taken into consideration when devices have similar computing power. A task distribution with low imbalance generally gives improved results with reduced execution time and high throughput by considering resource-aware task allocation. The proposed algorithm divides the tasks into chunks also known as packets and maps them into different devices having multi-core architectures. The number of packets to each node is adjusted based on the current load on the system as well as computing efficiency of the device. The algorithm is dynamic in the sense that the decisions are taken on fly and the number of packets is decided during run time. In a more detail way, the contribution of the proposed algorithm is described as under:

- The device utilization is improved with the concept of traditional scheduling approach
- Higher throughput and lesser makespan are obtained for compute-intensive applications
- Computation power of each device is monitored to further minimize the makespan
- Implementation of resource-aware allocation is carried out and load is allocated with different chunk sizes.

There are different criterions in different scheduling algorithms which decide the sequence of execution. In the proposed scheme, the tasks are divided into chunks to obtain optimal performance. Processors having high computing power and under-loaded are considered as acceptors, whereas those over-loaded and having lesser computing power are considered as donors. Processors with similar computational capabilities are scheduled using conventional scheduling algorithms. A number of parameters are used to evaluate the computational capabilities. The algorithm is described in pseudocode and is given below:

```
MULTICORE (Output: Configured Network with balance
load)
Load generation at all multi-core processors
/* let Pr is number of nodes and N is the cores in the
system*/
For central node obtain all connected nodes
Identify and count the number of donors and acceptors
Calculate computing power based on number of cores
Calculate load (sum), I_Load for all nodes
perform Migration based on I_Load Load(central_Node)
==Ideal
If Load(central_Node)  >   I_lLoad) and Total Number of
cores < N
Migrate excess load to other nodes which have
connectivity
else if(Load(Central_Node) < I_Load)and Number of cores >
N
```

```
Accept excess load from remaining nodes which are
connected
Else
If Number of cores == N, Migrate load based IdealLoad
For(i=0; i<N; i++)
Adjust the donors and acceptors
If (Load <= IL) N_acceptors++; acceptors[N_acceptors]=I;
Else N_donors++; donors[N_donors]=I;
Adjust the packet size
If load < ideal load and Cores < N, packet size <= 4*N
Migrate_Load(donor[i], acceptor[j], Pkt_size)
Evaluate Imbalance of load (LIF %)
/*identify all connected nodes*/
For all nodes check their connectivity
if (connectivity[central_node][node]==1)
node is connected to Central_node
For all nodes at level one find out second level nodes
which are also connected to these nodes and apply
migrations
Evaluate makespan
```

The given algorithm is designed to minimize the execution time by dividing tasks into packets and deciding the donors/acceptors depending upon the cores. Similarly, tasks are balanced using the same criterion of migration. Load imbalance % is evaluated for systems having different processing capabilities. Set of processing nodes at every level is decided in which task migrations could be carried out. As there is greater number of tasks, the scheduling parameters also increase. For uniform tasks, the scheduling decision is taken based on ideal load. The results obtained are also evaluated by using a conventional approach where tasks migration decisions are taken based on the load and processing elements available. In particular, the standard scheduling approach for multiprocessor system, namely Two-Round Scheduling (TRS) scheme is taken into consideration [13]. The given algorithm does not require offline partitioning and decides the mapping of load during run time. Therefore, a comparison of results is also carried out to evaluate the enhanced performance.

4   Results Analysis

Performance of the given algorithm and existing scheduling algorithms is evaluated and obtained results are compared. To evaluate the performance of the mentioned algorithms, the two different systems having six-core and eight-core processors are taken into consideration [14–16]. The considered systems are simple interconnection networks that could be extended with an enhanced number of cores depending upon the volume of task structures. The proposed scheme is applied on these networks with six cores and eight cores, respectively, and results are evaluated to check load imbalance and makespan in terms of total execution time. To compare the results, standard scheduling algorithm namely Two-Round Scheduling (TRS) scheme is also implemented on these systems. The simulation is carried out on Intel Core i3 processor with 4 GB RAM under Linux environment. For comparative study, the percentage of load imbalance with number of tasks is evaluated; curves are drawn and shown in Figs. 1 and 2.

The results shown in curves clearly indicate that the initial value of load imbalance is lesser when proposed scheme is implemented on six-core and eight-core systems as compared to standard scheduling scheme. The lower value of imbalance depicts that the system is having more balanced execution. Tasks are generated randomly with increasing pattern. With the increase in tasks volume, the behavior of load imbalance is showing similar pattern. In particular, for six-core processor, the load imbalance is improved with 20%. The improvement in eight-core processor is improved in the initial stage of task structure and approximately equal to 14%. In eight-core system, however, when a number of tasks are increased, the improvement is not significant and similar results are obtained with both the scheduling approaches.

The execution time is evaluated which demonstrates the total execution time consumed in execution of tasks including the time involved in balancing. The smaller is the execution time, the better is the performance. Figures 3 and 4 show the average execution time against number of tasks for considered systems. With the similar task behavior, the execution time for six-core processor is reduced considerably and approximately equal to 55%. The best speedup may be achieved by adjusting the packet size during mapping of tasks. Tasks are increased randomly and behavior is obtained while evaluating the execution time. With the increasing input size, the

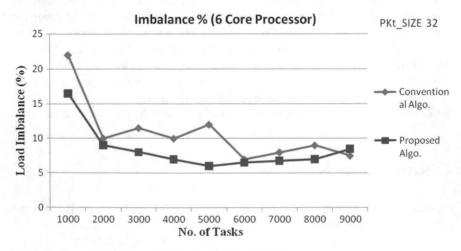

Fig. 1 Performance of given algorithm based on load imbalance

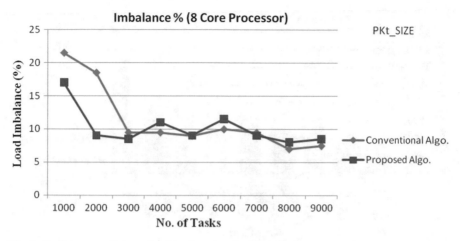

Fig. 2 Performance of given algorithm based on load imbalance

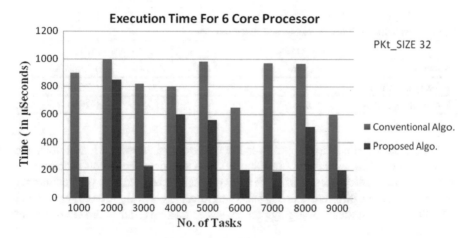

Fig. 3 Performance of given algorithm based on execution time

execution time with the given algorithm when implemented on eight-core processors is reduced up to 35% as compared to existing scheduling algorithm.

The reduction in execution time is obtained because of the consideration of task behavior and chunk size. Results obtained for both the systems, i.e., six cores and eight cores considered chunk size of 32. The reduction of execution time may further be improved with higher values of chunk sizes. However, greater value of chunk size may lead to performance degradation in terms of load imbalance. Therefore, an optimal value of chunk size is desirable for better performance of the system. For chunk size of 32, the initial value of load imbalance and execution time is depicted

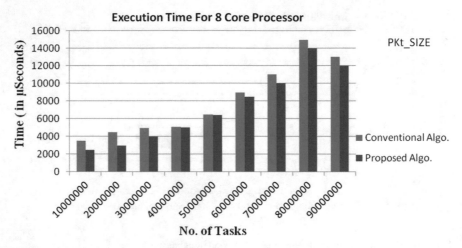

Fig. 4 Performance of given algorithm based on execution time

Table 1 Evaluation of load imbalance and execution time

Networks	Load imbalance (%)		Execution time (μs)	
	Six core	Eight core	Six core	Eight core
Existing algorithm	22.0	21.5	900	3500
Proposed algorithm	16.5	17.0	150	2500

in Table 1. Each value shows that there are improved results when load imbalance is obtained as well as the proposed algorithm is showing significant reduction in execution time with chunk size of 32. Similar results are obtained with greater number of tasks and having different chunk sizes. Therefore, the given algorithm is best suited to fulfill the requirements of highly parallel systems having multi-core environment under heavy load.

5 Conclusions

The present work focuses on the novel scheduling algorithm which ensures the task balancing of compute-intensive applications on multi-core systems. The algorithm performs assignment of tasks based on computing power of resources and computes load imbalance and total execution time of tasks on different systems. The experimental evaluation revealed that the proposed algorithm has significant role to balance the load on different computing units and employed reduced makespan by considering computational power of the machines. The resource utilization is efficient which makes the algorithm better for multi-core systems. The simulation results show that the performance of the system by applying the given scheme is improved as compared to when traditional state-of-the-art scheduling algorithms are applied.

In particular, the load imbalance is improved by 20–30% and execution time is reduced by 35–55% as compared to traditional algorithms. Therefore, it can be concluded that the given algorithm may be applied to any parallel system that employed multi-core systems.

References

1. Owens, J.D., M. Houston, D. Luebke, S. Green, J.E. Stone, and J.C., Phillips. 2008. GPU computing, In *IEEE proceedings*, 879–899.
2. Aleem, M., R. Prodan, and T. Fahringer. 2011. Scheduling java symphony applications on many-core parallel computers. In *Euro-Par 2011 parallel processing*, 167–179. Springer.
3. Lösch, A., T. Beisel, T. Kenter, C. Plessl, and M. Platzner. 2016. Performance-centric scheduling with task migration for a heterogeneous compute node in the data center. In *Proceedings of the 2016 conference on design, automation and test in Europe. EDA consortium*, 912–917.
4. Wen, Y., Z. Wang, and M.E.P. O'boyle. 2014. Smart multi-task scheduling for OpenCL programs on CPU/GPU heterogeneous platforms. In *2014 21st International conference on high performance computing (HiPC)*. IEEE, 1–10.
5. Wen, Y., and M.F. O'Boyle. 2017. Merge or separate multi-job scheduling for OpenCL kernels on CPU/GPU platforms. In *Proceedings of the general purpose GPUs*, 22–31. ACM. https://doi.org/10.1145/3038228.3038235.
6. Khalid, Y.N., M. Aleem, R. Prodan, et al. 2018. E-OSched: a load balancing scheduler for heterogeneous multicores. *Journal of Supercomputing* 74: 5399. https://doi.org/10.1007/s11227-018-2435-1.
7. Prasad, N., P. Mukkherjee, S. Chattopadhyay, and I. Chakrabarti. 2008. Design and evaluation of ZMesh topology for on-chip interconnection networks. *Journal of Parallel and Distributed Computing*, 17–36.
8. Grewe, D., and M.F. O'Boyle. 2011. A static task partitioning approach for heterogeneous systems using OpenCL. In *International conference on compiler construction*, 286–305. Springer.
9. Ghose, A., S. Dey, P. Mitra, and M. Chaudhuri. 2016. Divergence aware automated partitioning of OpenCL workloads. In: *Proceedings of the 9th India software engineering conference*, 131–135. ACM. https://doi.org/10.1145/2856636.2856639.
10. Albayrak, O.E., I. Akturk, O. Ozturk. 2012. Effective kernel mapping for OpenCL applications in heterogeneous platforms. In *Proceedings of international conference on parallel processing work*, 81–88. https://doi.org/10.1109/ICPPW.2012.14.
11. Belviranli, M.E., L.N. Bhuyan, and R. Gupta. 2013. A dynamic self-scheduling scheme for heterogeneous multiprocessor architectures. *ACM Transaction on Architecture Code Optimization* 9: 1–20. https://doi.org/10.1145/2400682.2400716.
12. Boyer, M., K. Skadron, S. Che, and N. Jayasena. 2013. Load balancing in a changing world: dealing with heterogeneous and performance variability. In *Proceedings of ACM international conference on computing frontiers*, 21.
13. Samad, A., M.Q. Rafiq, and O. Farooq. 2012. Two round scheduling (TRS) scheme for linearly extensible multiprocessor systems. *International Journal of Computer Applications* 38 (10): 34–40.
14. Khan, Z.A., J. Siddiqui, and A. Samad. 2016. Properties and Performance of Cube-based Multiprocessor Architectures. *International Journal of Applied Evolutionary Computation (IJCNIS)* 7 (1): 67–82.

15. Mohammad, S.B., and I. Ababneh. 2018. Improving system performance in non-contiguous processor allocation for mesh interconnection networks. *Journal of Simulation Modeling and Practice*, *80*, 19–31.
16. Manju, K., K. Raghvendra, N.D. Le., and M.C. Jyotir. 2017. Interconnect network on chip topology in multi-core processors: a comparative study. *International Journal of Computer Network and Information Security*, *11*, 52–62.

Almost Self-centered Index
of Some Graphs

Priyanka Singh and Pratima Panigrahi

Abstract For a simple connected graph G, center $C(G)$ and periphery $P(G)$ are subgraphs induced on vertices of G with minimum and maximum eccentricity, respectively. An n-vertex graph G is said to be an almost self-centered (ASC) graph if it contains $n - 2$ central vertices and two peripheral (diametral) vertices. An ASC graph with radius r is known as an r-ASC graph. The r-ASC index of any graph G is defined as the minimum number of new vertices, and required edges, to be introduced to G such that the resulting graph is r-ASC graph in which G is induced. For $r = 2, 3$, r-ASC index of few graphs is calculated by Klavžar et al. (Acta Mathematica Sinica, 27:2343–2350, 2011 [1]), Xu et al. (J Comb Optim 36(4):1388–1410, 2017 [2]), respectively. Here we give bounds to r-ASC index of diameter two graphs and determine the exact value of this index for paths and cycles.

Keywords Radius · Diameter · Almost self-centered graphs · r-ASC embedding index

1 Introduction

In the paper, we consider simple and connected graphs unless otherwise stated. For vertices u and v in a graph G, the length of a shortest u–v path is the distance $d(u, v)$ between them. The *eccentricity* of a vertex u in G, denoted by $e_G(u)$ (or simply $e(u)$), is the distance from u to a farthest vertex in G. The minimum and maximum eccentricity of vertices in G are the radius and diameter ($diam(G)$) of the graph G, respectively. If G is a graph in which graph H is induced, then we say that H is embedded in G. The induced subgraph $C(G)$ (respectively $P(G)$) on the vertices of minimum (respectively maximum) eccentricity in a graph G is called

P. Singh (✉) · P. Panigrahi
Indian Institute of Technology Kharagpur, Kharagpur, India

P. Panigrahi
e-mail: pratima@maths.iitkgp.ernet.in

© Springer Nature Singapore Pte Ltd. 2020
Y.-C. Hu et al. (eds.), *Ambient Communications and Computer Systems*, Advances in Intelligent Systems and Computing 1097,
https://doi.org/10.1007/978-981-15-1518-7_15

the *center* (respectively *periphery*) of G. Moreover, vertices of $C(G)$ and $P(G)$ are
called *central vertices* and *peripheral vertices*, respectively. Graph G of order n in
which there are exactly two peripheral vertices and $n - 2$ central vertices is called an
almost self-centered (ASC) graph. An r-ASC graph is ASC with radius r. We note
that the diameter of an r-ASC graph is $r + 1$.

The r-ASC index [1] of a graph G, denoted by $\theta_r(G)$, is defined as the minimum
number of vertices, and required edges, to be added to G such that the new graph is
an r-ASC graph in which graph G is induced.

Graph embedding is an area of interest of many researchers in the last few decades.
For many interesting problems on embedding of graphs, one can refer to [3–9].
Klavžar, Narayankar, and Walikar [1] were the first to introduce and study ASC
graphs, where the authors gave several results on ASC graphs. They discussed embedding of any graph into some ASC graph and proposed constructions for embedding of any graph into an ASC graph of radius two. Also, they calculated exact
values of 2-ASC index (or $\theta_2(G)$) of graphs like complete bipartite graphs, cycles,
and paths. The following two results are due to Klavžar et al. [1].

Theorem 1.1 *[1] For any graph G and positive integer $r \geq 2$, $0 \leq \theta_r(G) \leq 2r + 1$.*

Theorem 1.2 *[1] Suppose graph G is obtained by adding a pendant vertex to C_{2r},
and graph H is obtained by making a new vertex adjacent to two adjacent vertices
of C_{2r+1}. Then both G and H are r-ASC graphs.*

The upper bound in above theorem was improved by Xu et al. [2] which is stated
below.

Theorem 1.3 *[2] For every graph G and positive integer $r \geq 2$, $\theta_r(G) \leq 2r$.*

Also, in [2], authors determined 3-ASC index of complete graphs, paths, cycles, and
some trees. Further, they provided lower and upper bounds to $\theta_3(G)$ for graphs with
diameter two as stated below.

Theorem 1.4 *[2] For a diameter two graph G, $3 \leq \theta_3(G) \leq 4$.*

In this paper, we study r-ASC index of graphs for $r \geq 4$. In Sect. 2, we generalize the
result of [2]; that is, we provide lower and upper bounds to r-ASC index of diameter
two graphs, for $r \geq 4$. We also construct graphs satisfying upper and lower bounds
of r-ASC index of graphs with diameter two. In Sect. 3, we determine the exact value
of r-ASC index of cycles and paths for $r \geq 4$.

2 R-ASC Index of Graphs with Diameter Two

The following lemma is very basic and will be used in the sequel.

Lemma 2.1 *Let $P_1 : x, v_1, v_2, \ldots, v_k$ and $P_2 : x, u_1, u_2, \ldots, u_r$ be shortest x–v_k and
x–u_r paths in a graph G, respectively. If P_1 and P_2 intersects at v_i and u_j, i.e., $v_i = u_j$,
then $i = j$.*

Proof On the contrary assume that $i \neq j$. Then either $i < j$ or $i > j$. If $i < j$, then we get x–u_r path $P : x, v_1, v_2, \ldots, v_i = u_j, u_{j+1}, u_{j+2}, \ldots, u_r$ of length $r + i - j - 1 < r$ (since $i < j$). Similarly, if $i > j$, then we obtain a x–v_k path $P' : x, u_1, u_2, \ldots, u_j = v_i, v_{i+1}, \ldots, v_k$, where the length of P' is $k - (i - j) - 1 < k$ (since $i - j \geq 1$). In both the cases, we obtain x–v_k and x–u_r paths of shorter length, a contradiction, and hence, $i = j$.

In the rest of the paper, if vertices x and y are adjacent in a graph, then we use the notation $x \sim y$.

Theorem 2.1 *Let G be a simple connected diameter two graph and $r \geq 4$ be an integer. Then $2r - 3 \leq \theta_r(G) \leq 2r - 2$.*

Proof We have $diam(G) = 2$. Let H be an r-ΛSC graph which embeds G as an induced subgraph. For every vertex $x \in V(H)$, we have $e(x) = r$ or $r + 1$. Moreover, H contains exactly two vertices with eccentricity $r + 1$ and these two vertices are called diametral vertices of H. G cannot contain two diametral vertices of H because $diam(G) = 2$ and $r \geq 4$. So, G contains at most one diametral vertex, and we will have cases (a) and (b) below according to G contains a diametral vertex or not.

1. G contains a diametral vertex, say x, $e_H(x) = r + 1$, i.e., x is an end vertex of diametral path $P : x, y, z, x_1, x_2, \ldots, x_{r-1}$. Since $diam(G) = 2$, at the most the vertices y and z can be in G. So, P contains at least $r - 1$ vertices which are not in G. Since $e_H(z) = r$, there exists a vertex w in H such that $d_H(z, w) = r$, and a shortest path $P_1 : z, w_1, w_2, \ldots, w_r = w$ such that $d_H(z, w) = r = l(P_1)$. Obviously, $w \notin P$.

   First, let w be adjacent to x_{r-1}.

   Since there are only two diametral vertices with eccentricity $r + 1$ (i.e., x and x_{r-1}), we have $d(x, w) \leq r$. Now, if $d(x, w) < r$, then $d(x, x_{r-1})$ will be at the most r because $w \sim x_{r-1}$ and there is a path of length at the most $r - 1$ between x and w. So, $d(x, w) = r$ and there exists a path $P_2 : x, u_1, u_2, \ldots, u_r = w$ of length r. Now, no x_i, $i = 1, 2, \ldots, r - 1$ belong to P_2, otherwise $d(z, w)$ or $d(x, x_{r-1})$ will reduce.

   Since P_2 is of length r, P_2 contains at the most three vertices of G and so at least $r + 1 - 3 = r - 2$ new vertices which are not in G and different from

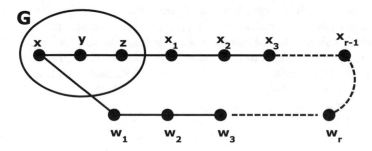

Fig. 1 Case when $w \sim x_{r-1}$

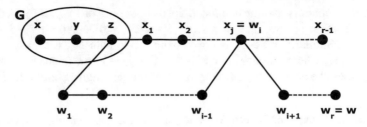

Fig. 2 P_1 intersects $\{z, x_1, x_2, \ldots, x_{r-1}\}$ but not $G - \{z\}$

$x_1, x_2, \ldots, x_{r-1}$. Hence, $|V(H) - V(G)| \geq r - 1 + r - 2 = 2r - 3$. This case is illustrated in Fig. 1.

Next we take that w is not adjacent to x_{r-1}. We have following subcases depending on intersection of shortest z–w path P_1 with G and its complement.

a. P_1 intersects $G - \{z\}$ but not $\{x_1, x_2, \ldots, x_{r-1}\}$.
   Since diameter of G is two, in this subcase, P_1 contains at least $r - 2$ vertices not in $G \cup \{x_1, \ldots, x_{r-2}\}$, and hence, we get $\theta_r(G) \geq (r - 1) + (r - 2) = 2r - 3$.

b. P_1 intersects both $G - \{z\}$ and $\{x_1, x_2, \ldots, x_{r-1}\}$.
   If P_1 contains vertices in $\{x, y\}$ and $\{x_1, \ldots, x_{r-1}\}$, then either $d(x, x_{r-1})$ or $d(x, w)$ reduces, which is not possible. So next we consider the situation where P_1 intersects $G - \{x, y, z\}$ and $\{x_1, x_2, \ldots, x_{r-1}\}$ both. Since $diam(G) = 2$, P_1 can intersect $G - \{x, y, z\}$ at the most two vertices.
   Let $P_1 \cap (G - \{x, y, z\}) = \{u\}$. Since $diam(G)$ is 2, we have $d_G(z, u)$ is 1 or 2. Let $d_G(z, u)$ be 1. We have $d_G(x, u) = 1$ or 2. Let $d_G(x, u) = 1$. Since P_1 also intersects $\{x_1, \ldots, x_{r-1}\}$, say $x_i = w_j$, then by Lemma 2.1, we get $i = j$. But then we obtain a x–x_{r-1} path of length less than $r + 1$ which is a contradiction. Next, let $d_G(x, u) = 2$. In this case also, we obtain x–x_{r-1} path of length less than $r + 1$ if P_1 intersects $\{x_1, \ldots, x_{r-1}\}$. Thus, P_1 has at least $r - 2$ (not in G) vertices and different from $\{x_1, \ldots, x_{r-1}\}$. Hence, we get $\theta_r(G) \geq r - 1 + r - 2 = 2r - 3$.
   Next, let $d_G(z, u) = 2$ and $d_G(x, u) = 1$. The path P_1 also intersects the vertex set $\{x_1, \ldots, x_{r-1}\}$. Since u lies on P_1 and $u \sim x$, then the length of x–x_{r-1} is reduced which is a contradiction. Again, let $d_G(z, u) = 2$ and $d_G(x, u) = 2$. Since P_1 intersects the vertex set $\{x_1, \ldots, x_{r-1}\}$, we obtain another x-x_{r-1} path which reduces the distance between x and x_{r-1} as new path passes through vertex u. So, P_1 contains at the most two vertices of G other than the vertex z. Hence, P_1 contains at least $r + 1 - 3 = r - 2$ vertices from G and other than $\{x_1, \ldots, x_{r-1}\}$. This proves that $\theta_r(G) \geq 2r - 3$ in this case.

c. P_1 intersects $\{x_1, x_2, \ldots, x_{r-1}\}$ but not $G - \{z\}$.
   If P_1 intersects $P \setminus \{x, y, z\}$ at x_j, i.e., $w_i = x_j$, then by Lemma 2.1, $i = j$. So we get $d(x, w) = r + 1$, which is a contradiction because $e(w) = r$. This clearly shows that P_1 contains $r - 2$ vertices which are different from $\{x_1, \ldots, x_{r-1}\}$

and not in G. In this case also, we obtain $\theta_r(G) \geq 2r - 3$. The construction for this case is shown in Fig. 2.

d. P_1 does not intersect any of $G - \{z\}$ and $\{x_1, x_2, \ldots, x_{r-1}\}$.
This case is not possible because in this case we get $d(x, w) > r + 1$ again contradicting the fact that $e(w) = r$. The case is referred and explained in the Fig. 3.

2. G does not contain any diametral vertex.
Since $diam(G)$ is 2, here also P contains at least $r - 1$ new vertices. Since $r \geq 4$ and $r + 1 \geq 5$, P contains at least six vertices. If x and y are end vertices of P which are not in G, then there exists $z \notin V(G)$ and $x \sim z$ or $y \sim z$. Without loss of generality, let $x \sim z$. Since $e(z) = r$, there exists a z–w path $P_2 : z, w_1, w_2, \ldots, w_r = w$ such that $l(P_2) = r = d(z, w)$. Now, the path P cannot be extended, otherwise $d(x, y) > r + 1$. Now, the P_2 contains at most three vertices from G since $diam(G) = 2$. Further, if P_2 intersect vertices of the P then $d(z, w) < r$, which is a contradiction because $e(z) = r$. This follows that at least $r - 2$ vertices in P_2 are not in G. This proves that we need to add at least $(r - 1) + (r - 2) = 2r - 3$ new vertices to G to get an r-ASC graph. The construction is illustrated with the help of an example in Fig. 4, where P_2 contains three vertices of G.

For the upper bound it is sufficient to construct an r-ASC graph H in which G is induced and $|V(H)| = |V(G)| + 2r - 2$. We give construction of an r-ASC graph H in the below. Let $x, y, z \in V(G)$, where x and y are non-adjacent vertices and z is a common neighbor of them. Let $V(H) = V(G) \bigcup \{x_1, x_2, \ldots, x_{2r-2}\}$ such that y–x_1–x_2–$\ldots$–x_{2r-2} forms a path and x_{2r-2} is adjacent to all the vertices of G except x and y. The construction is shown in Fig. 5.

Next we shall show that H is an r-ASC graph in which G is induced. In Fig. 5, we can see that $d_H(x, x_{r-1}) = d_H(x, y) + d_H(y, x_1) + d_H(x_1, x_{r-1}) = r + 1$. We note that in graph H, we obtain a cycle $C : g, y, x_1, x_2, \ldots, x_{2r-2}, g$ of length $2r$, where $g \in V(G) \setminus \{x, y\}$. Now any vertex lying on C has the eccentricity r except x_{r-1} whose eccentricity is $r + 1$. Next let $g \in V(G)$, where $g \neq x, y$. Now,

$$d_H(g, x_{r-1}) = d_H(g, x_{2r-2}) + d_H(x_{2r-2}, x_{r-1}) = 1 + 2r - 2 - r + 1 = r,$$

and $d_H(g, x_k) < r$ for $1 \leq k \leq 2r - 2$ for $k \neq r - 1$. This implies that $e_H(g) = r$, where $g \in V(G)$. Thus, every vertex of H has the eccentricity r except x and x_{r-1}, where $e(x) = e(x_{r-1}) = r + 1$. Thus, H is r-ASC graph where G is induced.

By Theorem 2.1, the r-ASC index of G is either $2r - 3$ or $2r - 2$. Next we construct graphs for which lower and upper bounds of Theorem 2.1 are achieved.

Theorem 2.2 *Let G be the graph obtained by attaching a pendant vertex x to a complete graph K_n, $n \geq 3$. Then $\theta_r(G) = 2r - 2$ for $r \geq 4$.*

Proof It is clear that $diam(G) = 2$. By Theorem 2.1, it is sufficient to show that $\theta_r(G) \geq 2r - 2$. Let x be the pendant vertex attached to K_n and $vx \in E(G)$ for some

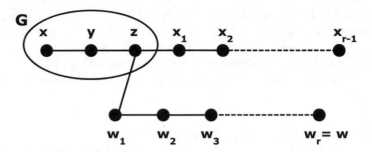

Fig. 3 P_1 intersect neither $G - \{z\}$ nor $\{x_1, x_2, \ldots, x_{r-1}\}$

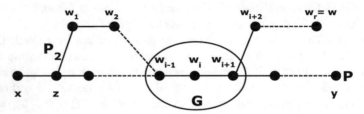

Fig. 4 Graph H : when G does not contain any diametral vertex

Fig. 5 Embedding of diameter two graph G in an r-ASC graph

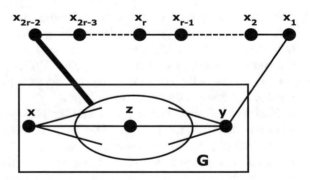

$v \in V(K_n)$. Let H be an r-ASC embedding graph in which graph G is induced. We have following two cases here depending upon eccentricity of vertex x in H.

Case (a) x is a diametral vertex of H, i.e., $e_H(x) = r + 1$.

Then there exists a vertex y and a diametral path $P : x = x_0, x_1, \ldots, x_{r+1} = y$ in H such that $d(x, y) = r + 1 = l(P)$. The path P contains at the most three vertices of G and thus at least $r + 2 - 3 = r - 1$ new vertices. Let x_2 be a vertex of K_n such that $d_G(x, x_2) = 2$. Since $e_H(x_2) = r$ and $diam(K_n) = 1$, if we proceed like Theorem 2.1, we need to have at least $r - 1$ new vertices. Hence finally, we will have at least $2r - 2$ new vertices and thus get the upper bound.

Case (b) x is not a diametral vertex in H, i.e., $e_H(x) = r$.

In this case, we have following subcases depending whether K_n contains a diametral vertex or not.

Fig. 6 Construction of
graph with $\theta_r(G) = 2r - 2$

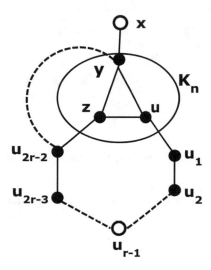

1. K_n contains a diametral vertex, say y. Then there exists vertex z in H such that $d_H(y, z) = r + 1$. Let $P_1 : y = y_0, y_1, \ldots, y_{r+1} = z$ be a diametral path. Since $diam(G) = 2$, then P_1 contains at the most three vertices from G, and hence, $r + 2 - 3 = r - 1$ vertices. Let z be a vertex adjacent to x. Then $e(z) = r$. There exists a vertex z' and a path $P_2 : z = u_0, u_1, \ldots, u_r = z'$, where each u_i is different from y_i. Since $e_G(y) = 1$, P_2 can at the most contain two vertices of G. This follows that at least $r + 1 - 2 = r - 1$ vertices of P_2 are not in G and different from y_i. Thus, in this case, we need at least $2r - 2$ number of new vertices and hence the upper bound.

2. K_n does not contain a diametral vertex.

To finish the proof, we need to prove that $\theta_r(G) \leq 2r - 2$. Let y and u be vertices in G such that $d_G(x, y) = 1$ and $d_G(x, u) = 2$. Let H be a graph obtained by attaching a path $P : u_1, u_2, \ldots, u_{2r-2}$ to G such that $uu_1, yu_{2r-2}, zu_{2r-2} \in E(H)$ for some vertex $z \in G$ and $z \neq x, y, u$. Then x and u_{r-1} are diametral vertices with eccentricity $r + 1$. The construction is explained in Fig. 6.

Theorem 2.3 *Let G be a graph obtained by attaching two or more pendant vertices to the same vertex of a complete graph K_n, $n \geq 3$. Then for $r \geq 4$ we have $\theta_r(G) = 2r - 3$.*

Proof It is clear that $diam(G)$ is 2. From Theorem 2.1, $\theta_r(G) \geq 2r - 3$. To finish the proof, we need to prove that $\theta_r(G) \leq 2r - 3$ with the help of a construction. We have following two cases.

1. Exactly two pendant vertices in G.

   Let x, y be pendant vertices of G and z be a vertex in K_n adjacent to both x and y in G. Consider a vertex u so that $d_G(x, u)$ is 2. Let H be the graph constructed from G by adding vertices $u_1, u_2, \ldots, u_{2r-3}$ so that $uu_1, zx \in E(H)$

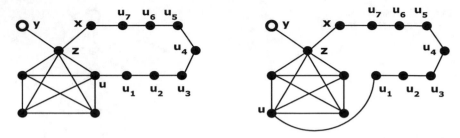

Fig. 7 Embedding of K_5 into 5-ASC graph

Fig. 8 Construction of
graph with $\theta_r(G) = 2r - 3$

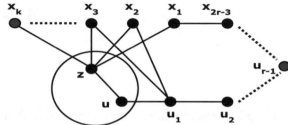

and $C : x, z, u, u_1, u_2, \ldots, u_{2r-3}, x$ forms a cycle. It is easy to evaluate that $d_H(y, u_{r-1}) = r + 1$ and rest of the vertices of H have eccentricity r. Thus, H is an r-ASC graph where G is induced. This case is illustrated with the help of an example in Fig. 7, where we considered different u in each case. In Fig. 7, we embed K_5 into an 5-ASC graph, where vertices y and u_4 are diametral vertices with eccentricity six.

2. More than two pendant vertices in G.

   Let $k \geq 3$ and $x_1, x_2, \ldots, x_k$ be pendant vertices attached to the vertex z, where $z \in V(K_n)$. Let $u \in V(G)$ so that $d_G(u, x_1) = 2$. Let H be a graph we obtain by adding $2r - 3$ vertices $u_1, u_2, \ldots, u_{2r-3}$ to G such that $C : x_1, z, u, u_1, u_2, \ldots, u_{2r-3}, x_1$ induces a cycle in G. Also draw the edges $x_i u_1$ for every $2 \leq i \leq k - 1$. It can be noted that eccentricity of vertices x_k and u_{r-1} is $r + 1$ and hence are diametral vertices. Also, rest of the vertices in H have eccentricity r and hence the result. The construction is given in Fig. 8.

3 R-ASC Index of Paths and Cycles

In this section, we compute r-ASC index of cycles and paths. In the the first theorem, we calculate r-ASC index of paths.

Theorem 3.1 *Let P_n be a path on $n \geq 2r + 1$ vertices for some positive integer $r \geq 4$. Then we have the following.*

$$\theta_r(P_n) = \begin{cases} 1, & r = \lfloor \frac{n-1}{2} \rfloor, \\ 2, & 4 \leq r < \lfloor \frac{n-1}{2} \rfloor. \end{cases} \tag{1}$$

Proof It is known that paths are not almost self-centered graphs and thus $\theta_r(P_n) > 0$. Let $P_n : v_1, v_2, v_3, \ldots, v_n$. We have following two cases now.

1. Let $r = \lfloor \frac{n-1}{2} \rfloor$.

   When n is even, we construct a graph G with $V(G) = V(P_n) \cup \{x\}$ and $E(G) = E(P_n) \cup \{xv_1, xv_2, xv_3\} \cup \{xv_{n-1}, xv_n\}$. Then v_1 and v_{r+2} are diametral vertices with eccentricity $r + 1$. The vertex x lie on a cycle $C : x, v_3, v_4, \ldots, v_{n-2}, x$, and we get $e(x) = r$. Eccentricity of rest of the vertices in G is r. So G is an r-ASC graph. Similarly, when n is odd, graph H is constructed as below. $V(H) = V(P_n) \cup \{x\}$ and $E(H) = E(P_n) \cup \{xv_1, xv_2, xv_n\}$. Obviously, H is the graph C_{2r+1}^*. Therefore, H is an r-ASC graphs (by Remark 1.2). This proves that $\theta_r(P_n) = 1$ for $r = \lfloor \frac{n-1}{2} \rfloor$.

2. Let $4 \leq r < \lfloor \frac{n-1}{2} \rfloor$.

   To prove the result, it is sufficient to show that $\theta_r(P_n) \geq 2$. If possible, let $\theta_r(P_n) = 1$. Then there exists an r-ASC graph G with $V(G) = V(P_n) \cup \{x\}$.

   First, let $e_G(x) = r + 1$. Let $d(x, v_i) = r + 1$ for some vertex v_i in P_n, $1 \leq i \leq n$. As $r < \frac{n}{2}$, we have $n \geq 2r + 1$. Since $d(x, v_i) = r + 1$, vertex x cannot be adjacent to $v_{i+1}, v_{i+2}, \ldots, v_{i+r}, v_{i-1}, v_{i-2}, \ldots, v_{i-r}$. Note that depending on values of i, at most one of v_{i+r} and v_{i-r} may not be a vertex in P_n. We have following subcases now.

   a. Either v_{i+r} or v_{i-r} is a vertex in P_n.

      Without loss of generality, let v_{i+r} be a vertex in P_n. Since $n \geq 2r + 1$, either the vertex v_{i-1} or v_{i+r+1} (at least one of these have to be a vertex in P_n) has eccentricity $r + 1$ or more which is a contradiction.

   b. Both v_{i+r} and v_{i-r} are vertices in P_n.

      Since $n \geq 2r + 1$, either v_{i-r-1} or v_{i+r+1} (at least one of them exists in P_n) will have eccentricity $r + 1$, which contradicts to the fact that there are exactly two diametral vertices.

   Next, let x is not a diametral vertex, i.e., $e_G(x) = r$. Then P_n contains both the diametral vertices. Let these vertices be v_i and v_j, $i < j$. Since $n \geq 2r + 1$ and $d_G(v_i, v_j) = r + 1$, there are some vertices of P_n which appear before v_i or after v_j. Eccentricity of these vertices in G has to be r, and for this, only possibility is that they have to be adjacent with x. Since $e_G(x) = r \geq 6$, x will have a neighbor in between v_i and v_j. But in this case, $d(v_i, v_j)$ reduces and we get a contradiction. Hence, we have $\theta_r(P_n) \geq 2$.

   To finish the proof, we need to show that $\theta_r(P_n) \leq 2$. To prove this, we construct an r-ASC graph G in which P_n is induced, where $V(G) = V(P_n) \cup \{x, y\}$. Now vertex x is made adjacent to vertices v_2, y, and v_i for every $i \geq 2r$. Also, y is attached to vertices v_j for every $2r - 1 \leq j \leq n$. This gives a cycle $C : v_2, v_3, v_4, \ldots, v_{2r}, x, v_2$ of length $2r$ to which a pendant v_1 is attached ($v_1v_2 \in E(H)$). We see that G is an r-ASC graph in which P_n is induced, where v_1 and v_{r+2} are diametral vertices. The construction is demonstrated in Fig. 9, where we embed path P_{15} into a 6-ASC graph.

   From the above two cases, we conclude that $\theta_r(P_n) = 2$ and hence the result.

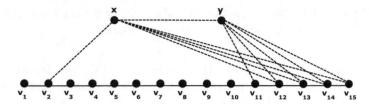

Fig. 9 Embedding of path P_{15} into 6-ASC graph H

In the following theorem, we calculate r-ASC index for cycles C_n for particular range of r.

Theorem 3.2 *Let C_n be a cycle on $n \geq 2r + 2$ vertices for some positive integer $r \geq 2$. Then*

1. *For $n = 2r + 2$ and $r = \lfloor \frac{n}{2} \rfloor - 1$ we have $\theta_r(C_n) = 1$.*
2. *For $n > 2r + 2$ and $2 < r < \lfloor \frac{n}{2} \rfloor - 1$, $\theta_r(C_n) = 2$.*

Proof Let $C_n : v_1, v_2, v_3, \ldots, v_n, v_1$ be a cycle.

1. We prove this part of the theorem with the help of a construction. Let $G = C_n = C_{2r+2}$ and $k = \frac{n}{2} + 1$. We check that $d(v_1, v_k) = \frac{n}{2}$ in cycle C_n. Let H be the graph constructed by adding a vertex x to G such that x is adjacent to $v_{k-2}, v_{k-1}, v_{k+1}$, and v_{k+2}. We shall prove that H is an r-ASC graph and C_n is induced in H. The construction is shown in Fig. 10. We see that v_k is the eccentric vertex of v_1 at a distance $r + 1$ and $d_H(v_1, v_i) \leq r$ for every $v_i \in V(C_n) \setminus \{v_k\}$. This gives $e_H(v_1) = e_H(v_k) = r + 1$. In the graph H, we obtain a cycle $C : v_1, v_2, \ldots, v_{k-2}, x, v_{k+2}, v_{k+3}, \ldots, v_n, v_1$ of length $2r$. Since x is adjacent to $v_{k-2}, v_{k-1}, v_{k+1}, v_{k+2}$, and x lies on C, we get $e_H(x) = r$. With the help of the construction, we can easily conclude that $e_H(v_i) = r$ for every $v_i \in V(C_n) \setminus \{v_1, v_k\}$. Thus, H is r-ASC graph in which C_n is induced, and this proves the result.

2. First, we shall show that $\theta_r(C_n) \geq 2$. Suppose G is an r-ASC graph with $V(G) = V(C_n) \cup \{x\}$. Let $e(x) = r + 1$. Then we get a v_i–v_j path P of length $2r$ such that x is adjacent to both v_i, v_j, and there is a vertex $v_k \in V(P)$ with $d_G(x, v_k) = r + 1$. Now, for any $v_s \in V(C_n) \setminus V(P)$, $d_G(v_k, v_s) \geq r + 1$, a contradiction for G to be an r-ASC graph.

    Further, if $e(x) = r$, then there exists v_i–v_j path P_1 of length $2r - 1$ such that x is adjacent to both v_i, v_j, and for some $v_t \in V(P_1)$, we have $d_G(x, v_t) = r$. It follows that if $e(v_t) = r$, then $d_G(v_t, v_i) = r = d_G(v_t, v_j)$. Then every vertex in $X = V(C_n) \setminus V(P_1)$ must be adjacent to v_{i+1} or v_{j-1} which is impossible as C_n contains vertices with degree two only. Finally consider $e(v_t) = r + 1$. Since $n \geq 2r + 2$, we have $|X| \geq 2$. So we get a vertex v_k in X so that $d_G(v_t, v_k) = r + 1$. However, for some $v_{s_1} \in P_1$ there is a vertex v_{s_2} with $d(v_{s_1}, v_{s_2}) \geq r + 1$, which is not possible. Thus $\theta_r(C_n) \geq 2$.

    To finish the proof, we need to prove $\theta_r(C_n) \leq 2$. For this, we construct an

Fig. 10 Embedding of cycle
C_{2r+2} into r-ASC graph for
$r = \lfloor \frac{n}{2} \rfloor - 1$

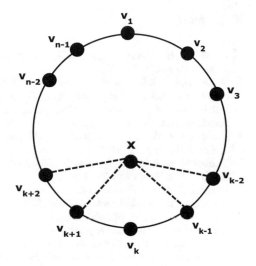

r-ASC graph H, whose vertex set is $V(C_n) \bigcup \{x, y\}$ and edge set is $E(C_n) \cup \{x_1 x, xy, yx_{2r-2}\} \cup \{xx_i : 2r-1 \leq i \leq n\} \cup \{yx_j : 2r-1 \leq j \leq n-2\}$, see Fig. 11. Since $n \geq 2r+2$, we get a path $P_1 : x_1, x_2, x_3, \ldots, x_{2r-2}$ of length $2r-3$ which gives a cycle $C : x_1, x_2, x_3, \ldots, x_{2r-2}, y, x, x_1$ of length $2r$. From the construction, it can be easily seen that $d_H(x, x_r) = r$, and hence, $d_H(x_{n-1}, x_r) = r+1$. As x is adjacent to vertices x_{n-1} and x_i, $2r-1 \leq i \leq n$, we get $d(x_{n-1}, x_i) = 2$. This proves that x_{n-1} and x_r are diametral vertices with eccentricity $r+1$. Let $n-2 \leq j < 2r-1$. Since vertices x_j are adjacent to x and y, the vertices farthest from x_j lie on C at a distance r. Hence, $e(x_j) = r$. Next for $x_k \in V(C) \setminus \{x_r\}$, the vertex farthest from x_k is at a distance r (since C is of length $2r$). Thus, the constructed graph is an r-ASC graph and hence the result.

Fig. 11 Embedding of cycle
C_n into r-ASC graph

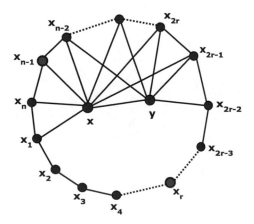

References

1. Klavžar, S., K.P. Narayankar, and H.B. Walikar. 2011. Almost self-centered graphs. *Acta Mathematica Sinica* 27: 2343–2350.
2. Xu, K., H. Liu, K.C. Das, and S. Klavžar. 2017. Embeddings into almost self-centered graphs of given radius. *Journal of Combinatorial Optimization* 36 (4): 1388–1410.
3. Balakrishnan, K., B. Bresar, M. Changat, S. Klavzar, M. Kovse, and A. Subhamathi. 2010. Simultaneous embeddings of graphs as median and antimedian subgraphs. *Networks* 56: 90–94.
4. Buckley, F., Z. Miller, and P.J. Slater. 1981. On graphs containing a given graph as center. *Journal of Graph Theory* 5: 427–434.
5. Dankelmanna, P., and G. Sabidussib. 2008. Embedding graphs as isometric medians. *Discrete Applied Mathematics* 156: 2420–2422.
6. Graham, R.L., and P.M. Winkler, On isometric embeddings of graphs. *Transactions of the American Mathematical Society* 288 (2), 527–536
7. Janakiraman, T.N., M. Bhanumathi, and S. Muthammai. 2008. Self-centered super graph of a graph and center number of a graph. *Ars Combinatoria* 87: 271–290.
8. Klavžar, S., H. Liu, P. Singh, and K. Xu. 2017. Constructing almost peripheral and almost self-centered graphs revisited. *Taiwanese Journal of Mathematics* 21 (4): 705–717.
9. Ostrand, P.A. 1973. Graphs with specified radius and diameter. *Discrete Mathematics* 4: 71–75.

Computationally Efficient Scheme for Simulation of Ring Oscillator Model

Satyavir Singh, Mohammad Abid Bazaz and Shahkar Ahmad Nahvi

Abstract This work addresses the computational difficulties involved with the simulation of the nonlinear ring oscillator system in the offline and online phase. Conventional POD has offline basis extraction burden along with online computation in nonlinear systems. The computation of the nonlinear term can be improved with a discrete empirical interpolation method (DEIM). However, the issue of offline computation in POD still persists. This work proposed approximate snapshot ensemble generation for basis extraction in nonlinear model order reduction to improve offline computation. Hence, proposed approach reduces offline basis computation, and DEIM reduces the online computational burden. This approach is tested on a large ring oscillator model with two different sets of inputs.

Keywords Ring oscillator · Proper orthogonal decomposition · Approximate snapshot ensemble basis · Discrete empirical interpolation method

1 Introduction

Two nonlinear models, ring oscillator and cross-coupled oscillators are most commonly used in electronic circuits. Application area of ring oscillator is clock recovery circuits, disk drive channels and function generators and frequency synthesizers [1]. In [2], the authors have applied linear system techniques for the study of frequency oscillation and magnitude of nonlinear feedback oscillators.

S. Singh (✉) · M. A. Bazaz
Department of Electrical Engineering, National Institute of Technology, Srinagar, India

M. A. Bazaz
e-mail: abid@nitsri.net

S. A. Nahvi
Department of Electrical Engineering, IUST, Awantipora, India
e-mail: shahkar.nahvi@islamicuniversity.edu.in

© Springer Nature Singapore Pte Ltd. 2020
Y.-C. Hu et al. (eds.), *Ambient Communications and Computer Systems*, Advances in Intelligent Systems and Computing 1097,
https://doi.org/10.1007/978-981-15-1518-7_16

A ring oscillator has various delay stages with response to the excitation of the primary stage from the last stage. To achieve oscillation frequency, the ring should give a phase shift of 2π and unit gain of voltage at that frequency. The individual stage of delay provides a phase shift of π/n, where n is the delay stage number. The remainder of the phase shift is a DC inversion. This means that an odd number of stages in the DC inversion case are important for an oscillator with single-ended stages. When different stages are used, the ring has an even number of stages. Eventually, the feedback lines are swapped, and each oscillator in the ring is controlled by the previous one [3]. Every component contains a negative gain for an odd number n; therefore, interconnection behaves as a negative feedback system.

This paper describes a method for constructing proper orthogonal decomposition (POD)-based reduced order models (ROMs) from simulations of transient full order models (FOMs), ignoring parameter dependence. The proposed method reduces the offline costs of ROM snapshot selection and basis computation in both memory and simulation time relative to existing methods in the literature [4]. It is intended to be used in constructing ROMs from large-scale FOM simulations. This approach required snapshots of FOM and imposed a priori heavy computational burden in extracting snapshots [5]. Approximate snapshot ensemble basis generation scheme is proposed to reduce aforementioned limitation of the offline basis extraction implications in POD, called APOD. However, this procedural scheme will not reduce the computational burden of nonlinear function in POD. Discrete empirical interpolation method (DEIM) is a motivating scheme to reduce the computational burden of the nonlinear function and proposed for the POD [6]. Hence, a hybridized APOD-DEIM scheme is developed for the numerical computation of a nonlinear system.

This paper is organized as follows: Mathematical model of nonlinear ring oscillator system is presented in Sect. 2. In Sect. 3, the nonlinear reduction strategies are reviewed, and subsequent modifications are proposed. This is accompanied with a numerical validation of hybridized APOD-DEIM scheme on a nonlinear ring oscillator model in Sect. 4.

2 Mathematical Model of a Ring Oscillator system

A nonlinear oscillator is selected with an odd number of digital inverters connected to a feedback loop. Figure 1 shows the proposed model for a cascaded - tanh(.) nonlinearity and a low-pass filter (LPF) in digital inverters. The model of nonlinear term is expressed as $V_n = V_0 \tanh(V_i/V_s)$, where V_0 is a saturation voltage, V_i input

Fig. 1 Low-pass filter with nonlinearity

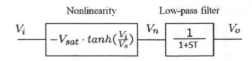

Fig. 2 n-stage ring oscillator model with feedback

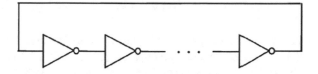

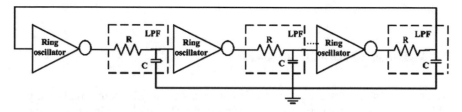

Fig. 3 Low-pass filter with ring oscillator model

voltage, and V_s is used to the internal slope adjustment [7]. An inverter's behavior is described by the following differential equation.

$$V_o + T\dot{V}_o = -V_{sat}\tan(V_i/V_s)$$

The model is formulated in state space as follows:

$$\dot{V}_o = -1/T * V_0 - V_{sat} * V_i/T * \tanh(V_i/V_s)$$

$$\dot{x} = -\eta x - \beta\tanh(\alpha x)$$

where $\eta = 1/T, \alpha = 1/V_s, \beta = V_{sat} * V_i/T$ In the form of feedback loop (as shown in Figs. 2 and 3), system model can be expressed as follows:

$$
\begin{aligned}
\dot{x}_1 &= -\eta_1 x_1 - \beta_1\tanh(\alpha_1 x_n)\\
\dot{x}_2 &= -\eta_2 x_2 - \beta_2\tanh(\alpha_2 x_1)\\
\dot{x}_3 &= -\eta_3 x_1 - \beta_3\tanh(\alpha_3 x_2)\\
\dot{x}_n &= -\eta_n x_n - \beta_n\tanh(\alpha_n x_{n-1})
\end{aligned}
\tag{1}
$$

Since all n blocks are having gain with negative sign. Therefore, the interconnection is a negative feedback system, provided n is an odd number [8]. The condition of global stability is important under negative feedback system because it serves as a necessary condition for the existence of limit cycles [9]. Under the odd number n, the similarity transformation $D = diag(d_1,\ldots d_n)$ with $d_i = (-1)^{i+1}$ results in compact form. This work addressed n identical characteristic models of oscillators. The input $u(t)$ is representing the coupling between the oscillator models. The stages of ring oscillator are similar, $C_1 = C_2 = \cdots = 2\mu F$ and $R_1 = R_2 = \cdots = 1k$. The time constant of circuit, $T = RC$ and inverse of internal slope factor , $\alpha_1 = \alpha_2 = \cdots = 5$. The oscillator model formulated as follows:

$$\dot{x}(t) = -\eta x(t) - \beta \tanh(\alpha x(t)) + u(t) \tag{2}$$

A linearized form applies to cyclic systems and states that the ω-limit describes the solutions of a sustained point or a periodic orbit [3]. The simulation of the nonlinear ring oscillator model is challenging as due to associated nonlinear term. Interpolation method is introduced to evaluate it in addition to improved basis extraction scheme in the offline computation.

3 Methods of Nonlinear MOR

Given the high computational costs, it makes sense to minimize simulation time while improving the design criteria. The unpredictability of nonlinearity should be taken care of by some resource for numerical computation. The nonlinear model order reduction is an infancy field of research in ROM. Along these lines, various strategies have been proposed to compute nonlinearity effectively. The standard formulation of the model is given as follows:

$$\begin{aligned} \dot{x}(t) &= f(x(t)) + Bu(t) \\ y(t) &= Cx(t) \end{aligned} \tag{3}$$

In (3) nonlinear term, $f(x) = -\eta x(t) + \beta * \tanh(\alpha x(t))$. The approximation error and its accuracy are associated with the adopted methodology of MOR [10]. The MOR methodologies are described in the next sections.

3.1 Proper Orthogonal Decomposition

The economic reasons motivate to adapt snapshots-based technique [10]. It can be collected by some physical experiments or short time simulation stacked at regular intervals.

$$\chi = [x_{(t_1)}, \dots, x_{(t_{n_s})}] \in R^{n \times n_s}, \tag{4}$$

where n_s are the number of snapshots and should be captured as many as so that system dynamics is addressed.

Now performing the singular value decomposition (SVD) on snapshot matrix χ

$$\chi = U \Sigma W^T \tag{5}$$

The aim is to approximate the state vector $x(t) \in R^n$ by $V x_k(t)$, where $x_k(t) \in R^k$ with $k \ll n$ is the state vector of ROM, and $V = U(:, 1 : k)$ is the orthogonal projection matrix, obtained by truncating k dominant singular values of Σ, i.e., $\sigma_1 \geq \sigma_2 \cdots \geq \sigma_k$, on main diagonal and other elements are zero. Dynamical behavior of full order model (FOM) lies in k-dimensional space of approximated high fidelity model.

The error in the approximation of high fidelity model (5) in reduced state vector $(\hat{\chi} = V_k \Sigma_k W_k^T)$ is given as follows:

$$||\chi - \hat{\chi}||_F^2 = \sum_{i=k+1}^{n} \sigma_i^2 \tag{6}$$

which is optimal 2-induced norm [10]. Finally, constructing ROM for nonlinear dynamical system given in (3) by applying Galerkin projection $x(t) = V_k x_k(t)$, the reduced system is of the form as follows:

$$\dot{\tilde{x}}(t) = V_k^T f(V_k \tilde{x}(t)) + V_k^T B u(t) \tag{7}$$

Here, $\tilde{x}(t) \in R^k$ and $x(t) \approx V_k \tilde{x}(t)$ and $k \ll n$. The formulated ROM with the choice of dominant SVD with Galerkin projection reside in k-dimensional reduced subspace; whereas, dimension of nonlinear term will not reduced [10].

3.2 Approximate Snapshot Ensemble Basis Extraction

The techniques used for the simulation of nonlinear models are not well developed. Most of the techniques are based on linearization schemes. In (3), let m be linearization points, $(x_0, \ldots, x_i, \ldots, x_{m-1})$ are selected along a training trajectory, and the Taylor series first-order approximation of $f(x)$ about x_i is given by

$$\tilde{f}(x) = f(x_i) + A_i(x - x_i) \tag{8}$$

where A_i is the Jacobian of $f(x)$ evaluated at x_i. The linear system at x_i is hence given by

$$\left. \begin{array}{l} \dot{x} = f(x_i) + A_i(x - x_i) + Bu \\ y = Cx \end{array} \right\} \tag{9}$$

The dynamics is restricted around a point of linearization only. The objective is to extract a subsequent linearization model along the trajectory of the nonlinear model based on Euclidean distance $|x - x_i|$, where x_i is the point of linearization. This process creates subsequent localized linear models. These models are expected to be a good approximation of the nonlinear system for that part of the state space which has been covered by the training trajectory. It is also expected that the subsequent models would remain a valid approximation for any other input called the evaluation input. This method of subsequent linearization would naturally create new linear models at places where the behavior of previous ones starts becoming changed from particular selected distance [11]. Pseudo-code is given in Algorithm 1.

Algorithm 1 Approximate snapshot ensemble generation and basis extraction

---
1: $i \leftarrow 0$, $j \leftarrow 0$, $X = [x_j]$. x_0: initial state, T: number of simulation time steps, δ is an appropri-
   ately selected constant.
2: Linearize the nonlinear system at x_i

$$\begin{cases} \dot{x} = A_i x + (f(x_i) - A_i x_i)) + Bu \\ y = Cx \end{cases} \tag{10}$$

3: Simulate (10) with initial condition $x = x_j$ for one time step
4: $X = [X \quad x_{j+1}]$;   ▷ *Store the state-snapshots*
5: **if** $min_{(0 \leq k \leq i)} \frac{||x_{j+1} - x_k||}{||x_k||} > \delta$ **then**
6:     $i \leftarrow i + 1$, $j \leftarrow j + 1$. go to (2)
7: **else**
8:     $j \leftarrow j + 1$ go to (3)
9: Snapshot matrix=$X = [x_1 \quad x_2 \dots \quad x_T]$
10: Find the SVD of X, $X = U \Sigma Q^T$
11: Retain columns of U corresponding to singular values larger than some ϵ.
12: The projection matrix is $V_{n \times k} = \{u\}_{i=1}^{k}$.

3.3 Discrete Empirical Interpolation Method (DEIM)

The DEIM is a discrete variation of the empirical interpolation method (EIM) proposed by Barrault et al. [12] in regard to dimensional reduction in nonlinear function efficiently [12]. In the previous section, we have addressed ROM via the snapshots of the states. In this section, the ROM is addressed with the snapshots of the nonlinear term. Nonlinear function of (7) can be written as follows:

$$\tilde{N}(\tilde{x}(t)) = V_k^T f(V_k \tilde{x}(t)) \tag{11}$$

The function $\tilde{N}(\tilde{x})$ has dimension n, which is not reduced. However, the state vector size is reduced to k from n. Hence, to address this issue, DEIM is proposed to reduce the computational burden of nonlinear function. The nonlinear function in (7) is approximated by projecting it onto a subspace came from snapshots of the nonlinear function. Its POD modes are selected corresponding to m dominant singular values, where $m \ll n$. From [13], the nonlinear function in (7) can be approximated as follows:

$$\tilde{f}(V_k \tilde{x}(t)) \approx U_m (P^T U_m)^{-1} P^T f(V_k \tilde{x}(t)) \tag{12}$$

where the nonlinear function is first projected onto the basis vectors spanned by the columns of $U_m \in R^{n \times m}$, then m distinguished rows are selected by premultiplying the whole system with P^T, $P = [e_{\rho_1}, \ldots, e_{\rho_m}] \in R^{n \times m}$, and e_{ρ_i} is the ρ_ith column of $I_n \in R^{n \times n}$. The basis vectors $U_m = [u_1, \ldots, u_m]$ for the nonlinear term $f(V_k \tilde{x}(t))$ can be constructed by performing the SVD of the snapshots matrix. The nonlinear function snapshots are given as follows:

$$F = [f(x_1(t)), \ldots f(x_{n_s}(t))]$$

The interpolation indices $\rho_1, \ldots, \rho_m$ are coming from the basis $[u_1, \ldots u_m]$ using the DEIM algorithm.

The matrix P for nonlinear function is given by minimizing function $f - Uc$ at each iteration.

Substituting Eq. (12) in Eq. (11), the nonlinear term can be written as follows:

$$\tilde{N}(\tilde{x}(t)) = V_k^T U_m (P^T U_m)^{-1} f(P^T V_k \tilde{x}(t)) \tag{13}$$

where $P^T f(V_k \tilde{x}(t))$ in (11) has been replaced by $f(P^T V_k \tilde{x}(t))$ in (13) if the function f evaluates component-wise at $x(t)$ (see [13]). Here, subspace matrix V_k is obtained from Algorithm 1. Further, the part $\Psi = V_k^T U_m (P^T U_m)^{-1}$ is constant and can be pre-computed cheaply. From (7), (11) and (13), the final reduced order model with APOD-DEIM is in the following form:

$$\dot{\tilde{x}}(t) = \Psi f(P^T V_k \tilde{x}(t)) + V_k^T Bu(t) \tag{14}$$

Algorithm 2 Reduced model using *DEIM* Algorithm

1: Input $U_m = \{u_i\}_{i=1}^m$
2: $[|p|, \rho_1] = max|u_1|$
3: $U_m = [u_1]$, $P = [e_{\rho_1}]$, $\rho = [\rho_1]$
4: **for** l = 2 to m **do**
5: Solve $(P^T U_m)c = P^T u_l$ for c
6: $r = u_l - U_m c$
7: $[|p|, \rho_l] = max|r|$
8: $U_m \leftarrow [U_m \quad u_l]$, $P \leftarrow [P \quad e_{\rho_l}]$
9: **end for**
10: Pre-compute offline $\Psi = V_k^T U_m (P^T U_m)^{-1}$
11: Reduced model, $\dot{\tilde{x}}(t) = \Psi f(P^T V_k \tilde{x}(t)) + V_k^T Bu(t)$

---

4 Numerical Validation

Consider the model given by (2) with system size, $n = 10000$ [3]. The size of reduced order model is having number of POD modes and DEIM modes, $k = 25$ and $m = 23$, respectively.

Case-1: Cosine input $u(t) = 7 \cos \omega t$, $\omega = 2\pi/3 \times 10^5$ Hz, $\epsilon = 0.001$. Computational burden reduces in basis extraction with proposed approach as compared to conventional POD. The orthonormal basis is extracted to reduce offline basis extraction time in a reduced order model as given in Algorithm 1. The computational burden of nonlinear term is reduced by Algorithm 2. It is observed that the simulation time for the basis significantly reduced with APOD-DEIM as compared to conventional POD-DEIM, as summarized in Table 1. The computational results are presented in Fig. 4. In Fig. 4, synchronizations of output voltages (limit cycle behavior) and error profile are shown. It can be seen that the output voltage is perfectly synchronized with insignificant errors in the reduced order modeling.

Case-2: Consider step input $u(t) = 9H(t)$, $\epsilon = 0.001$. Computational burden reduces in basis extraction with proposed approach as compared to conventional POD. The orthonormal basis is extracted to reduce offline basis extraction time in a reduced order model as given in Algorithm 1. The computational burden of nonlinear term is reduced by Algorithm 2. It is observed that the simulation time for the basis significantly reduced with APOD-DEIM as compared to conventional POD-DEIM, as summarized in Table 2.

The computational results are presented in Fig. 5. In Fig. 5, synchronizations of output voltages (limit cycle behavior) and error profile are shown. It can be seen that

Table 1 Comparison of computational savings and errors in APOD and APOD-DEIM

–	FOM	MOR approaches	
		APOD	APOD-DEIM
Sim. time	441.02	131.54	81.90
Error (in states) %	–	1.24×10^{-10}	4.51×10^{-11}

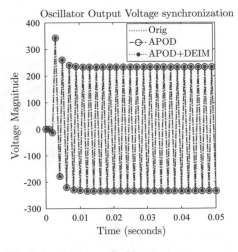
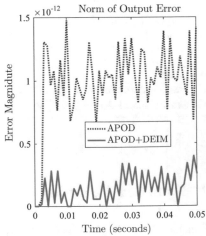

Fig. 4 Output voltage for input $7 \cos \omega t$

Table 2 Comparison of computational savings and errors in APOD and APOD-DEIM

–	FOM	MOR approaches	
		APOD	APOD-DEIM
Sim. time	366.89	120.40	74.59
Error (in states) %	–	3.10×10^{-10}	2.81×10^{-10}

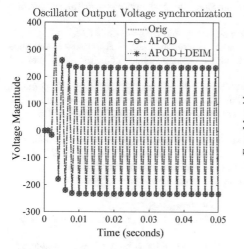
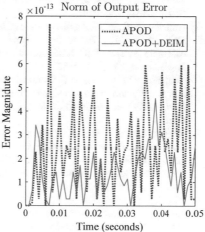

Fig. 5 Output voltage for input $9H(t)$

the output voltage is perfectly synchronized with insignificant errors in the reduced order modeling. From the above comparative analysis, it is clear that the basis with a nonlinear term approximation gives an appropriate ROM model. Generally, this model has an error that quickly settles to the steady-state values. This shows that the limit cycle behavior still persists. The stability preserved and savings in computation are significant with a reduced basis extraction time.

5 Conclusion

This work presents a successful attempt to implement approximate snapshot ensemble bases extraction and DEIM for POD scheme. The approximate snapshot ensemble algorithm is addressed to improve the offline simulation time of the conventional POD technique, and DEIM improves the computation of the nonlinear function. Hence, overall computational savings in the hybridized APOD-DEIM scheme are verified with a nonlinear ring oscillator model. The proposed model is preserved system stability and accuracy in all test cases. The approach can be extended to improve the simulation time of the available MOR techniques with an insignificant error.

References

1. Hajimiri, A., S. Limotyrakis, and T.H. Lee. 1999. Jitter and phase noise in ring oscillators. *IEEE Journal of Solid-State Circuits* 34 (6): 790–804.
2. Borys, A. 1987. Elementary deterministic theories of frequency and amplitude stability in feedback oscillators. *IEEE Transactions on Circuits and Systems* 34 (3): 254–258.
3. Singh, S. 2017. Efficient simulation of nonlinear ring oscillator via deim with approximate basis. In *2017 International conference on computing, communication, control and automation (ICCUBEA)*, 1–5. IEEE.
4. Antoulas, A.C. 2005. *Approximation of large-scale dynamical systems*. SIAM.
5. Nahvi, S.A., M.A. Bazaz, M.-U. Nabi, and S. Janardhanan. 2014. Approximate snapshot-ensemble generation for basis extraction in proper orthogonal decomposition. *IFAC Proceedings Volumes* 47 (1): 917–921.
6. Chaturantabut, S., and D.C. Sorensen. 2009. Discrete empirical interpolation for nonlinear model reduction. In *Proceedings of the 48th IEEE Conference on decision and control, 2009 held jointly with the 2009 28th Chinese control conference. CDC/CCC 2009.*, 4316–4321. IEEE.
7. Roychowdhury, J., et al. 2009. Numerical simulation and modelling of electronic and biochemical systems. *Foundations and Trends® in Electronic Design Automation* 3 (2–3): 97–303.
8. Ge, X., M. Arcak, and K.N. Salama. 2010. *Nonlinear analysis of ring oscillator and cross-coupled oscillator circuits. Dynamics of continuous, discrete and impulsive systems series B: applications and algorithms.*
9. Khalil, H.K. 1996. *Noninear systems*. New Jersey: Prentice-Hall.
10. Kutz, J.N. 2013. *Data-driven modeling & scientific computation: methods for complex systems & big data*. Oxford University Press.
11. Singh, S., M.A. Bazaz, and S.A. Nahvi. 2018. A scheme for comprehensive computational cost reduction in proper orthogonal decomposition. *Journal of Electrical Engineering* 69 (4): 279–285.
12. Barrault, M., Y. Maday, N.C. Nguyen, and A.T. Patera. 2004. An empirical interpolation method: application to efficient reduced-basis discretization of partial differential equations. *Comptes Rendus Mathematique* 339 (9): 667–672.
13. Chaturantabut, S., and D.C. Sorensen. 2010. Nonlinear model reduction via discrete empirical interpolation. *SIAM Journal on Scientific Computing* 32 (5): 2737–2764.

Web and Informatics

Comparative Analysis of Consensus Algorithms of Blockchain Technology

Ashok Kumar Yadav and Karan Singh

Abstract In today's era of big data and machine learning, IoT is playing a very crucial role in nearly all areas like social, economic, political, education, health care. This drastic increase in data creates security, privacy, and trust issues in the era of the Internet. The responsibility of IT is to ensure the privacy and security for huge incoming information and data due to the drastic evolution of the IoT in the coming years. The blockchain has emerged as one of the major technologies that have the potential to transform the way of sharing the huge information and increase trust among. Building trust in a distributed and decentralized environment without the call for a trusted third party is a technological challenge for researchers. Due to the emergence of IoT, the huge and critical information is available over the Internet. The trust over the information is reduced drastically, causing an increase in security and privacy concern day by day. Blockchain is one of the best-emerging technologies for ensuring privacy and security by using cryptographic algorithms and hashing. We will discuss the basics of blockchain technology, consensus algorithms, comparison of important consensus algorithms, and areas of application.

Keywords Consensus algorithm · Merkle tree · Hashing · DLT · Hash cash

1 Introduction

Recent advancements of wireless communication, computing power, Internet, big data, cloud computing increase the data day by day. The drastic increase in data creates a lot of problems like security, privacy, trust, and authentication. The responsibility of IT is to ensure the privacy and security for huge incoming information and data due to the drastic evolution of the IoT in the coming years. The blockchain has emerged as one of the major technologies that have the potential to transform the way of sharing the huge information and trust to another. Building trust in the distributed

A. K. Yadav (✉) · K. Singh
School of Computer and Systems Sciences, Jawaharlal Nehru University, New Delhi, India
e-mail: ashokyadav88.jnu@gmail.com

K. Singh
e-mail: karan@mail.jnu.ac.in

© Springer Nature Singapore Pte Ltd. 2020
Y.-C. Hu et al. (eds.), *Ambient Communications and Computer Systems*, Advances in Intelligent Systems and Computing 1097,
https://doi.org/10.1007/978-981-15-1518-7_17

and the decentralized environment without a trusted third party is a technological advancement that has the potential to change upcoming scenarios of the society, industries, and organizations. In today's era of big data and machine learning, IoT is playing a very crucial role in nearly all areas like social, economic, political, education, health care. Disruptive technologies such as big data and cloud computing have been benefited by IoT. Due to the emergence of IoT, the huge and critical information is available over the Internet. The trust over the information is reduced drastically, causing an increase in security and privacy concern day by day. The blockchain is one of the best-emerging technologies for ensuring privacy and security by using cryptographic algorithms.

Blockchain technology has turned out from the concept of timestamping of digital document published by Stuart Haber and W. Scott Stornetta in 1991. Time stamping of a digital document is used to maintain the dignity and integrity of the digital document by a particular node [1]. Cryptocurrency Bitcoin has acquired so much fame implemented by Satoshi Nakamoto in the year 2009 [2]. There are many cryptocurrencies that exist, but no one gets the same as Bitcoin. It has emerged as a decentralized system. Blockchain technology considered and regarded as a public ledger. "Blockchain is an incorruptible decentralized digital public ledger of economic transactions that can be programmed to record not just financial transactions but also virtually everything of values to facilitate data decentralization, transparency, the immutability of digital ledger, security and privacy provenance, trust and finality in peer-to-peer network." Blockchain implemented as a digital ledger on top of the Web which can be seen as an analogy to SMTP, HTTP, or FTP running on top of TCP/IP. Blockchain is append-only, immutable and only updatable with the consent of peers within the network is possible, which can be performed using the built-in consensus mechanism [3].

Blockchain technology is the effective application of existing technology such as decentralization, hashcash, public ledger, consensus, Merkle tree, public-key encryption, and hashing algorithm. Decentralization can be considered as the first most important perspective of blockchain technology. Basically, decentralization is a platform where various peers can participate to generate block having the same authority and to cooperate. Every peer connected will have the same authority to make changes in public ledger if applicable. Network failure during the execution of the transaction does not affect the transaction too much because every peer makes their own separate network. Public ledger is documentation of every successful transaction which is available and sharable to all peers (in peer-to-peer network) (Fig. 1).

Structure and the size of the block are implementation-dependent. The maximum number of transactions that a block can contain depends upon the block size and the size of each transaction. Blockchain cannot guarantee the transaction privacy since the values of all transactions and balance for each public key are publicly visible. A block has block header, and block body block header contains block version, Merkle tree root hash, timestamp, N bits target threshold of a valid block hash, nonce and parent block hash. Block body contains transaction counter and transactions. Blockchain is a chain of blocks. It is considered longer the chain of blocks more will be prioritized to add a new block to provide privacy and security. All the blocks

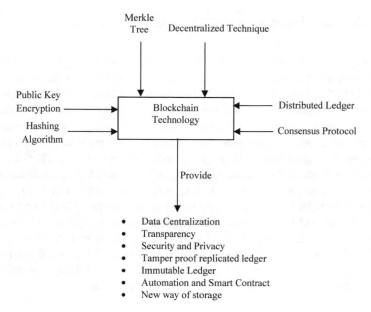

Fig. 1 Component of blockchain and output

connected in the chain can avail security, decentralization and permissionless facility in systems where any users can take participate without providing their identity [4]. It will lead to taking care of all malicious activity in the transaction phase. Mining is the solution to such a problem. Miners will decide the block size and transaction probability, and whether it will add to blockchain or not. If the answer is yes, then which chain will be used to add a block also decided by miners. Finally, after investigation new block will add at the longest chain. Making a bit change of nonce will affect the whole hash of all successor blocks. It is very difficult to identify the real hash value. Miners get some incentives to maintain their honesty with block size and transaction. Change in a transaction will provide a replica of that particular block to every peer connected with that transaction. Variation in transactions can be done by miners. It leads to an honesty problem and failing to design a secure and privacy system. Miner having high-power computing machine will get more incentive. The high-power computing machine consumes a huge amount of electricity. It is a major concern for the miner. The solution to this is to use an effective consensus algorithm. There are several consensus algorithms have proposed. We will investigate some most important consensus algorithms and do a comparative analysis [5].

2 Background and Types of Consensus Protocol

The concept of hashcash was suggested by Adam Back in 1997 [6]. Hashcash is a mining algorithm used as proof-of-work consensus algorithm (used for permissionless blockchain technology, i.e., Bitcoin). It is used to restrain e-mail and save such system from denial of attacks. Brute force method is the only way to implement the hashcash. The consensus algorithm is the heart of blockchain technology. The consensus is considered as the pillar of the blockchain network. Many consensus algorithms have been suggested to get system safe from any malicious activity in blockchain technology: Proof of work (PoW), proof of stake (PoS), delegated proof of stake (DPoS), practical byzantine fault tolerance (PBFT), etc., are some of them. Basically, consensus ensures the attainment of logical decision so every peer should agree whether a transaction should be committed in the database or not [7]. Blockchain uses the technique of hash function, Merkle tree, nonce (to make hash function harder to retrace) and others to provide data centralization, transparency, security and privacy, tamper proof replicated ledger, immutable ledger non-repudiation, irreversibility of records, automation and smart contract, a new way of storing.

2.1 Proof-of-Work (PoW) Consensus Algorithm

Proof of work was invented by Dwork and Moni Naor 1993 and formalized by Markus Jakobsson and Ari Juels in 1999. It ensures economic measures to prevent denial-of-service attacks. DoS attacks to prevent legitimate users from using the service. It is the asymmetry, i.e., hard on the requester side, but easy to check for the service provider. The proof of work prevents fraud nodes to get a grip of true nodes. The concept of PoW is used beyond blockchain. Ideally, the concept is to produce a challenge to a user, and the user has to produce a solution which should show some proof of work being done against that challenge. Once it is validated, the user accepted it. It eliminates the entity that is slow or not capable enough to generate PoW. In blockchain, PoW is used to generate a value that is difficult to generate and easy to verify. To generate block hash, there are n leading 0. It will help in solution and known as a nonce (Fig. 2).

The brute force method is applied to find the value of the nonce. The combination of the nonce and the block data which has generated, including the hash value of the previous block comes out with required leading 0. More is the value of n, more the complexity. PoW in the context of blockchain signifies that the computation required is exponential to the number of leading 0 required in PoW. As the blocks are chained, redoing will require an entire chain to be redone. It also signifies that some amount of computation and effort has been invested in finding the solution to the problem [5].

Since many of the miners working in the same permissionless network, it will be difficult to identify which miner will commit the block and verify the committed

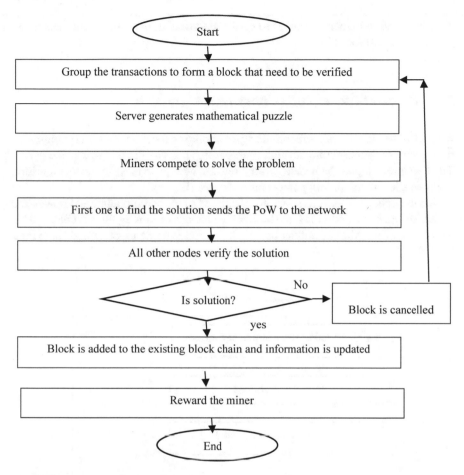

Fig. 2 Flowchart of PoW algorithm

transaction block. In PoW, miners are taking around 10 min to gather all the commit-ted transactions and generate a new block for them. So now what can be metadata contain in a block that should be previous block hash, block hash, Merkle tree, nonce, and it makes the attackers very hopeless unless attacker should not be mined that is why miners are awarded some incentives in form of cryptocurrency when they generate a new block [8]. PoW is also difficult for the miners to propose a new block (i.e., to find a nonce that will not affect the previous block hash), and miners must show their prior done work which he had generated before proposing a new block to other nodes. Timestamping should also be a factor of the block so that later peer cannot disagree on its transaction made. The major problem for the miners is to get how many zeros the hashcode should be generated. In PoW, node having high-power machine will perform more transactions to be committed and generate a new block.

This will lead to higher incentives toward node utilizing more powerful machine in generating new blocks [9].

2.2 Proof-of-Stake (PoS) Consensus Algorithm

In proof of work, miners are required to give a solution for the complex cryptographic hash problem. Miners compete with each other to become the first to find the nonce. The first miner solves the puzzle gets the reward. Mining in proof-of-work algorithm requires a lot of computing power and resources (Fig. 3).

All the energy is used to solve the puzzle. Higher the computational power, the higher the hash rate, and thus, the higher the chances of mining the next block. This leads to the formation of mining pools where miners come together and share their

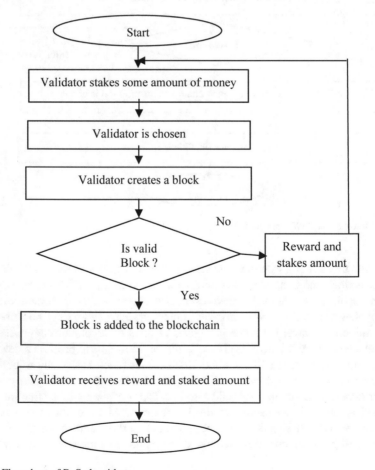

Fig. 3 Flowchart of PoS algorithm

computational power to solve the puzzle and share the reward among themselves. Proof of work uses a huge amount of electricity and encourages mining pools which take the blockchain toward centralization [10]. To solve these issues, a new consensus algorithm was proposed called proof of stake. A validator is chosen randomly to validate the next block. To become a validator, a node has to deposit a certain amount of coins in the network as a stake. This process is called staking/minting/forging. The chance of becoming the validator is proportional to the stake. The bigger the stake is, the higher the chances validate the block. This algorithm favors the rich stake. When a validator tries to approve an invalid block, he/she loses a part of the stake. When a validator approves a valid block, he/she gets the transaction fees and the stake is returned. Therefore, the amount of the stake should be higher than the total transaction fee to avoid any fraudulent block to be added. A fraud validator loses more coins than he/she gets. If a node does not wish to be a validator anymore, his/her stake, as well as transaction fees, is released after a certain period of time (not immediately as network needs to punish the node if he/she is involved in fraudulent block). Proof of stake does not ask for huge amounts of electrical power, and it is more decentralized.

It is more environmentally friendly than proof of work. 51% attack is less likely to happen with proof of stake as the validator should have at least 51% of all the coins which is a very huge amount. Proof of stake is performing for the more protective way and to use less usage of power to execute the transaction. Sometimes, a person having the more cryptocurrency (i.e., Bitcoin) will have more probability to mine new block, but again, it was arising the problem of dominance when a person having 50% or more and then it will have the highest probability to mine the block so the solution has been made in terms of some randomization protocol in which random nodes are selected to mine new block. Since it was also found that nodes are priory starting with PoW, they move to PoS for better and smoother usage.

In PoW, miners can mine only one block and choosing a wrong fork is costly for miners. If a miner chooses the wrong branch, then later, another branch ends up being the longest chain, and the miner's resources for mining the block are wasted. In PoS, the validators can forge multiple forks and choosing a wrong fork is not costly as miners did not spend expensive resources. Every other validator can work on multiple branches. A fraud validator can double spend with the money. A node can include a fraudulent block in one branch and wait for it to be confirmed by the service; once it is confirmed, the node can double-spend the money by including the block in the other branch [11]. A malicious validator can approve a "bad" block in one fork and a "good" block in the other. In case if the same validator again gets the chance to validate the blocks, he/she might work in "bad" branch, making the "bad" branch longer than the "good" one. Hence, other validators, too, may start working on the longest chain that includes a fraudulent block. In PoS, validators should have some amount of money for the stake. The problem is how the validators would manage to acquire money at the beginning when the PoS was at its initial stage. Proof of stake needs the coins to be distributed initially as the coins are needed for forging. In Pos, the attacker can go back to the previous blocks and rewrite the history. The attacker may buy some old private keys from the old validators who have lost interest in forging. A node staking a larger amount of money than the other nodes has more chances of becoming the validator.

2.3 Proof-of-Activity (PoA) Consensus Algorithm

Since PoA can be considered as the combination of proof of stack (PoS) and proof of work (PoW), PoA modifies the solution for PoS. If any miner wants to commit some transactions in a block as to mine a new block, then if that miner wants to commit that mined block into the database, and then most of every node sign the block for validation [12] (Fig. 4).

2.4 Proof-of-Elapsed Time Consensus Algorithm (PoET)

PoET algorithm suggests some common steps to select the miner that which would mine a new block. Each miner that had mined the prior block had waited for random time quantum to do so. Any miner which is proposing any new block to mine should wait for the random moment of time, and it will be easy to determine whether any miner which is proposed for the new block to mine has waited for some time or not by making a determination that a miner has utilized a special CPU instruction set (Fig. 5).

2.5 Practical Byzantine Fault Tolerance (PBFT)

PBFT algorithm concerns when one or more node in any network becomes faulty and behave maliciously that result in improper communication among all nodes connected to that network. Such things result in a delay in functioning, whereas time is a very serious concern as we already are working in an asynchronous system where if at least one fault occurs, then it would be impossible to solve the consensus problem. It will also generate discrimination in responses of various nodes. PBFT works for the permissioned model. In practical byzantine fault tolerance, state machine replication occurs at multiple nodes and the client will wait for an n + 1 response from all nodes where n is the number of faulty nodes, but it isn't giving the proper solution for this because $n + 1$ cannot determine the majority vote for the client. PBFT applies to the asynchronous system [13].

Generally, PBFT was attained after PAXOS and RAFT that both have maximum fault tolerance of $n/2 - 1$ among all nodes where n seems to be the number of faulty nodes [14]. However, PBFT is getting around $3n + 1$ response among all non-faulty nodes where n is determined as the faulty nodes. As we are discussing the state machine replication, then it is important to understand it (Fig. 6).

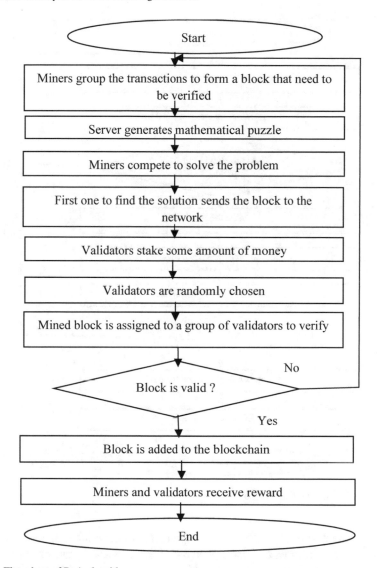

Fig. 4 Flowchart of PoA algorithm

2.6 Delegated Proof-of-Stake (DPoS) Consensus Algorithm

Delegated proof of stake is similar to PoS algorithm. It refers to more decentralized fashion in blockchain network, and it also modifies the way by which energy can be utilized often very less in executing the proper manipulation. Delegated proof of stake generally offers the chances to stockholders to give their votes to those want to mine further coming block should be committed in the database. Cryptocurrency holders will also have the opportunity to select the miner to mine a further block. Stockholders

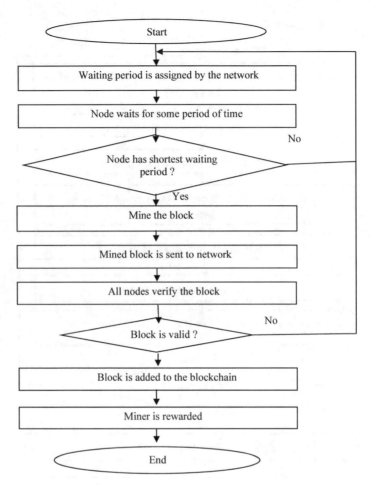

Fig. 5 Flowchart of PoET algorithm

will choose the delegates which will be responsible for the mining of new block, and somehow some witnesses are also selected on election basis by currency holders to perform proper manipulation like searching of nonce and validation of block and what the delegates need to do is that they will decide how much incentives to be given to witnesses, and they will also decide the factors like block size, power and final decision will be made by stakeholders to what delegates will have proposed to them. Witnesses will change within some time duration or within a week. Witnesses should perform the transaction allotted within the given time duration. It is all about the reputation of witnesses; more they perform the transaction efficiently within the given time duration, more will be their chances to get selection again in the mining process by selectors (i.e., cryptocurrency holders). DPoS is also increasing more decentralized fashion as what proposed in PoS was more in a centralized fashion to whom will have the higher amount of currency will have the more dominating

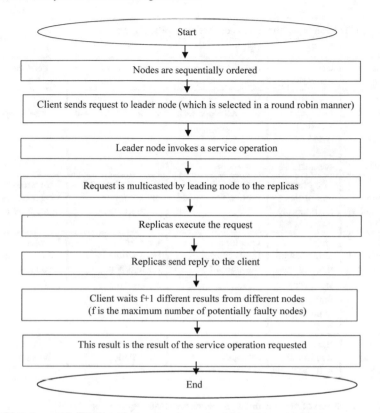

Fig. 6 Flowchart of PoET algorithm

effect in the whole network; but in DPoS, it has been modified and made a system something distributed that is removing the centralization process [15].

3   Comparison of Consensus Algorithms

See Tables 1 and 2.

4   Applications Area

In the recent, blockchain-based application is not only limited financial domain, but have grown in accounting, voting, energy supply, quality assurance, self-sovereign, identity (KYC), health care, logistics, agriculture and food, law enforcement, industrial data space, digital identifications, and authentications, gaming and gambling,

Table 1 Comparison of permissioned network consensus algorithms

BFT	RBFT	PBFT	PAXOS	RAFT
Closed network	Closed network	Synchronous	Synchronous	Synchronous
Used for business working based on smart contracts	Used for business working based on smart contracts	Smart contract-dependent	Smart contract-dependent	Smart contract-dependent
State machine replication is used	The proper authorities are handling proper work	State machine replication is used	Sender, proposer, and acceptor jointly work	Collecting selected ledger on some agreement to work
Good transactional throughput	Good transactional throughput	Greater transactional throughput	Greater transactional throughput	Greater transactional throughput
Byzantine faults	Byzantine faults (i.e., hyperledger body)	Byzantine faults	Crash fault	Crash fault
Based on traditional notions	Based on traditional notions	It can bear f-1 tolerance	PAXOS can bear f/2-1 faults	RAFT can bear f/2-1 faults

Table 2 Comparison of permissionless network consensus algorithms

Proof of Work (PoW)	Proof of Stake (PoS)	Proof of Burn (PoB)	PoET
Used for industries working on financial level	Used for industries working on financial level	Used for industries working on financial level	Used for industries working on financial level
Using public-key encryption (i.e., Bitcoin)	Using RSA algorithm for encryption	RSA algorithm for encryption	RSA algorithm for encryption
Miners having higher work done after investing higher power will have a higher probability to mine the new block	It is some election type selection of miners for next block to be mined	PoB acquires some cryptocurrencies (wealth) to mine new block using virtual resource	Person spends some time and power to mine new block who finishes first the prior task will be the next miner
Power inefficient	Power efficient	Power efficient	Power efficient
Open environment	Open environment	Open environment	Open environment
Bitcoin script is used	Mostly, Golong is used	Mostly, Golong is used	

government, and organizational governance, job market, market forecasting, media, and content distribution, network infrastructure, philanthropy transparency and community services, real state reputation verification and ranking ride, sharing service, social network, supply chain certification in the food industry. Blockchain has a problem with scalability and security, which must be tackled.

5 Technical Challenges

Blockchain technology implementation has issues like scalability, block size, number of transactions per second. It may not apply to high-frequency transactions. The trade-off between block size and security leads to selfish mining strategy. Miners can hide their mined block for more revenue in the future. There is a chance of privacy leakage even when users only make transactions with their public key and the private key. Users' real IP addresses could be tracked and a number of blocks are mined per unit time cannot fulfill the requirement of process of millions of transactions in a real-time fashion. What will the maximum chain length and the maximum number of miners be a question? Is there any possibility to go for a centralized system using blockchain? Larger block size could slow down the propagation speed and lend blockchain branch is a challenge for many application. There is a possibility of small transactions can be delayed due to miners give more performance to a high transaction fee technical.

6 Conclusion

Today, due to the emergence of IoT and big data, the huge, diverse, and critical information is available over the Internet. The trust over the information is reduced drastically, causing an increase in security and privacy concern day by day of the industry and organizations. Blockchain technology has the potential to change upcoming scenarios of society, industries, and organizations. The blockchain is one of the best-emerging technologies for ensuring privacy and security by using cryptographic algorithms and hashing. We have discussed the basics of blockchain technology, consensus algorithms, comparison and analysis of important consensus algorithms, and area of application in this paper. In the future, we will cover the different implementation platform such as Ethereum and Hyperledger.

References

1. Haber, S., and W. Scott Stornetta. 1991.How to time-stamp a digital document. *Journal of Cryptology* 3 (2): 99–111.
2. Nakamoto, S. 2008. *Bitcoin: A peer-to-peer electronic cash system*, Self-published Paper, May 2008 (Online). Available: https://bitcoin.org/bitcoin.pdf.
3. Bentov, I., A. Gabizon, and A. Mizrahi. 2016. Cryptocurrencies without proof of work. In *International conference on financial cryptography and data security*, Christ Church, Barbados, 142–157,Feb 2016.
4. Castro, M., and B. Liskov. 2002. Practical byzantine fault tolerance and proactive recovery. *ACM Transactions on Computer Systems* 20 (4): 398–461.
5. Dai, H., Z. Zheng, and Y. Zhang. Blockchain for internet of things: A survey. *IEEE Internet of Things Journal*. https://doi.org/10.1109/jiot.2019.2920987.
6. Back, Adam. *Hashcash—A denial of service counter-measure*. http://www.cypherspace.org/hashcash/.
7. Vukoli, M. 2016. The quest for scalable blockchain fabric: Proof-of-work vs. BFT replication. *IFIP WG 11.4 international workshop on open problems in network security iNetSec 2015*, 112–125, 29 Oct 2015–29 Oct 2015.
8. Jaag, C., and C. Bach. 2017. Blockchain technology and cryptocurrencies opportunities for postal financial services. In *The changing postal and delivery sector*, 205–221. Springer: Cham. https://doi.org/10.1007/978-3-319-46046-8_13.
9. Bentov, I., A. Gabizon, and A. Mizrahi. 2016. Cryptocurrencies without proof of work. Paper presented at the international conference on financial cryptography and data security, 142–157. Heidelberg: Springer, February. https://doi.org/10.1007/978-3-662-53357-4_10.
10. Zheng, Z., S. Xie, H.-N. Dai, and H. Wang. 2016. Blockchain challenges and opportunities: Survey. School of Data and Computer Science, Sun Yat-sen University, Tech. Rep.
11. Sankar, L.S., M. Sindhu, and M. Sethumadhavan. 2017. Survey of consensus protocols on blockchain applications. In *2017 4th international conference on advanced computing and communication systems (ICACCS)*.
12. Larimer, D. 2018. *DPOS consensus algorithm—The missing Whitepaper*, Steemit (Online). Available: https://steemit.com/dpos/dantheman/dpos-consensus-algorithm-this-missingwhite-paper Accessed 03 Feb 2018.
13. Correia, M., G. Veronese, and L. Lung. 2010. Asynchronous Byzantine consensus with 2f + 1 processes. In *Proceedings of the 2010 ACM symposium on applied computing—SAC '10*.
14. Theng, Z., S. Xie, H. Dai, X. Chen, and H. Wang. 2017. An overview of blockchain technology: Architecture, consensus, and future trends. In *2017 IEEE international congress on big data (BigData Congress)*, Honolulu, 557–564, IEEE.
15. Zheng, Zibin, Shaoan Xie, Hong-Ning Dai, and Xiangping Chen. 2018. Blockchain challenges and opportunities: A survey. *Huaimin Wang International Journal of Web and Grid Services (IJWGS)* 14 (4).

A Roadmap to Realization Approaches in Natural Language Generation

Lakshmi Kurup and Meera Narvekar

Abstract Text realization is the most significant step involved in natural language generation. It involves the approaches used to generate syntactically and semantically valid text, given an abstract linguistic representation. Based on the data and nature of data, a typical task generation includes text-to-text generation, database-to-text generation, concept-to-text generation, and speech-to-text generation. There are many approaches of natural language generation to generate texts, usually from non-linguistic structured data, which varies from a canned text approach to methods of learning from a text corpus and generating text based on the characters, content, keywords, size of the text, context, etc. Much work has also been done in learning and generating text mimicking a writing style. For applications like tutoring systems, wherein the text has to be manipulated and validated, we have to rely more on a template-based approach. Machine learning and other probabilistic-based statistical approaches can generate text for applications like report generation, summarization. This paper presents a roadmap and a comparative analysis of various text realization approaches.

Keywords Canned text · Context-free grammar · Template-based systems · Tree adjoining grammar · Lexical functional grammar

1 Introduction

Text generation has been an important aspect in the area of natural language generation, dated from 1970s which uses the concepts of artificial intelligence, computational linguistics, syntactic and semantic interpretations. Auto-text generation has its significance across a wide range of applications like weather forecasting, news headline generation, medical/military report generation, summarization, auto-story generation, explanation of mathematical proofs, descriptions of museum artifacts, intelligent tutoring systems, auto-completion, auto-paraphrasing, dialog systems, machine translation, question answering. Any NLG system follows three basic processing steps to generate the text which is structured, natural and semantically and

L. Kurup (✉) · M. Narvekar
DJ Sanghvi College of Engineering, Vileparle-W, Mumbai, India

© Springer Nature Singapore Pte Ltd. 2020
Y.-C. Hu et al. (eds.), *Ambient Communications and Computer Systems*, Advances in Intelligent Systems and Computing 1097,
https://doi.org/10.1007/978-981-15-1518-7_18

syntactically correct. The first one being document planning, the content and information needed to generate the text has to be decided. The next step is microplanning, wherein the linguistic structure of the chosen content, the structure of the knowledge base, the syntactic and semantics of the content, etc., has to be planned. The last and most important step is realization wherein syntactically, semantically, and morphologically valid text adhering to the grammatical correctness like syntax, word order, agreement is generated from an abstract linguistic representation. In this paper, we are concentrating on the microplanning and realization approaches in detail.

To generate a text, adaptation of a linguistically justified grammar is needed. In addition, we need a knowledge representation, wherein the information is encoded in diverse forms. Each domain has its own domain representation, decision rules, and variables. Appropriate modeling of domain knowledge plays an important role in text generation. Moreover, the type of text that has to be generated varies based on the intended audience or application. It could be a single phrase, multi-phrases, or a paragraph or an entire report generation. The discourse model should be adapted in such a way that it should generate sentences adhering to the knowledge base and target application [1].

The paper is structured as follows. Section 2 includes a brief analysis of various approaches involved in text generation and a comparative analysis. Experimentation results are also shown in the paper for statistical modeling. The paper ends with a conclusion and a direction to applications, wherein each approaches are fits in.

2 Approaches Used for Auto-Text Generation

The process of transforming a meaning-based representation into a target string is termed as language generation [2]. As per the level of sophistication and the power of expressiveness, a typical NLG can be performed using the following techniques.

2.1 Canned Text

It is a pre-authored set of statements or expressions which is used for generating a single sentence or multi-sentences. As it has a predefined structure, it is inflexible or difficult to generate a dynamic text. But it remains the simpler technique of text generation. It is widely used in generating error messages or warnings. A representation of a canned text approach is as follows [3]: "This is a [x] generated warning." Here [x] takes in values from the knowledge base.

2.2 Template-Based Systems

It can use in the generation of multi-sentences. This technique is considered much more sophisticated than a canned text; though the template structure is predefined, it can have a few open fields, or fill in the gaps, which gets populated from the underlying knowledge base. It is used in places where the same type of sentences has to be repetitively generated, but with slight alterations like in a business report or formal letters. A template structure can be enhanced by adding more control expression and a feature-based grammar-based mechanism to fill in the missing gaps or slots [4]. An augmented template structure used in YAG is given as follows [4]. The statement "John walks" is represented as

```
(template clause)
(agent "John")
(process "walk")
```

This definition says that a clause is to be realized as the value of the agent, followed by a verb- form followed by the object, with a period concatenated to its left end.

2.3 Feature-Based Systems

An English sentence can be IMPERATIVE, a QUESTION or a TAGGED sentence with its TENSE being PRESENT, PRESENT CONTINUOUS, PAST or FUTURE. Representation of these diverse features in the domain knowledge is very difficult. Feature-based systems generate single sentences based on these features. Though it has very high expressive power, it is very difficult to maintain the coherence between the linguistic constituents. Most of the systems currently use a hybrid of the above-mentioned systems.

2.4 Rule-Based Systems

The rule-based systems are dependent on the handmade rules and are less flexible if number of coherent variables increases.

2.5 Example-Based Approaches

Most of the approaches, though they adhered to grammatical rules, it was far from semantic modeling. Example-based systems [5] are used to infer semantic and syntactic representations from examples. Here the need for hand-authored rules was eliminated. These approaches can generate a dynamic output with more productive output and less grammatical inferences.

2.6 Grammar-Based Approaches

To meet the needs of the target system, grammar can be hand authored. A grammar defines the syntactic structure of the text to be generated. We can use a regular grammar, a context-free grammar (CFG), a context-sensitive grammar, or a regular grammar, depending on the context and semantic representations.

2.6.1 CFG-Based Text Generation

A context-free grammar (CFG) is a set of rules that define the sequence of words in a language. As the name implies, it is context-free and hence not sufficient in generating meaningful sentences, although it can create syntactically correct sentences. A CFG-based system, TGEN [5] defines its generation rules in CFG form where RHS consists of a frame and LHS consists of a frame type.

[frame_student] → [studname] ([sex]) [vp_studId]. [studentname] [vp_DOB], [vp_cityborn]. [studentname] [vp_classId]

Research has been carried out to learn the production rules by using a recurrent model and generate valid sentences. The paper [6] evaluates the ability of LSTM to learn context-free grammars. Since LSTMs are inherently sequential models, the question arises whether they can learn the correlation between the hierarchical structures inherent in any human language or can learn the syntactic units in a language. The disadvantage of a CFG is that it is ambiguous. It can generate more than one parse tree for the same sentence. To disambiguate, CFGs are augmented with a probabilistic context-free grammar [7]. For the below-mentioned CFG

S → NP VP
NP → Article ADJP
ADJP → Adj SN
VP → V ADVP
ADVP → slowly | quickly
Article → A | An|The
ADJP → old |young
SN → lady | man| child
V → walked | ran

A sentence generated from above CFG is "An old man walked slowly." There could be different variants of the sentences with different substitutions of lexicons. So the maximum edit distance for the sentences generated with above CFG accounts to 5. More the lexicons, more the edit distance.

2.6.2 Probabilistic Context-Free Grammar (PCFG)

Each production in PCFG is assigned a probability. The most probable parse trees can be identified by learning the estimation of probabilities by parsing the text corpus and counting the rule probabilities with the help of Expectation Maximization algorithm, Inside–Outside algorithms, etc. But the disadvantage with PCFGs is that the calculation of the probability is based on the count of occurrence of production rules, no matter where they occur or rather it is context-independent. Another drawback is that there is no lexical conditioning. Specific words may lead to different probability calculations. One way to improve the situation is to annotate the nodes with the name of the parent or have some tagging system. But then with more annotation or tags, it needs more training data.

2.6.3 Lexical Functional Grammar (LFG)

LFG is a framework for constraint-based grammar. It has two different levels of representation: C (constituent)-structure and F (functional)-structure. C-structure is represented by phrase structure trees, depicting the structure of syntactic constituents; while F-structure is represented by grammatical functions, depicting the abstract functional relationships such as subject and object. First-order predicate logic is used to represent meaning representation [8, 9]. Though LFG generates syntactically correct sentences, the variable values in PRED in F-Section have to be predefined and would be very specific to the domain. So playing around F-Section would be a tedious task. Figure 1 depicts a C-structure, and Fig. 2 depicts a F-structure for the sentence "John picked a small stone."

2.6.4 Head-Driven Phrase Structure Grammar (HPSG)

Here the feature descriptions are described by attribute value matrix (AVM). Such AVMs consist of feature value pairs. It is a generalized formalism which helps in recursive feature passing. A feature type is ordered in hierarchies with the most general type at the head and the most specific types or subcategories at the bottom. For example, Fig. 3 shows how feature and "case" is ordered in hierarchies.

Phrase-based systems also can use a predefined template, but a more structured and generalized one. It relies on the lexical grammar rules. A typical English phrase can be determined in the form of a phrase structure grammar (PSG). For example, S → NP VP where a sentence is first matched with a noun phrase pattern followed by

Fig. 1 C-structure [8]

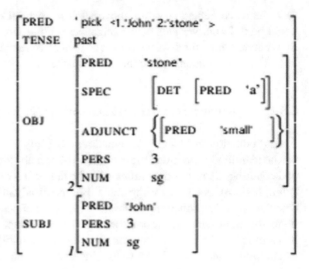

Fig. 2 F-structure [8]

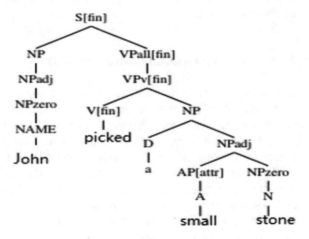

Fig. 3 A feature-based HPSG

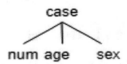

a verb phrase pattern. A generation process comes to an end when each grammatical rule is recursively matched and replaced by one or more words. It is much powerful and robust, but is not feasible for very large text generation.

Fig. 4 Schematic parse tree: One possible LTAG derivation tree: C1 uses elementary trees A1 and B1 [8]

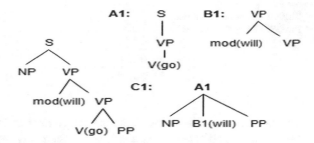

2.6.5 Lexicalized Tree-Adjoining Grammar (LTAG)

LTAG is a formal grammar mechanism, wherein it consists of a set of formal trees and wherein each tree is associated with one lexical item and operation for combining those lexical items [10]. The formal elementary trees can be either depicted as argument–predicate-type relations or recursion. It is an extension of TAG formalism. It can generate more complex trees by substitution and adjunction [11]. TAG formalism is not that very powerful enough to reach the objectives of generating a computationally efficient text. But due to its simplicity of operation and consideration of lexicalization rules, it can be considered as much sophisticated approach than mere context-free grammars. Figure 4 shows a parse tree where there could be more than one derivation trees. In the figure, only one derivation tree is shown for reference.

2.6.6 Statistical-Based Approach

Statistical-based systems typically require a test corpus of a specific domain where training is done using statistical models to generate similar text. Mostly used statistical models are Markovian model and recurrent neural network. These models learn from a given text corpus and generate a sample text with some logic and sense. A RNN based on character level cannot be considered as a good approach, as they would not create syntactically correct sentences; though with proper training, they would be able to generate correctly spelled words. A Markovian model, with more depth of training, can generate syntactically correct text though the probability of generation of such a text is unpredictable. In recent years, research has been concentrated more on text generation using neural networks that work on dense vector representations with the help of word embeddings and other deep learning methods.

Language Modeling for Text Generation

Text generation requires a trained language model which can be operated at character level, n-gram level, paragraph level, or sentence level or which can learn the likelihood of occurrence of word based on previous occurrences. For this representation of

Table 1 Evaluation metrics
for GPT-2, BERT, and GloVe

Language models	BLEU-1	BLEU-2
GPT-2	0.764706	0.692364
BERT	0.77431	0.64321
GloVe	0.661337	0.524324

vocabularies in a document, depicting the context of the word in that document, the syntactic and semantic similarities between each word are needed. There are different ways in which these representations are created. Count vectorization, TF-IDF and co-occurrence matrix calculations can be used to create a frequency-based word embedding. Most widely used word embedding is a predication-based vector representation called as Word2vec which works in combination with continuous bag of words (CBOW) and skip-gram model. Skip-gram is considered as a suitable approach with a small data and work well for rare words, whereas CBOW works well with frequent words. Character-level embeddings are bound to outperform the word-level embeddings, on languages with rich morphology (Arabic, Czech, French, German, Spanish and Russian [12]. Vector representation of words using GloVe has gained much attention recently. Here the similarities are identified using the Euclidian distance. ELMo, a pre-trained word representation, proposes a deep contextualized word representation. A bidirectional LSTM is used to generate vectors in ELMo [13]. Research shows that generative adversarial networks (GANs) generate more valid and semantically correct sentences. BERT [14] is also considered as an efficient transformer-based bidirectional language model for context-based text generation. A very recent model GPT-2 [15], released by OpenAI released in 2019, outperformed BERT and was found more efficient in applications like auto-story generation, auto-essay generations. Table 1 shows the evaluation metrics for the language models GPT-2, BERT, and GloVe. Experiments were run on a dataset of Donald Trump speeches. It is observed that GPT-2 outperformed the other two models.

3 Comparative Analysis of Text Generation Approaches

The table mentioned below shows a brief analysis of pros and cons of the above-discussed approaches (Table 2).

4 Conclusion

Much advancement has been achieved in the area of text generation. Though many of the approaches date back to 1970s, neither of the approaches can be disregarded. Each of the above approaches has got its own significance based on the application it builds. In a simple application like auto-generation of small email responses, we

Table 2 Comparative analysis of text generation approaches

S. No.	Approach	Advantage	Disadvantage	Applications
1	Canned Text	• The simpler approach to implement. • Have a predefined structure	Cannot be used for dynamic text	Can be used only for small warning texts and error display messages
2	Template-based	It can be used for iterative generation of multi-sentences of similar structure	Cannot be used for multiple variants of the sentence. Each variant should have a template defined	Can be used to generate a paraphrase with some predefined slots or fill in the blanks with some intelligence
3	Example-based	From the examples, inference on semantic and grammatical rules is done. Less importance on hand-authored rules	More productive output with less development costs	Sentences similar to underlying examples
4	Feature-based	It can use any of the generating approaches based on the feature of the sentence to be generated	Very difficult to maintain the relationships between the features	Sentences based on specific features
5	Grammar-based	It can generate syntactically correct sentences	Ambiguous nature of the production rules have to be handled and not very large text can be generated	Sentences adhering to lexical grammatical rules are generated

(continued)

Table 2 (continued)

S. No.	Approach	Advantage	Disadvantage	Applications
6	TAG	It can improve semantic role labeling and simplicity	More variations difficult to achieve	Sentence that adheres more to semantic capabilities are generated
7	Statistical-based	More flexibility in generating data based on the corpus	Not always well-defined syntactic structures generated, though latest models like GPT-2 and MT-DNN have achieved to generate most probable valid sentences	TEXT generation/summarization

can use a canned text approach. But if we want to incorporate text generation for the use of tutoring systems where repetition and genuineness have to be ensured, a template-based or a grammar-based approach would be most appropriate. Statistical-based generation can be used to auto-summarize document with the help of a trained corpus. Recent years have seen advancements in language models using RNNs like LSTM and GANs with the use of vector embeddings like Word2vec, GloVe, BERT, ELMo, GPT-2. We have shown BLEU-I and BLEU-2 metrics on some of the above-said language models. It is clearly evident that GPT-2 outperforms all the models. But these pre-trained models are highly beneficial due to the fact that it can be fine-tuned to solve any specific task, thereby eliminating the need to train the model from scratch, which in turn reduces the computational load and time. This paper is an attempt to review various NLG approaches used till now showcasing the pros and cons of each in their unique form.

References

1. Mann, W. Text generation. *American Journal of Computational Linguistics* 8(2) April–June 1982.
2. Baptist, L., and S. Seneff. 2000. Genesis-II: A versatile system for language generation in conversational system applications. In *Sixth international conference on spoken language processing*.
3. Busemann, Stephan, and Helmut Horacek. 1998. In *9th international workshop on natural language generation*, Niagara-on-the-Lake, Canada, August 1998, 238–247. arXiv:cs/9812018 [cs.CL].
4. McIoy, S., S. Channarukul, and S. Ali. 2003. An augmented template-based approach to text realization. *Natural Language Engineering* 9 (4): 381–420. https://doi.org/10.1017/S1351324903003188.
5. Le, H.T. 2007. A frame-based approach to text generation. In *Proceedings of the 21st Pacific Asia conference on language, information and computation*, 192–201.
6. Sennhauser, L., R. Berwick. Evaluating the ability of LSTMs to Learn context-free grammars. In *Proceedings of the 2018 EMNLP workshop blackbox NLP: Analyzing and interpreting neural networks for NLP*, 115–124, Brussels, Belgium, 1 Nov 2018.
7. Sundararajan, S. 2001. *Probabilistic context-free grammars in natural language processing*, Joan Bresnan. Lexical-Functional Syntax. Oxford, UK: Blackwell.
8. Lareau, F., M. Dras, B. Borschinger, and R. Dale. 2011. Collocations in multilingual natural language generation: Lexical functions meet lexical functional grammar.
9. Kahane, S. 2000. How to solve some failures of LTAG. In *Proceedings of the fifth international workshop on tree adjoining grammar and related frameworks* (TAG + 5), 123–128.
10. Forbes-Riley, K., B. Webber, and A. Joshi. 2005. Computing discourse semantics: The predicate-argument semantics of discourse connectives in D-LTAG. *Journal of Semantics* 23 (1): 55–106.
11. Kim, Y., Y. Jernite, D. Sontag, and A.M. Rush. 2016. Character-aware neural language models, In *Thirtieth AAAI conference on artificial intelligence*.
12. DeVault, David, David Traum, and Ron Artstein. 2008. *Practical grammar-based NLG from examples*, INLG 2008.
13. Mikolov, Tomas, Ilya Sutskever, Kai Chen, Greg Corrado, and Jeffrey Dean. 2013. Distributed representations of words and phrases and their compositionality. In *Advances in neural information processing systems*.

14. Wang, Alex, and Kyunghyun Cho. 2019. *BERT has a mouth, and it must speak: BERT as a Markov random field language model.* arXiv preprint arXiv:1902.04094. Apr 2019.
15. Radford, Alec, Karthik Narasimhan, Tim Salimans, and Ilya Sutskever. 2018. *Improving language understanding with unsupervised learning.* Technical report, OpenAI.

A Case Study of Structural Health Monitoring by Using Wireless Sensor Networks

Der-Cherng Liaw, Chiung-Ren Huang, Hung-Tse Lee and Sushil Kumar

Abstract In this paper, an intelligent structural health monitoring platform is proposed with an example study of vibration signal detection on a multi-floor building. The platform is consisted of self-designed hardware devices and graphical user monitoring interface. In addition, the proposed framework is a three-layered architecture. Among them, the first layer is a wireless sensor network client which is deployed on the top three floors of a nine-floor building to collect vibrational data. The second layer is laptop computer which is connected with a master WSN node to collect the data sensed by slave nodes. All data packets will then be transmitted to master computer by laptop computer. The master computer plays the role of the third network layer and is responsible for post-data processing and signal monitoring. A testing experiment is also given to demonstrate the function of the proposed platform.

Keywords WSN · Structural health monitoring · Intelligent platform

1 Introduction

In recent years, the study of structure and building health monitoring has attracted lots of attention (e.g., [1–4]). The reliability and safety of a building will be considered as the most important issue for those people who have lived in the same building for several years. In addition, there are many cities located at the intersection of plates and/or topographic fault areas. Those areas are known to be easier to have earthquakes caused by the movement of plates. Due to the occurrence of unexpected natural disasters, people might be shocked and the crisis may cause the loss of hundreds of people's lives if the critical emergency messages are not transmitted to resident immediately. Thus, how to obtain an early notification like an impulse signal for the possible occurrence of earthquakes and minimize the damage becomes an essential subject.

D.-C. Liaw (✉) · C.-R. Huang · H.-T. Lee · S. Kumar
Institute of Electrical and Control Engineering, National Chiao Tung University,
Hsinchu, Taiwan, ROC
e-mail: ldc@cn.nctu.edu.tw

© Springer Nature Singapore Pte Ltd. 2020
Y.-C. Hu et al. (eds.), *Ambient Communications and Computer Systems*, Advances in
Intelligent Systems and Computing 1097,
https://doi.org/10.1007/978-981-15-1518-7_19

Due to the rapid development on the Internet of Things (IoT), lots of new technology devices such as wireless sensors, RFID tags, electric appliances and mobile phones are able to communicate with each other and cooperate with their neighbors for certain specific objectives (e.g., [5–12]). Among those recently developed technologies for IoT, the wireless sensor networks (WSNs) are known especially for its possible application to the development of a real-time, low cost, high stability and robustness structural health monitoring system. It is known to be useful in managing crisis situations via providing a good disaster management and emergency response information. One of the most useful functions of WSN is environmental monitoring. There are many related schemes that have been developed for structural health monitoring (SHM). For instance, WSN is usually deployed on bridge to avoid collapsing incident [1–3], radiation monitoring [10] and environment monitoring [12]. Besides, it can be used for the alignment of the tracks as well as the detection of the corrosion and deformation of tracks [13]. Furthermore, health monitoring of famous architectural heritage and civil engineering are also the common applications [14]. From those application designs, it is clear that the WSNs seem to be feasible and practicable for building health monitoring.

In this paper, we will propose an intelligent building health monitoring platform which can detect the unusual vibration signal from each of multi-floor building simultaneously. It is consisted of the self-designed hardware devices and graphical user monitoring interface. The proposed framework is three-layered architecture. Among them, the top layer is a wireless sensor network client which is deployed in every floor of the building to collect vibration data. The middle layer is laptop computer which is connected with a master WSN node to collect the data sensed by slave nodes and performs the preliminary data processing. The data will then be transmitted to master computer for post-data processing and signal monitoring. The master computer will play the role as the third network layer.

The paper is organized as follows. The architecture of the proposed intelligent vibration monitoring platform is introduced in Sect. 2. Detailed design for software and hardware of the proposed platform is described in Sect. 3. It is followed by the experimental results for the verification of the proposed design. Finally, concluding remarks are also given in Sect. 4.

2 Design of the Intelligent Vibration Monitoring Platform

In the design of the structural health monitoring (SHM) platform, people might use two-layer scheme. The first layer is to construct the sensor node for collecting the information of SHM and those data will then be passed to the server by wireless communication for analysis. Here, the server plays the role of the second layer of the platform. Such an approach has been used in several existing applications (e.g., [4, 11, 12]). It is known that the wireless communication might not be able to pass through the ceiling between two floors of a building. Thus, the two-layer design might not be feasible for the SHM on a multi-floor building. In order to relax such a burden, in this study we propose to insert a second layer as intermediary to solve the wireless communication problem on a multi-floor environment. This is achieved by adding a laptop computer to pass the data from sensor node to the server.

One of the main goals of this paper is to construct a building health monitoring platform for monitoring unusual vibration and offering essential information to the resident. In this study, two different network architectures will be considered for the design of a three-layer network. One is sensor network topology and the other is heterogeneous network architecture. The design and implementation of both hardware and software for the proposed platform will also be discussed. Details are given as follows.

2.1 Sensor Network Architecture

First, we consider the development of the architecture for sensor node. In order to guarantee time synchronization of sensor data fetching, here we adopt a star shape network topology as depicted in Fig. 1 [15]. The star topology has a master node to manage and maintain the architecture of the network. Other sensor nodes cannot talk to each other and all the data communication should pass through the master node. The biggest advantage of the master–slave architecture as shown in Fig. 1 is that all nodes can be managed in a unified manner to improve the efficiency of execution. In addition, we can design an algorithm to prevent the collision of data transmission. The major task of the master node is focusing on efficient execution which includes data transfer, system management and security protection.

Next, we consider the hardware scheme for the proposed vibration sensor node. As shown in Fig. 2, we adopt our previous self-designed hardware used in [11, 12] to add an extra three-axis acceleration sensing module for detecting the vibrational signal. The proposed WSN module is consisted of a TI CC2530 microcontroller,

Fig. 1 Star network topology [15]

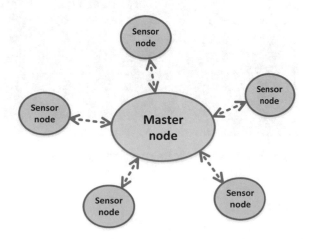

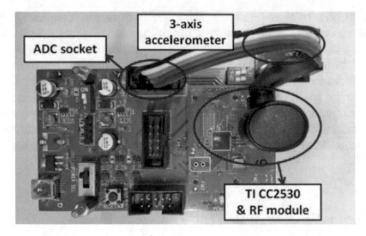

Fig. 2 The self-designed WSN sensor node

ADC socket, SPI socket, a three-axis accelerometer and power module. Among them, the microcontroller is mainly responsible for data transmission and processing. The three-axis accelerometer module is mounted on the main board and connected to the main board via the ADC socket and/or SPI socket by industry standard interface bus. After fetching the data from the accelerometer, the microcontroller will process the raw data and send the processed data packet to the master node by its built-in RF module. In order to the moving of vibrational signal, we will set up five sets of the proposed WSN nodes on each floor of the building as shown in Fig. 3.

Fig. 3 Five set of sensor nodes are deployed in each floor

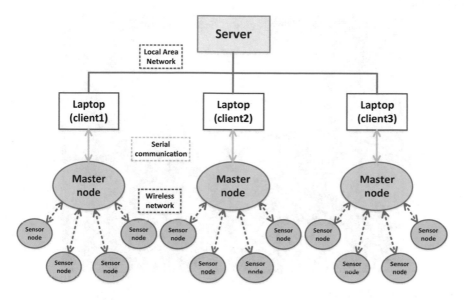

Fig. 4 Master–slave cluster network architecture

2.2 Heterogeneous Network Architecture

Now, we propose the second part of the proposed design. It is consisted of the second layer and the third layer of the design. The second layer is a laptop computer connected with the master WSN node defined in Sect. 2.1 and the third layer is the server of the proposed platform as shown in Fig. 4. Here, in the first layer, the sensor node and the master node are wirelessly communicated with each other by 2.4 GHz radio frequency. In the second layer, the master WSN node and laptop are connected by serial transmission while the laptop and the server are connected through the wired area network. The computer architecture among the laptop computer and the server is observed to be a star shape network topology, which will make the whole platform be easy to achieve the function of time synchronization.

In the proposed design, the laptop is responsible for receiving the data collected by the master node and then transmits the data packet to the server for analysis. Besides, it will also execute the tasks assigned by the server to manage the operation of the master node. The practical hardware of the second layer design is shown in Fig. 5.

2.3 Software: Sensing and Monitoring Interface

To facilitate the design, a graphical user interface (GUI) is developed on the server to provide the functions for system testing and results displayed as shown in Fig. 6.

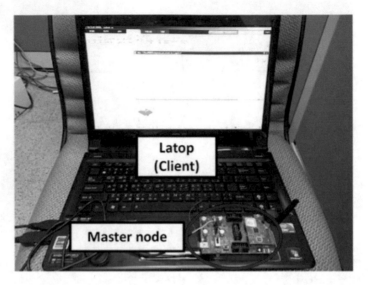

Fig. 5 Laptop and master node

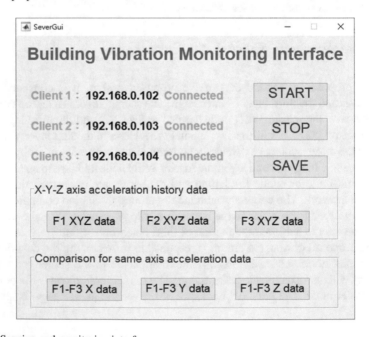

Fig. 6 Sensing and monitoring interface

Here, for simplicity and without loss of generality, we only consider three clients as an example. It is not difficult to extend the proposed design to more general cases with lots of clients. The proposed functions include area network address setting, connection status showing, sensing data graphs displaying and recording. After setting of the area network IP, each client can connect to the server through local area network and the connection status of each group of star networks will be displayed on the GUI. Then the user can press the "START" button to send a command to each client to execute the sensing task synchronously. Next, the sensing data will be stored in the database automatically. There are six buttons given in GUI, and the vibrational signal can be monitored instantly by the users after clicking. By clicking the buttons in the upper group, the user can obtain the X-Y-Z axis vibrational data of the five sensor nodes from each floor. In addition, by clicking the buttons in another group, only the X-axis (Y-axis or Z-axis) data between different floors can be compared with each other simultaneously. Moreover, the user can suspend monitoring by pressing the "STOP" button. After clicking "SAVE" button, the graphical timing diagram of vibrational data can then be recorded and stored into a file.

3   Experimental Study

In this section, we will present the experimental results for the verification of the proposed design given in Sect. 2. The goal of this study is to read the impact of earthquakes or other external vibrations on each floor of the building simultaneously. It will be achieved by the proposed design for the synchronous sampling of each star network through the establishment of an area network. Since the entire network architecture is synchronized, the real-time vibrational signal can then be judged by the timing chart and the direction of seismic wave transmission. Two types of experiments will be carried out. One is the testing of the proposed functions by putting five WSN nodes on the same table. The other is a real-time detection of the vibrational signal which is similar to seismic wave. It is carried out by putting the sensors on the top three floors of a nine-floor building. Details are given as follows.

3.1   Testing of the Proposed Functions

First, five sets of WSN nodes are deployed on the experimental table for the testing of the proposed functions as shown in Fig. 7. In order to verify the performance of each vibration sensor, we hit the table from the first node to the fifth node to simulate a seismic wave and observe the vibration data generated from the Z-axis movement. From the results given in Fig. 8, it is clear to see that the maxima impulse signal first appears in timing response at the first node. Then it appears at the second node to the fifth node sequentially. Hence, it can be concluded that the vibrational signal moves from the first node to the fifth node. Moreover, the coupling effect is also

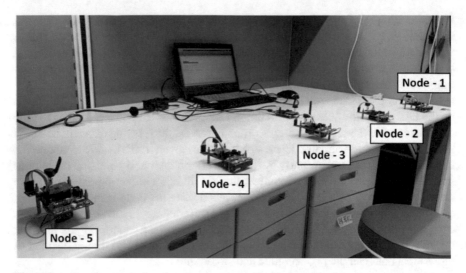

Fig. 7 Sensor nodes are deployed on the experimental table for testing

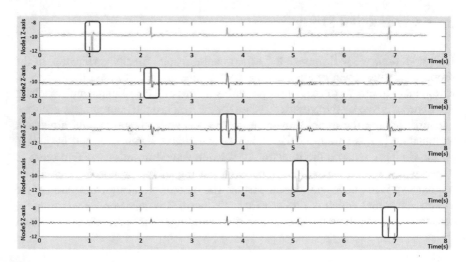

Fig. 8 Z-axis vibration data from five sensors on the table

observed in the experiment. For instance, all remaining four sensors will also detect the vibrational signal with smaller magnitude when we hit the table nearby each node.

Fig. 9 Configuration diagram of wireless sensor network nodes on a nine-floor building

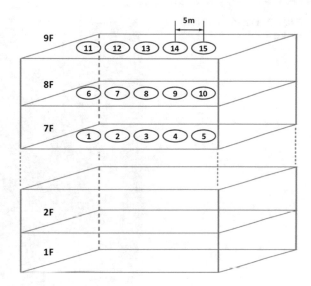

3.2 Real-Time Detection of the Vibrational Signal

Now, we consider the real-time detection of the vibrational signal on the floors of a building. There are fifteen sets of WSN nodes are placed on the floor of the top three of a nine-floor building as depicted in Figs. 9 and 10. The reason why we choose the top three floors for testing is that the vibrational signal will be easily damped out on the lower floor of a building. Based on the proposed architecture presented in Sect. 2, we can obtain the experimental results as shown in Figs. 11, 12 and 13. Here, we choose the distance between WSN nodes is 5 m and use people's walking signal to simulate the seismic wave.

The vibrational signal created by people's walking will go through from node 1 to node 5 on the seventh floor, from node 6 to node 10 on the eighth floor and from node 11 to node 15 on the ninth floor, respectively. It is clear from Figs. 11, 12 and 13 that we can observe traveling waves on the WSN nodes placed on each floor. However, the coupling effect can not be observed as those in Sect. 3.1. The possible reason might be the structure of the building has larger damping effect than that of the table.

4 Conclusions

In this study, we have developed an intelligent building health monitoring platform which can simultaneously detect the unusual vibrational signal from each floor of a multi-floor building. A second layer is added in the design and inserted into a general two-layer design for SHM problem to prevent the signal block of wireless

Fig. 10 Placement of sensor nodes

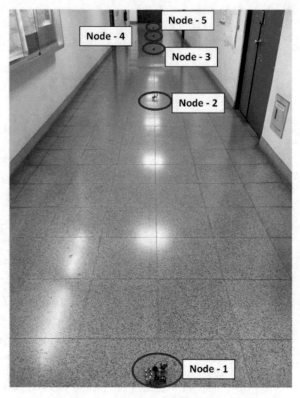

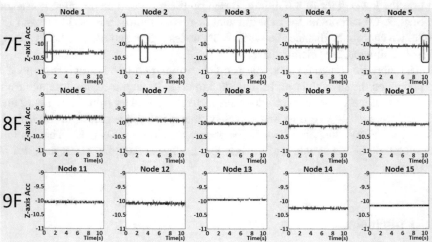

Fig. 11 Comparison of the Z-axis data among different floors (walking at 7F)

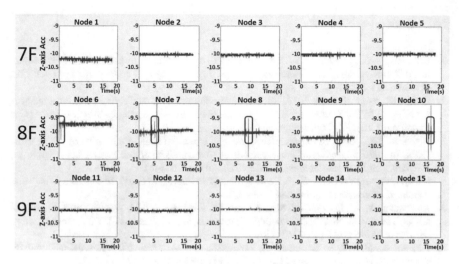

Fig. 12 Comparison of the Z-axis data between different floors (walking at 8F)

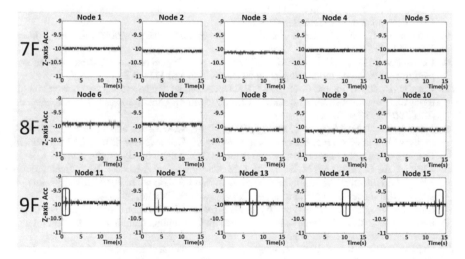

Fig. 13 Comparison of the Z-axis data among different floors (walking at 9F)

data transmission. The experimental results from the testing of the top three floors of a nine-floor building have demonstrated the success of the proposed design. Although in this paper we only consider the top three floors of a building as an example for the SHM study, it will not be difficult to extend the results to more general case of covering every floor of a multi-floor building.

References

1. Patil, P.K., and S.R. Patil. 2017. Structural health monitoring system using WSN for bridges. In *2017 International conference on intelligent computing and control systems (ICICCS)*, Madurai, 371–375.
2. Patil, P.K., and S.R. Patil. 2017. Review on structural health monitoring system using WSN for bridges. In *2017 International conference of electronics, communication and aerospace technology (ICECA)*, Coimbatore, 628–631.
3. Risodkar, Y.R., and A.S. Pawar. 2016. A survey: structural health monitoring of bridge using WSN. In *2016 International conference on global trends in signal processing, information computing and communication (ICGTSPICC)*, Jalgaon, 615–618.
4. Liaw, D.-C., and T.-H. Lin. 2013. Development of an internet of civil infrastructure framework. *Advanced Materials Research* 684: 583–587.
5. Al-Fuqaha, A., M. Guizani, M. Mohammadi, M. Aledhari, and M. Ayyash. 2015. Internet of things: a survey on enabling technologies, protocols, and applications. *IEEE Communications Surveys & Tutorials* 17 (4): 2347–2376.
6. Zanella, A., N. Bui, A. Castellani, L. Vangelista, and M. Zorzi. 2014. Internet of things for smart cities. *IEEE Internet of Things Journal* 1 (1): 22–32.
7. Xu, L.D., W. He, and S. Li. 2014. Internet of things in industries: a survey. *IEEE Transactions on Industrial Informatics* 10 (4): 2233–2243.
8. Ganti, R.K., F. Ye, and H. Lei. 2011. Mobile crowd sensing: current state and future challenges. *IEEE Communications Magazine* 49 (11): 32–39.
9. Stankovic, J.A. 2014. Research directions for the internet of things. *IEEE Internet of Things Journal* 1 (1): 3–9.
10. Lin, T.-H., and D.-C. Liaw. 2015. Development of an intelligent disaster information integrated platform for radiation monitoring. *Natural Hazards* 76 (3): 1711–1725.
11. Liaw, D.-C., S.-M. Lo, and C.-H. Tsai. 2015. Development of a wireless sensor network monitoring platform for positioning systems. In *The 2015 international automatic control conference (CACS2015)*, Yilan, Taiwan.
12. Liaw, D.-C., J.-H. Liao, Y.-C. Chang, C.-W. Tung, C.-C. Kuo, and L.-F. Tsai. 2016. A design of IoT platform for environment monitoring. In *Proceedings of 2016 16th* International Conference on Control, Automation and Systems (ICCAS 2016), Gyeongju, Korea.
13. Abrardo, A., L. Balucanti, M. Belleschi, C.M. Carretti, and A. Mecocci. 2010. Health monitoring of architectural heritage: the case study of San Gimignano. In *2010 IEEE workshop on environmental energy and structural monitoring systems*, Taranto, 98–102.
14. Rajendra Dhage, M., and S. Vemuru. 2017. Structural health monitoring of railway tracks using WSN. In *2017 International* Conference on Computing, Communication, Control and Automation (ICCUBEA), Pune, 1–5.
15. Callaway, Edgar H., Jr. 2003. *Wireless sensor networks: architectures and protocols.* CRC Press.

TBSAC: Token-Based Secured Access Control for Cloud Data

Pankaj Upadhyay and Rupa G. Mehta

Abstract Growing digital world has led to enormous data. With the growing data, its availability and processing have become a tedious task which is rescued by cloud computing. The cloud computing and data integration pose critical security concern to protect the data from unauthorized access. The current literature suggests various data access control schemes and models for cloud, but all provide permanent access to the resources, i.e., once a user is authenticated, either he is required or is not to be authenticated or authorized for subsequent requests. The data industry nowadays works on auto-expiry access tokens with traditional access control mechanisms which the current literature lacks. This paper contains a framework for data privacy using contextual information, attribute-based encryption, time-based assured data deletion, JSON Web Token (JWT) for token-based authorization, challenge response-based authentication, policy and context update with hidden policy for attribute-based encryption.

Keywords Access control · Token-based authorization · Cloud data · Attribute-based encryption

1 Introduction

Cloud computing is used to manage, store, and process enormous data making it independently shareable and reliable. A cloud provides natural point for (i) aggregated data analysis, (ii) system and service management, control, and coordination, (iii) resource management benefits, (iv) meet scaling demands for storage and computation. Storing data to a remote site is continuous process in organizations which liberates the users from data storage and retrieval issues. The administration, management, and operation of cloud by certain vendors require a high trust level and security. Cloud serves various users through virtual machines and shared resources with

P. Upadhyay (✉) · R. G. Mehta (✉)
Sardar Vallabhbhai National Institute of Technology, Surat, Gujarat 395007, India
e-mail: rgm@coed.svnit.ac.in

© Springer Nature Singapore Pte Ltd. 2020
Y.-C. Hu et al. (eds.), *Ambient Communications and Computer Systems*, Advances in
Intelligent Systems and Computing 1097,
https://doi.org/10.1007/978-981-15-1518-7_20

multitenancy and virtualization risks. A general security mechanism should identify who is accessing what, when, and with which conditions. Data stored in cloud has various security requirements [1], viz. secure communication, access control, sensitive data identification, internal cloud data protection and sharing, entities' identification and encryption, data combination, malicious entity recognition, and trustworthiness of services. In this research, access control is extensively surveyed which emphasizes some limitations, viz. key manager dependency, semi-trusted third party, complex key management, data deletion, permanent access allocation, and policy enlightening based on which this research proposes a framework on access control security requirement for data stored in cloud which is based on assured data deletion, auto-expiry authentication, hidden access policy for attribute-based encryption, hidden recipients' credentials, and extracted data context.

This paper is organized as follows: Sect. 2 contains extensive literature survey, Sect. 3 contains theoretical background, Sect. 4 contains TBSAC details, Sect. 5 contains key features of TBSAC, and Sect. 6 contains future scope and conclusions.

2 Literature Survey

Cloud storage implementation to mitigate data privacy and security risks is a topic of rigorous work. A lot of research is being done on access control for cloud data. Tang et al. [2] propose an overlay approach for privacy and assured data deletion with policy-based deletion. It encrypts data with data key which is further encrypted with RSA-based control key maintained by a key manager. The authors make it mandatory to have control key available to decrypt the data key required for its retrieval and demands the key manager to be fully trusted. Nusrat et al. [3] modify the above-mentioned approach using independent key managers with encryption key sharing to users. AES is used for file encryption, and AES key is encrypted using a RSA key. The proposal also removes the certification complexities by asking secret phrase at every time of file manipulation. All the files are encrypted under the same policy, and if the owner revokes a policy, all the files encrypted under that policy will be permanently unreadable. Ali et al. [4] propose an improved key management and authentication system using (k,n) threshold secret sharing scheme for key generation with semi-trusted third party. The system provides multiple key managers to overcome single point of failure, uses session-based keys with semi-trusted third-party system, station-to-station protocol, and digital signature for authentication process.

Cui et al. [5] propose a low-cost-efficient key management and derivation model. The client may be too lightweight to provide encryption/decryption and, as a result, possesses single password to derive all other keys. The client does not perform encryption–decryption; it directs all the sensitive data to a trusted server performs the encryption and decryption and then sends it to a semi-trusted server. The author also proposes dynamic key authorization update method. Zhang et al. [6] propose multi-replica associated deletion scheme in the cloud using replica directory to manage multiple replicas. The author proposes data storage, replica generation & deletion and feedback algorithms. Arfaoui et al. [7] propose a contextual dynamic authorization approach for patient data . The authors incorporate hybrid certificateless signcryption (H-CLSC) scheme for ciphertext verification without decryption. The data owner's private key is the mixture of the contribution of data owner and key generation center.

Dong et al. [8] propose an access control context state-aware model as a combination of role-based access control (RBAC) and attribute-based access control (ABAC) using semantic web technologies and SWRL. The authors clearly state that the implementation of the proposed model is very difficult and context state awareness is also not very well mentioned in the model. Malamateniou et al. [9] propose a context-aware capability based, role-centric access control model using semantic web technologies for connected medical devices for personal health records. Zhang et al. [10] propose a compressed cipher length polycentric scheme using multi-authority CP-ABE. The attribute binary tree representation of child nodes for each attribute authority decreased ciphertext length and computation. The scheme focuses on reducing do not care conditions used in attributes using polycentric hierarchical AND gate attribute encryption. Esposito et al. [11] propose interoperable dynamic access control for heterogeneous organizations using ontology-based decision-maker, identity claim extractor, and risk-aware authorization.

Wang et al. [12] propose blockchain-based framework for fine-grained access control to resolve data sharing and storage issues in centralized storage systems using Ethereum blockchain. The proposed scheme lacks an important feature of access policy update and user attribute revocation. Kapadia et al. [13] proposed a system called as privacy-enhanced attribute-based publishing (PEAPOD) with various properties, viz. secure data publishing to multiple possible recipients using attribute-based policies, hidden plaintext message and policies (see Appendix Fig. 7).

3 Theoretical Background

Access control in cloud data restricts what a user or a program, representing the user, can do with cloud data resource. Figure 2 illustrates a basic user authentication and authorization (UAA) entity which allows a user to access the data resources stored on the cloud. The user requests UAA for permission to access the data stored in cloud data storage, and UAA grants or revokes the permission based on user credentials. User then approaches to data storage with the permission which after validation with UAA, is granted or rejected by data storage. There are many security requirements in Fig. 1, viz. confidentiality in request and permission transmission, integrity of granted authorization, time-based access revocation and data user anonymity with data owner. The proposed framework solves above requirements using attribute-based encryption [14], context [15] for key generation, JWT-based authorization [16], homomorphic encryption [17], hidden credentials [18], assured deletion and partial dependency on key manager.

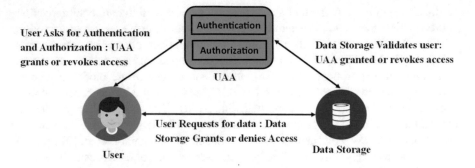

Fig. 1 UAA with cloud data

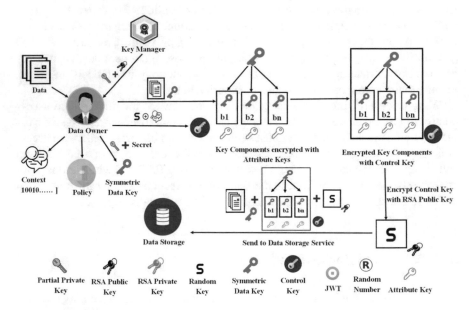

Fig. 2 Data storage

4 Proposed Framework

4.1 Introduction to Framework

TBSAC provides a data owner to securely store his data in the cloud which can be accessed only by the user satisfying the policy and having the same context as that of the owner. The owner is also provided with context and policy change features. It has following independent entities, viz. data owner which collects data into a data file, creates access control policy, extracts context, derives private key, and performs data encryption. Key manager acts as an authority to generate partial private key for data owner, issues JWT access token for data user, and performs challenge response-based authorization. Data storage service performs challenge-based authentication to the data owner, stores data provided by data owner, and supplies requested data to users after verifying JWT token. Data user requests data from the data storage service.

4.2 TBSAC Design

The proposed framework provides all data-related activities, viz. data storage, data retrieval, data policy, and context change. The data security maintained in these activities is as follows.

4.2.1 Data Storage

1. Data owner collects data from different sources and combines it into a data file F. The owner further creates an access control policy P required to be satisfied to decrypt the data and derives a context, C, from data. He then asks for a partial private key (used to drive the data encryption key) and RSA public key (used to encrypt control key, CK) from key manager. The data owner generates a random number S which is used to create control key, CK, as exclusive-OR operation between random number S and the derived context C (see Fig. 2).
2. Data encryption key K is derived as a combination of partial private key and a chosen secret. File F is then encrypted using key K. The owner breaks down the data encryption key K into n components and randomly generates b_i such that the product of all b_i is 1.

$$k_1 \cdot k_2 \cdot k_3 \ldots k_n = K \qquad (1)$$

$$b_1 \cdot b_2 \cdot b_3 \ldots b_n = 1 \qquad (2)$$

$$k_1 \cdot b_1 \cdot k_2 \cdot b_2 \cdot k_3 \cdot b_3 \ldots k_n \cdot b_n = K \qquad (3)$$

3. Each component of data encryption key is then encrypted with attribute specific encryption key and further with control key C using ElGamal cryptosystem.
4. The data owner encrypts random number S with RSA public key received from key manager and sends encrypted file F, encrypted key components, and encrypted random number S to data storage service.

4.2.2 Data Retrieval

1. User requests key manager for JWT to access data. Key manager performs challenge response-based authentication with the user and, on successful authentication, provides an auto-expiry access token (see Fig. 3).

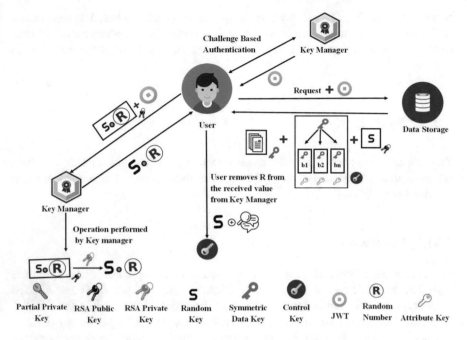

Fig. 3 Data retrieval

2. After successful JWT verification, data storage service sends key components, encrypted data file and encrypted random number S to the user.
3. The user decrypts random number S with the help of key manager. The key manager receives JWT and encrypted random number S attached with random number R from the user.
4. Key manager then decrypts encrypted random number with the help of private RSA key and then returns S.R. User then retrieves control key, CK, using context C:

$$CK = S \oplus C \qquad (4)$$

5. Control key CK is then used to decrypt key components further using which data is recovered (see Fig. 4).

$$[[k_1 \cdot b_1]_{Attribute\_1} \cdot k_2 \cdot b_2]_{Attribute\_2} \ldots [k_n.b_n]_{Attribute\_N}]_{ControlKey} \qquad (5)$$

$$[k_1 \cdots b_1]_{Attribute\_1} \cdot [k_2 \cdots b_2]_{Attribute\_2} \ldots [k_n.b_n]_{Attribute\_N} \qquad (6)$$

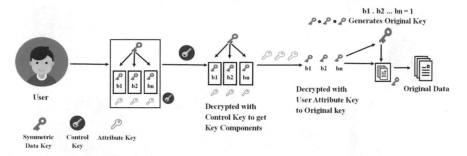

Fig. 4 Data decryption

4.2.3 Access Policy Change

1. Owner requests encrypted data key set and encrypted random number S from data storage service which performs challenge response-based authentication with owner and, if successful, returns encrypted key components and encrypted random number S to the data owner (see Fig. 5).
2. Data owner recovers random number S from the key manager and derives control key CK.
3. After recovering the control key CK, data owner recovers the original key components using CK and then encrypts them according to new access policy encryption attributes.
4. The newly encrypted components are then homomorphically encrypted with control key, and the resultant is then sent back to data storage service.

4.2.4 Context Change

1. Owner requests encrypted data key set and encrypted random number S from data storage service which performs challenge response-based authentication with owner and, if successful, returns encrypted key components and encrypted random number S to the data owner (see Fig. 6).
2. Data owner recovers random number S from the key manager and derives control keys CK_{old} with old context C_{old} and CK_{new} with new context C_{new} as shown in Fig. 6.

5   Key Features of TBSAC

- Proposed framework has **credential privacy**, i.e., hidden credentials if an adversary cannot decide non-negligibly better than random guessing of a recipient that it has a certain attribute in the system. The data owner may never know who has made an effort to decrypted his data.

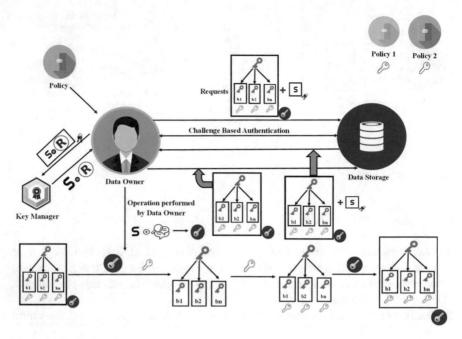

Fig. 5 Policy change

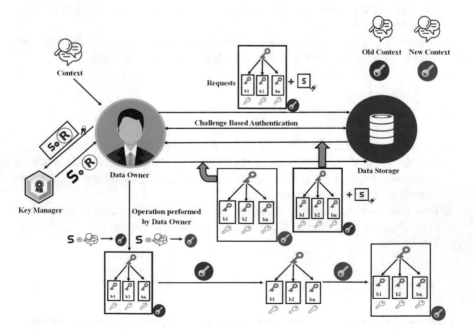

Fig. 6 Context change

- The data owner gets a partial private key from the key manager which is transformed into symmetric data key using a secret generated by himself. This procedure makes duplicate data encryption key generation impossible for key manager.
- The control key derivation is owner's **context dependent**. The key components are homomorphically encrypted using control key, and the origin control key is removed after encryption, which requires the same context to be used for control key retrieval.
- The data key is fragmented into different components, and then each component is homomorphically encrypted in accordance with the access control policy clauses. For the policy with multiple clauses, the key components may be homomorphically encrypted with different clause attributes which as a consequence would generate multiple data key component sets. A user is required to satisfy any of the clauses for data key generation.
- A user is required to get **JWT** token from the key manager for data retrieval from data storage service which auto-expires after a given amount of time as set by the key manager.
- The data can only be decrypted with homomorphically encrypted data key using control key, CK, which is RSA encrypted by the key generated by key manager. If the key manager is requested to remove the RSA decryption key for a particular RSA encryption key, in that case the data encrypting Key, K, would not be regenerated. Here, transitively the data file F will be permanently lost.

6 Future Scope and Conclusions

TBSAC works for fine-grained access control for cloud data. The framework provides context-based, hidden policy, and time-based expiry data publishing facility to a data owner. In TBSAC, the data owner breaks keys into separate components and then encrypts them as per policy attributes. The data user gains no knowledge about the policy satisfying attributes as the keys are also randomized using b1, b2, ...bn variables. The use of data owner and user context provides further access control. The context of user and data owner must be the same for control key generation which leads to context-based key generation which itself is a research area to be worked upon. There are various context models, representation, and extraction methods which can be aligned with the proposed framework. Improvements may be done to provide de-centralized key manager and data storage service. Only data owner is allowed to change the data context or the access policy. Attribute-based encryption techniques may further be worked upon for lightweight computation.

7 Appendix

See Appendix Fig. 7

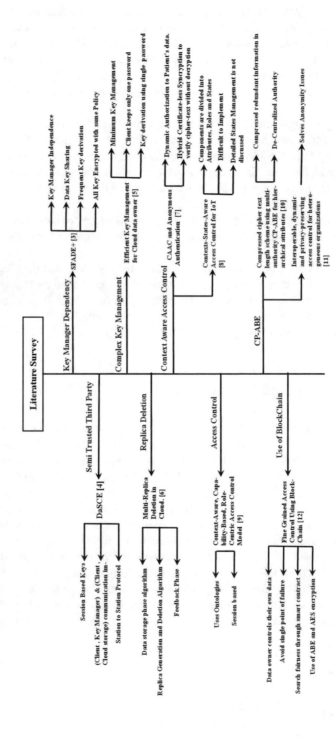

Fig. 7 Literature review

References

1. Singh, Jatinder, et al. 2015. Twenty security considerations for cloud-supported Internet of Things. *IEEE Internet of things Journal* 3 (3): 269–284. https://doi.org/10.1109/JIOT.2015. 2460333
2. Tang, Yang, et al. 2012. Secure overlay cloud storage with access control and assured deletion. *IEEE Transactions on dependable and secure computing* 9 (6): 903–916. https://doi.org/10. 1109/TDSC.2012.49
3. Nusrat, Raisa, and Rajesh Palit. 2017. Simplified FADE with sharing feature (SFADE+): A overlay approach for cloud storage system 2017. In *IEEE 7th annual computing and communication workshop and conference (CCWC)*. IEEE. https://doi.org/10.1109/CCWC.2017. 7868486.
4. Ali, Mazhar, Saif UR Malik, and Samee U. Khan. 2015. DaSCE: Data security for cloud environment with semi-trusted third party. *IEEE Transactions on Cloud Computing* 5 (4): 642–655. https://doi.org/10.1109/TCC.2015.2446458
5. Cui, Zongmin, et al. 2015. Efficient key management for IOT owner in the cloud. In *2015 IEEE fifth international conference on big data and cloud computing*. IEEE. https://doi.org/10.1109/ BDCloud.2015.40
6. Zhang, Yuanyuan, et al. 2016. A multi-replica associated deleting scheme in cloud. 2016 10th international conference on complex, intelligent, and software intensive systems (CISIS). IEEE 2016. https://doi.org/10.1109/CISIS.2016.68
7. Arfaoui, Amel, et al. 2019. Context-aware access control and anonymous authentication in WBAN. *Computers and Security*. https://doi.org/10.1016/j.cose.2019.03.017.
8. Dong, Yuji, et al. 2018. Contexts-states-aware access control for internet of things. In *2018 IEEE 22nd international conference on computer supported cooperative work in design (CSCWD)*. IEEE. https://doi.org/10.1109/CSCWD.2018.8465364.
9. Malamateniou, Flora, et al. 2016. A context-aware, capability-based, role-centric access control model for IoMT.In *International conference on wireless mobile communication and healthcare*. Cham: Springer. https://doi.org/10.1007/978-3-319-58877-3_16.
10. Zhang, Zhiyong, et al. 2018. Efficient compressed ciphertext length scheme using multi-authority CP-ABE for hierarchical attributes. *IEEE Access* 6: 38273–38284. https://doi.org/ 10.1109/ACCESS.2018.2854600.
11. Esposito, Christian. 2018. Interoperable, dynamic and privacy-preserving access control for cloud data storage when integrating heterogeneous organizations. *Journal of Network & Computer Applications* 108: 124–136. https://doi.org/10.1016/j.jnca.2018.01.017.
12. Wang, Shangping, Yinglong Zhang, and Yaling Zhang. 2018. A blockchain-based framework for data sharing with fine-grained access control in decentralized storage systems. *IEEE Access* 6: 38437–38450. https://doi.org/10.1109/ACCESS.2018.2851611.
13. Kapadia, Apu, Patrick P. Tsang, and Sean W. Smith. 2007. Attribute-based publishing with hidden credentials and hidden policies. *NDSS* 7.
14. Bethencourt, John, Amit Sahai, and Brent Waters. 2007. Ciphertext-policy attribute-based encryption. 2007. In *IEEE symposium on security and privacy (SP'07)*. IEEE. https://doi.org/ 10.1109/SP.2007.11.
15. Perera, Charith, et al. 2013. Context aware computing for the internet of things: A survey. *IEEE Communications Surveys & Tutorials* 16 (1): 414–454. https://doi.org/10.1109/SURV. 2013.042313.00197.
16. Bradley, John, Nat Sakimura, and Michael B. Jones. 2015. JSON web token (JWT).
17. Gentry, Craig, and Dan Boneh. 2009. A fully homomorphic encryption scheme. Vol. 20. No. 09. Stanford: Stanford University.
18. Holt, Jason E., et al. 2003. Hidden credentials. In *Proceedings of the 2003 ACM workshop on Privacy in the electronic society*. ACM. https://doi.org/10.1145/1005140.1005142.

2DBSCAN with Local Outlier Detection

Urja Pandya, Vidhi Mistry, Anjana Rathwa, Himani Kachroo
and Anjali Jivani

Abstract This research is related to designing a new algorithm which is based on the existing DBSCAN algorithm to improve the quality of clustering. DBSCAN algorithm categorizes each data object as either a core point, a border point or a noise point. These points are classified based on the density determined by the input parameters. However, in DBSCAN algorithm, a border point is designated the same cluster as its core point. This leads to a disadvantage of DBSCAN algorithm which is popularly known as the problem of transitivity. The proposed algorithm—two DBSCAN with local outlier detection (2DBSCAN-LOD), tries to address this problem. Average silhouette width score is used here to compare the quality of clusters formed by both algorithms. By testing 2DBSCAN-LOD on different artificial datasets, it is found that the average silhouette width score of clusters formed by DBSCAN-LOD is higher than that of the clusters formed by DBSCAN.

Keywords DBSCAN · Clustering · Border points · Local outliers · Global outliers

1 Introduction

Clustering is an unsupervised learning technique which groups a set of data objects into a number of clusters or classes based on the similarities and dissimilarities with one another. Data is categorized according to the clusters formed. Hence, an ideal approach to clustering must take into consideration, the quality of clusters. There are many algorithms designed in order to address the problem of clustering. Each algorithm has its own advantages and disadvantages, and hence, it is important to select a proper algorithm based on the dataset and the intended use of results. There are various approaches formulated for the process of clustering, for example, hierarchical clustering algorithms, partitioning algorithms, grid-based clustering algorithms, and density-based clustering algorithms [1]. Out of these, the focus here is density-based

U. Pandya (✉) · V. Mistry · A. Rathwa · H. Kachroo · A. Jivani
Department of Computer Science and Engineering, The Maharaja Sayajirao University of Baroda, Vadodara, India

A. Jivani
e-mail: anjali.jivani-cse@msubaroda.ac.in

© Springer Nature Singapore Pte Ltd. 2020
Y.-C. Hu et al. (eds.), *Ambient Communications and Computer Systems*, Advances in Intelligent Systems and Computing 1097,
https://doi.org/10.1007/978-981-15-1518-7_21

clustering algorithms and specifically the algorithm density-based spatial clustering of applications with noise (DBSCAN). Thus, the main objective of carrying out this research work is to design an algorithm based on existing ones, to perform better clustering qualitatively. Also, a statistical metric, known as the silhouette width index, is used in order to validate the quality of the clusters formed.

1.1 Density-based spatial clustering of applications with noise (DBSCAN)

One of the most widely used density-based clustering algorithms is the DBSCAN algorithm.

DBSCAN algorithm requires two input parameters:

1. **Epsilon radius (ε)**: Distance threshold which determines the neighborhood of a data object.
2. **Minimum number of Points (MinPts)**: Minimum number of points that must be present in the Epsilon (ε) neighborhood in order to satisfy the density required to classify the data objects as a separate cluster.

Some of the basic concepts can be expressed by the following definitions:

1. **Epsilon (ε) neighborhood**: For a data object Q, the set of data objects which are at most ε distance (The term distance, in this paper, refers to the Euclidean distance) away from Q constitute the Epsilon (ε) neighborhood of Q. Also, as this research is tested on 2D datasets, the Epsilon (ε) neighborhood of a point Q can be a circle of radius ε with Q as its center.
2. **Directly density reachable**: A data object P is directly density reachable from a data object Q if and only if P lies in the Epsilon (ε) neighborhood of Q.
3. **Density reachable**: A data object P is density reachable from data object Q if there is a path $q_1, q_2, \ldots, q_n$ with $q_1 = Q$ and $q_n = P$, where each $q_i + 1$ is directly density reachable from q_i. This means that all the data objects in the path must be core objects except the exception of object q_n.
4. **Density connected**: A data object P is density connected to a data object Q if there exists a data object V such that both P and Q are density reachable from V.

DBSCAN algorithm classifies each data object (data object and data point can be used interchangeably throughout the paper) into three types:

- Core point: A data object which has at least the minimum number of data objects (MinPts) in its Epsilon (ε) neighborhood. No data object is directly reachable from non-core data object.
- Border point: A data object which is density reachable from a core point, and which has less than the minimum number of data objects (MinPts) in its Epsilon (ε) neighborhood.

- Noise point (Outliers): A data object which is neither a border point nor a core point.

DBSCAN starts by checking Epsilon (ε) neighborhood of each individual data object in the dataset. If the Epsilon (ε) neighborhood of a data object contains more than the minimum number of neighbors, i.e., MinPts, a new cluster with that data object as a core point is created. The DBSCAN algorithm then collects directly density reachable data objects from these core points to merge all the density reachable clusters. This process terminates when there is no data object left to be added [2]. The data objects which do not fall in any cluster are considered to be noise or outliers [3].

2 The Problem of Transitivity

A border point in the DBSCAN algorithm is density reachable from a core point. There is a bridge of core points connecting a border point and a particular core point. However, it is to be noted that a border point does not actually cross the density threshold required to form a cluster [4].

Now, consider Fig. 1.

Here, points P, Q, and R are core points as they have more than or equal to the minimum number of points (MinPts) in their Epsilon (ε) neighborhood. The point S is a border point as it is density reachable from P, connected by a bridge of core points Q and R, and has less than the minimum number of points (MinPts) in its Epsilon (ε) neighborhood. Now, even though point S does not cross the required density threshold, the DBSCAN algorithm labels it in the same cluster as that of point P. Thus, it can be inferred that the DBSCAN algorithm considers point S and P to be as similar as points Q and P. Thus, practically even though data objects P and S differ by a greater degree than the points P and Q differ, this degree of difference is not depicted by the DBSCAN algorithm. This problem is popularly known as the problem of transitivity.

Fig. 1 Implementing the DBSCAN algorithm with Epsilon = ε and MinPts = 5

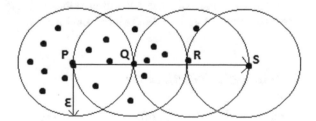

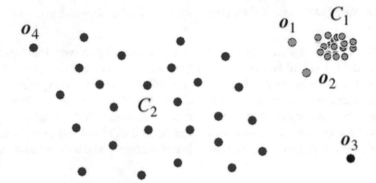

Fig. 2 Figure depicting local outliers and global outliers

3 Local Outliers and Global Outliers

Outliers or noise points are anomalous data objects. The data objects which are not directly density reachable or density reachable from any of the core points are classified as noise points. Generally, noise points are to be eliminated in the preprocessing step to identify a regular pattern in the data set. Consider Fig. 2 [5].

Here, the dataset is divided into two clusters, C1 and C2. Also, there is a data object O3 which is identified as a noise point by the DBSCAN algorithm. Here, it can be seen that cluster C2 is sparse with respect to cluster C1. Hence, to identify both the clusters properly, the value of MinPts and Epsilon would have to be kept such that the density threshold calculated is able to identify cluster C2.

Now, take a local point of view, i.e., with respect to cluster C1. Here, the relative density of the data objects O1 and O2 is lower than that of its neighboring data objects. Hence, data objects O1 and O2 can be outliers with respect to the data objects in cluster C1. However, globally, due to the relatively sparse cluster C2, the data objects O1 and O2 will not be identified as outliers but will be classified in the cluster C1. Thus, these points are known as local outliers. Also, it can be implied that the data object O3 is a global outlier.

Hence, to obtain better categorization of the data objects, the clustering algorithm implemented must identify both global and local outliers.

4 Silhouette Width Score

Validating a clustering algorithm can be hampered by our visual bias. Hence, some statistical measure is necessary to judge the quality of clusters formed. silhouette width index is a well-known validation tool for assessing clustering quality. It is to be evaluated for every individual data object. The average silhouette width score is used to compare the clustering algorithms for this research. The best advantage of using silhouette index for finding the number of clusters is that it allows for measuring the compactness or tightness and can be used to study the separation distance between the resulting clusters [6]. The silhouette width scores lie in the range of −1 to 1 and can be interpreted as follows:

- If the silhouette width score is close to +1, then the data object has a high probability of belonging to its designated cluster
- If the silhouette width score is close to 0, then the data object has low probability of belonging to its designated cluster
- If the silhouette width score is close to −1, then the data object has a high probability of not belonging to its designated cluster.

5 2DBSCAN with Local Outlier Detection (2DBSCAN-LOD)

This research proposes an enhancement for the existing DBSCAN algorithm, known as 2DBSCAN with local outlier detection (2DBSCAN-LOD). The basic idea of 2DBSCAN-LOD is that it considers a cluster to be a group of connected objects having densities greater than or equal to a given density. Thus, it attempts to solve the problem of transitivity by identifying border points as local outliers and eliminates them from the process of clustering. 2DBSCAN-LOD uses DBSCAN in the preprocessing step, to identify outliers and border points. It then forms a new data set by removing those points from the original data set. The DBSCAN algorithm is then again implemented on this newly formed data set.

The following steps are employed to implement 2DBSCAN-LOD:

1. Initialize the data set and implement the DBSCAN algorithm with suitable values of Epsilon (ε) and minimum number of points (MinPts) required for determining the density.
2. Identify border points and global outliers.
3. Form a new data set by eliminating the data objects identified in step 2.
4. Implement the DBSCAN algorithm on this new data set with the same values of Epsilon (ε) and minimum number of points (MinPts) as used in step 1.
5. On the clusters obtained, plot the border points (identified in step 2) as local outliers and noise points as global outliers (Fig. 3).

6 Empirical Results and Observations

The proposed algorithm is tested on 2D artificial datasets [7]. Three different types of data sets are demonstrated here, each showing positive change in the silhouette width score as compared to the original DBSCAN algorithm. Note that, in Figs. 4(ii), 5(ii),

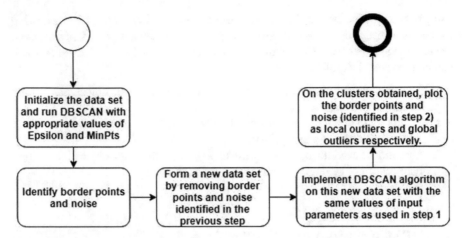

Fig. 3 Flow chart for 2DBSCAN-LOD

(1) **Test Case-1** (Number of attributes, Number of instances): 2 ,1000

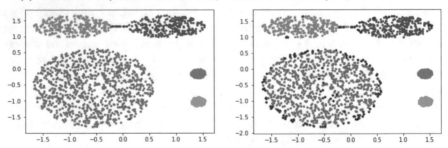

Fig. 4 Test case 1. Left figure shows implementation of DBSCAN algorithm, and right figure shows the implementation of 2DBSCAN-LOD

(2) **Test Case-2** (Number of attributes, Number of instances): 2 ,200

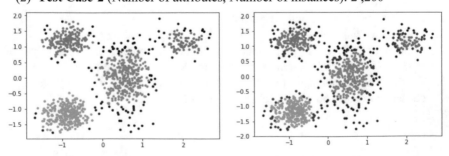

Fig. 5 Test case 2. Left figure shows implementation of DBSCAN algorithm. Right figure shows the implementation of 2DBSCAN-LOD

(3) **Test Case-3** (Number of attributes, Number of instances): 2, 250 [8]

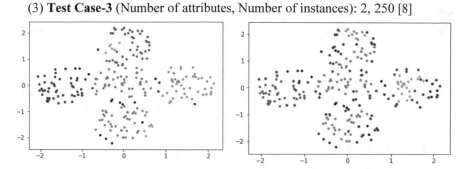

Fig. 6 Test case 3. Left figure shows implementation of DBSCAN algorithm. Right figure shows the implementation of 2DBSCAN-LOD

6(ii), the red-colored data objects are local outliers, and black colored data objects are global outliers.

(1) **Test Case-1** (Number of attributes, Number of instances): 2, 1000
(2) **Test Case-2** (Number of attributes, Number of instances): 2, 200
(3) **Test Case-3** (Number of attributes, Number of instances): 2, 250 [8].

Above three datasets are chosen due to their distinct characteristics.

Test case 1, i.e., Figure 4, has a bridge of data points connecting the upper two clusters. These data points are classified as a separate cluster according to the DBSCAN algorithm, whereas 2DBSCAN-LOD does not categorize those points as a separate cluster but as border points due to which the silhouette width index improves which thus indicates better clustering.

Test case 2, i.e., Figure 5, has four different clusters, where the data points located near the boundary of each cluster do not cross the density threshold, yet the DBSCAN algorithm classifies those data points in the same cluster. 2DBSCAN-LOD again results in better clustering, as indicated by the positive change in silhouette width score, as it identifies those data points as border points.

Test case 3, i.e., Figure 6, has sparsely distributed data points. 2DBSCAN-LOD results in better clustering here as well because it forms clusters of only those data points which cross the required density threshold.

Table 1 summarizes the change in silhouette width score in all the test cases considered. It can be clearly interpreted here, that the silhouette width score is nearer to $+1$ for clusters formed by 2DBSCAN-LOD algorithm, than the silhouette width score for clusters formed by the DBSCAN algorithm.

Table 1 Comparing DBSCAN with 2DBSCAN-LOD using silhouette width score

Sr. No.	Data set	DBSCAN		Avg. silhouette score of original DBSCAN	Avg. silhouette score of 2DBSCAN-LOD	Percentage change (%)
		Eps	MinPts			
1.	Test case one	0.2	20	0.42	0.55	31.67
2.	Test case two	0.2	15	0.60	0.74	22.80
3.	Test case three	0.3	7	0.42	0.48	13.78

7 Proof for Local Outliers

As stated earlier, the silhouette width score is calculated for each individual data object. Also, if the silhouette width score is close to zero, then that data object has a low probability of belonging to its designated cluster.

Hence, during the above steps, the silhouette width score was calculated for border points as well as the core points. And a histogram, as depicted in Fig. 7, for the core points and border points of the data set in test case one is plotted which clearly depicts that the silhouette width scores of most of the border points are close to zero and have a lower average than that of the core points. Thus, it facilitates the categorization of border points as local outliers.

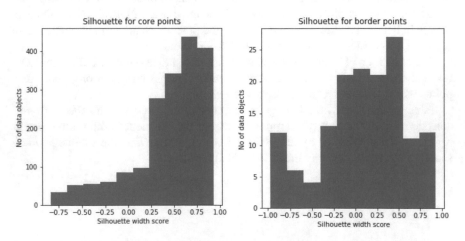

Fig. 7 Comparison of silhouette width score for core points (left) and border points (right)

Average silhouette width score for core points is 0.448
Average silhouette width score for border points is 0.080.

8 Conclusion

2DBSCAN-LOD can be further optimized to work faster on real data sets as well as high dimensional data. Moreover, 2DBSCAN-LOD fails to work well for datasets having less proportion of boundary points. Thus, in such cases, it is futile to implement DBSCAN twice. Hence, a preprocessing step can be added which would detect the proportion of boundary points beforehand and proceed with 2DBSCAN-LOD only if required. Hence, concluding, a new algorithm, 2DBSCAN-LOD, is proposed which solves the problem of transitivity and gives better outlier detection by considering the border points as local outliers. Furthermore, the quality of clustering is also improved as depicted by the positive change in the silhouette width score.

References

1. Ghuman, Sukhdev Singh. 2016. *Clustering techniques—a review.*
2. Khan, M.M.R., M.A.B. Siddique, R.B. Arif, and M.R. Oishe. 2018. ADBSCAN: adaptive density-based spatial clustering of applications with noise for identifying clusters with varying densities.
3. Martin, E., Hans-Peter Kriegel, Jörg Sander, and X. Xiaowei.1996. A density-based algorithm for discovering clusters a density-based algorithm for discovering clusters in large spatial databases with noise.
4. Campello, Ricardo J.G.B., Davoud Moulavi, Arthur Zimek, and Jörg Sander. 2015. Hierarchical density estimates for data clustering, visualization, and outlier detection.
5. Chepenko, D. 2018. A Density-based algorithm for outlier detection. *Medium.* https:// towardsdatascience.com/density-based-algorithm-for-outlier-detection-8f278d2f7983.
6. Rousseeuw, P.J. 1987. Silhouettes: a graphical aid to the interpretation and validation of cluster analysis.
7. Barton, T. 2015. *Clustering-benchmark.* Github. https://github.com/deric/clustering-benchmark.
8. Bandyopadhyay, S., and U. Maulik. 2002. Nonparametric genetic clustering: comparison validity indices.
9. Bhattacharyya, S. 2019. *DBSCAN algorithm: complete guide and application with python scikit-learn, clustering spatial database.* https://towardsdatascience.com/dbscan-algorithm-complete-guide-and-application-with-python-scikit-learn-d690cbae4c5d.
10. DBSCAN Wikipedia. https://en.wikipedia.org/wiki/DBSCAN.
11. Silhouette Width Wikipedia. https://en.wikipedia.org/wiki/Silhouette_(clustering).

Classification of Documents based on Local Binary Pattern Features through Age Analysis

Pushpalata Gonasagi, Rajmohan Pardeshi and Mallikarjun Hangarge

Abstract A method for identifying the age of a document using local binary pattern (LBP) features is presented here. The paper documents are the most important source of information which has been using for communication, record maintenance and proof for smoothing. A paper document is a sheet used for writing or printing something on it. We have been using a large amount of paper documents in day today life. Watermarking, embedding signatures and printed patterns have been using to secure legal documents; however, due to the misuse of digital technology, several challenges related to document security is raised. To address this problem, document age identification is one of the significant steps used to identify document's authenticity/originality. In this paper, 500 printed documents which are published during 1993–2013 are used for identifying as new or old documents based on their year of publication. We have considered whole document for study irrespective of the content like text, line, logo, noise, etc. Initially, we have segmented a document page into 512×512 blocks and retained those text blocks which are covered with full text. Later, applied LBP technique on these text blocks and extracted features. These features are fed to k-nearest neighbors (KNN) classifier and average classification accuracy achieved is 91.5%.

Keywords Local binary patterns (LBP) · K-nearest neighbor (KNN) · Document age identification

P. Gonasagi
Department of PG Studies and Research in Computer Science, Gulbarga University, Kalaburagi, Karnataka, India

R. Pardeshi · M. Hangarge (✉)
Department of PG Studies and Research in Computer Science, Karnatak Arts, Science and Commerce College, Bidar, Karnataka, India

© Springer Nature Singapore Pte Ltd. 2020
Y.-C. Hu et al. (eds.), *Ambient Communications and Computer Systems*, Advances in Intelligent Systems and Computing 1097,
https://doi.org/10.1007/978-981-15-1518-7_22

265

1 Introduction

In our daily life, paper documents are playing a vital role in keeping the records such as birth certificates, death certificates, identity cards, official documents, etc. But as digital technology has been growing, the concern about the authenticity of these paper documents is increasing. Even though watermarking system, printing pattern, logo, etc., are applied during production of documents but still security problems exist. Document classification based on its age is important task in forensic science. As a document becomes older or aging, color of the document changes. The color and background of the document depend on humidity, environment, storage conditions, temperature, frequently access, etc. Paper documents may have been generated in many centuries, decades, years, and months ago. In fact, paper manufacturing technologies, printing technologies, and paper quality have been changing with time. Besides, printed documents can be modified by using digital technology, such as Adobe Photoshop, Gimp, Color printer for malicious purposes. In this context, investigation for the document uniqueness is essential for several purposes. We present a method to deal with document age identification using LBP to support the forensic science experts in analyzing the authenticity of the documents.

The organization of the rest of the paper is as follows: Sect. 2 describes related work, Sect. 3 presents proposed method, Sect. 4 contains results and discussion, and Sect. 5 concludes the contributions of the proposed work.

2 Related Work

A number of research works have been carried out to classify the handwritten and printed documents as old and new based on their aging to support the investigation documents originality. Here, we have presented a brief review of the work pertaining to the documents classification based on their aging. Raghunandan et al. [1] explained identification of the documents as old and new based on the quality of the documents. They have used foreground and background information of handwritten documents as well as Fourier co-efficient features. The authors carried out the experimental study on their own dataset and achieved an accuracy of 78.5% for new documents and 77.5% for old documents. He et al. [2] proposed a method for historical document dating using scale-invariant features. Stroke shape elements of historical documents and evolutionary self-organizing map are discovering to identify the evolution of visual elements along the time line. This method achieved accuracy of 85.1%. Li et al. [3] presented a method to estimate publication dates of printed historical documents. They proposed hybrid model which employs image and text data both, they used Convolutional Neural Network (CNN) and Word Embedding. The same model further employed for Classification and Regression Tasks. They achieved text models are better than image models and combined models are better than both model. Khan et al. [4] presented a method which demonstrates the use of hyper-spectral imaging for fraud handwritten document analysis. Ink mismatches of handwritten documents are detected. But illicit purpose created same handwritten document using advance software which satisfied the characters of the original documents. Biswajit et al. [5] proposed a method to determine ink age of printed documents. This system is based on the color image analysis by finding the average intensities, pixel profile and kurtosis of the image. These features are used for ink age identification of the unknown samples. Google Life magazine cover pages were considered as dataset. Neural network is designed for ink age identification and achieved the recognition accuracy as 74.5%.

da Silva Barboza et al. [6] this article identifies the age of the document based on the RGB color components of documents background. The birth, wedding cards and other certificates of twentieth century were considered for experimentation. They have considered the document images for experimentation with a fifty-two years gap of each other and noticed encouraging results.

Gebhardt et al. [7] presented a system that differentiates the document from the bundle of documents which are printed with different printers. Different printers have different impact of printing the characters. With this clue, they have identified the forgery documents. They have noticed the changes in the edges of printed characters of laser printer and inkjet printer. They used unsupervised anomaly Grubb's test and global KNN algorithm to classify the documents printed using laser and inkjet printers.

Bertrand et al. [8] described a method that automatically detects the forgery of characters. The copy and paste techniques which are commonly used to make the documents forgery are identified. The copy and paste characters are detected through similarity checking between the characters. To detect the simulation forgery, they identified the unusual shapes of characters. They achieved the results as 0.77 recalls and 0.82 precision.

3 Proposed Method

The pre-processing, feature extraction and classification of the documents are three important steps performed in this method. The block diagram of the proposed method is depicted in Fig. 1.

3.1 Pre-processing

Basically, we have collected 100 document pages from five books which are printed in 1993, 1998, 2003, 2008 and 2013 (20 pages of each book). These documents are scanned by HP flatbed model LaserJetProM128fn scanner with resolution of 300 dpi. These scanned images are converted to gray-level images and then segmented them as

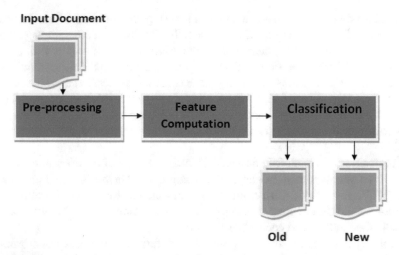

Fig. 1 Flow diagram of proposed method

Table 1 Details of the dataset developed for experiments

Year	No. of images	No. of text blocks extracted
2013	20	112
2008	20	111
2003	20	113
1998	20	140
1993	20	104
Total	100	580

512×512 pixels text blocks and retained only those blocks which are fully covered by the text. The details of dataset are presented in Table 1.

3.2 Feature Extraction

Local Binary Pattern: It is one of the efficient texture descriptors. LBP [9, 10] measures local spatial patterns and gray scale contrast of the underlying texture. It is robust for computing monotonic gray scale changes in the image. Rotation invariant is the basic property of the LBP. Thus, we choose LBP to obtain proficient texture descriptors of text blocks. In number of applications, it has been proved that LBP is more robust than statistical texture features. We extract the features of the text blocks using LBP ($P = 8$, $R = 1$). This yields a text block with 256 patterns. Among these, 58 patterns are uniform and each non-uniform are signed as single code which gives total 59 different patterns. Mathematically, LBP can be expressed as shown in Eq. 1.

$$\text{LBP}_{P,R} = \sum_{p=0}^{p-1} S(g_p - g_c)2^p \text{ where } S(x) \begin{cases} 1 & \text{if } x \geq 1 \\ 0 & \text{if } x < 0 \end{cases} \tag{1}$$

Where, P indicates sampling points on a circle of radius R ($P = 8, R = 1$), gc corresponds to the gray value of the center pixel and gp corresponds to the gray values of its neighbor pixel p.

3.3 Classification

K-nearest neighbor (KNN) [11] is simple and lazy classifier, even though we have employed it to identify the age of documents. Instead of sophisticated classifiers, we preferred it because LBP features are more discriminative in nature. This simple classifier has given encouraging results.

4 Results and Discussion

For experimentation, we have considered documents printed in 2013 as new and the documents which are printed in 2008, 2003, 1998 and 1993 are as old. We have designed two class problems and employed KNN to compute the distance between the feature vectors of documents printed in 2013 with feature vector of documents printed in 2008, 2003, 1998 and 1993, respectively. The classification accuracy is presented in Table 2.

The classification accuracy shown in Table 2 reveals interesting results that as the aging gap increases between the documents, accuracy also increases. This is because the aging of documents printed in 1993 is 20 years older than documents of 2013. Hence, the classification accuracy presented in Table 2 is 98.5%. Similarly, the aging gap between the documents printed in 2013 and 2003 is ten years. Therefore,

Table 2 Classification accuracy by considering documents of 2013 as new and others as old

Year (new/old document)	2008	2003	1998	1993
2013 (%)	87.5	92	95.5	98.5

Table 3 Classification accuracy by considering documents of 2013 as new and others as old

Accuracy	New documents (2013)	Old documents (1993–2008)
New documents (2013) (%)	87.5	12.5 (error)
Old documents (1993–2008) (%)	05 (Error)	95

Table 4 Classification rate of the proposed system without block-wise dataset compared to existing approach

Method	New documents (%)	Old documents (%)
Raghunandan et al. [1]	78.5	77.5
Proposed method	87.5	95

similarity between the documents of 2013 and 2008 is more; therefore, less accuracy is noticed that is 87.5%. The confusion rate of the classifier is shown in Table 3. The comparative analysis of the results with the reported work is not significant as there is no similarity in experimental environment (different dataset, method and number of features). Moreover, there is no age identification work reported on documents of Kannada script. However, a numeric comparative analysis is shown in Table 4 with the work cited at [1]. This work is carried out on handwritten documents age identification with different age gaps.

5 Conclusion

In this paper, we proposed a novel approach for classifying documents as new and old. The LBP has shown remarkable results in discriminating the text blocks as new and old. LBP efficiently captured texture properties of the documents and yielded high-discriminating features. Document age identification research is carried out to study the nature of the documents to validate its originality. In addition to this, we are extending this work to find the printing technologies used to produce the documents. Also we are extending it to identify the age of the handwritten documents. In the future, we validate the performance of the proposed algorithm when it is applied on the documents of different (printed and handwritten) scripts, font styles and font sizes.

References

1. Raghunandan, K.S., B.J. Palaiahnakote Shivakumara, B.J. Navya, G. Pooja, Navya Prakash, G. Hemantha Kumar, Umapada Pal, and Tong Lu. 2016. Fourier coefficients for fraud handwritten document classification through age analysis. In *2016 15th International conference on frontiers in handwriting recognition (ICFHR)*, 25–30. IEEE
2. He, Sheng, Petros Samara, Jan Burgers, and Lambert Schomaker. 2016. Discovering visual element evolutions for historical document dating. In *2016 15th International conference on frontiers in handwriting recognition (ICFHR)*, 7–12. IEEE.
3. Li, Yuanpeng, Dmitriy Genzel, Yasuhisa Fujii, and Ashok C. Popat. 2015. Publication date estimation for printed historical documents using convolutional neural networks. In *Proceedings of the 3rd international workshop on historical document imaging and processing*, 99–106. ACM.

4. Khan, Zohaib, Faisal Shafait, and Ajmal Mian. 2015. Automatic ink mismatch detection for forensic document analysis. *Pattern Recognition* 48 (11): 3615–3626.
5. Halder, Biswajit, and Utpal Garain. 2010. Color feature based approach for determining ink age in printed documents. In *2010 20th International conference on pattern recognition*, 3212–3215. IEEE.
6. da Silva Barboza, Ricardo, Rafael Dueire Lins, and Darlisson Marinho de Jesus. 2013. A color-based model to determine the age of documents for forensic purposes. In *2013 12th International conference on document analysis and recognition*, 1350–1354. IEEE.
7. Gebhardt, Johann, Markus Goldstein, Faisal Shafait, and Andreas Dengel. 2013. Document authentication using printing technique features and unsupervised anomaly detection. In *2013 12th International conference on document analysis and recognition*, 479–483. IEEE.
8. Bertrand, Romain, Petra Gomez-Krämer, Oriol Ramos Terrades, Patrick Franco, and Jean-Marc Ogier. 2013. A system based on intrinsic features for fraudulent document detection. In *2013 12th International conference on document analysis and recognition*, 106–110. IEEE.
9. Ojala, Timo, Matti Pietikäinen, and David Harwood. 1996. A comparative study of texture measures with classification based on featured distributions. *Pattern Recognition* 29 (1): 51–59.
10. Ojala, Timo, and Matti Pietikäinen. 1999. Unsupervised texture segmentation using feature distributions. *Pattern Recognition* 32 (3): 477–486.
11. https://machinelearningmastery.com/k-nearest-neighbors-for-machine.

An Efficient Task Scheduling Strategy for DAG in Cloud Computing Environment

Nidhi Rajak and Diwakar Shukla

Abstract Cloud computing is an active research topic in computer science and its popularity is increasing day-to-day due to the high demand of cloud in every field. Data center in cloud platform is having the number of computing resources which are interconnected with very high-speed network. These resources are accessed at the rapid speed so that minimum interaction with service provider. Task scheduling is a burning area of research in cloud environment. Here an application program is represented by directed acyclic graph (DAG). Major concerned of the task scheduling method is to reduce overall execution time. i.e., to minimize the makespan. This paper presents a new strategy for task scheduling in DAG which based on two well-known attributes critical path and static level. By using these attributes, we have developed new attributes CPS which is summation of critical path and static level. New strategy works on two phases such as task priority and resource selection. The proposed method is tested using two DAG models which shows outperformance as compared to heuristic algorithm HEFT. Comparisons have been done using some performance metrics which also gives good result of proposed method.

Keywords Cloud computing · DAG · Scheduling length · Critical path · Speedup

1 Introduction

Cloud computing is spreading in the current era of next generation of computing, and it is also called as Internet-based computing. The major principal of cloud computing is "pay-as-you-go-basis" [1] that is customer gives rent for using cloud resources. There are number of applications such as scientific computing, healthcare and DNA computing, where cloud computing uses for their high-speed computing and fast access of data. Two computing models [2] such as *data sharing* and *service sharing* are used in cloud computing and both models play very important role for cloud platform. The major objective of cloud computing is to maximize resource utilization and minimize the processing time.

N. Rajak · D. Shukla (✉)
Department of Computer Science and Applications, Dr. Harisingh Gour Central University,
Sagar, M.P., India

© Springer Nature Singapore Pte Ltd. 2020   273
Y.-C. Hu et al. (eds.), *Ambient Communications and Computer Systems*, Advances in
Intelligent Systems and Computing 1097,
https://doi.org/10.1007/978-981-15-1518-7_23

Scheduling is one of the research areas of cloud platform. It is mechanism to allocate the tasks onto the available resources by minimizing the scheduling length. The three main steps of any task scheduling algorithm that finds the priority of the tasks, sorting tasks as per their priority value either increasing or decreasing order and allocate the sorted tasks onto the available virtual machines. Task Scheduling is also known as NP-complete [3] problem. There are two basic types of task scheduling which are static and dynamic task scheduling. Static scheduling is scheduling in which information about the number of tasks, number of resources and their connectivity is known in advanced. It is also called as compile time scheduling, whereas dynamic scheduling method does not have pre-information that is only in execution time all information is known. Generally, task scheduling is represented by directed acyclic graph (DAG), i.e., an application program is represented by DAG in which each task represent by node and their communication link depicted by edges. To minimize the overall execution time, i.e., makespan is a major objective of the task scheduling algorithm.

This paper proposes an efficient method for task scheduling in cloud computing environment which is based on *critical path (CP) attribute* and *static level (SL) attribute*. There are two basic steps of the proposed algorithm. First step is to find priority of the tasks of given DAG using summation of CP and SL attributes. Second step is to resources selection, i.e., virtual machine selection which takes minimum execution time of the tasks. The performance of the new approach gives outstanding results in respect of scheduling length (makespan), speedup, efficiency and scheduling length ratio as compared to heuristics algorithm such as HEFT [4, 5].

The rest of the paper is arranged as follows: Sect. 2 addresses related work and 3 discusses the system model and scheduling attributes are discussed in Sect. 4. The performance metrics and proposed method are discussed in Sect. 5 and 6, respectively. Performance analysis of proposed algorithm and heuristic algorithms is discussed in Sect. 7 and finally, we conclude this paper.

2 Related Work

There are number of literature where task scheduling algorithm for DAG scheduling has been proposed to deal with optimized scheduling length. Heterogeneous Earliest Finished Time (HEFT) [5] algorithm which compute upward rank for priority of the tasks, critical path on a processor (CPOP) [4] uses upward and downward rank for computing the rank of the tasks. In Performance Effective Task Scheduling (PETS) algorithm, it uses [12] summation of mean of computation cost and communication time as rank attribute for the tasks. Similarly, Simulate annealing [6] is one of the methods to optimize the scheduling length in both environment grid and cloud computing. Some algorithms such as delay-based workflow scheduling [4] and particle swarm optimization-[13] based workflow compute cost parameter. Moreover, we use two well-known attributes, static level and critical path, for find the rank or priority of the tasks. The proposed algorithm is compared with HEFT algorithm in respect of scheduling length and other performance metrics which gives better results.

3 System Model

To discuss the task scheduling problem in this section, we present the formal definition of application model and computing resource.

An application model is represented by a directed acyclic graph (DAG), and it can be formally defined as a graph $G = (T, E, C_t)$, where $T = \{T_1, T_2, T_3, ..., T_n\}$ is a finite collection of nth tasks, $E = \{E_{i,j}\}$ is set of edges between the tasks T_i and T_j and which should maintain the precedence constrain between the tasks. C_t is communication time between the tasks T_i and T_j. In given DAG, there are two important tasks such as entry is a task which has no any parents and similarly, an exit task is task which has no any children. Figure 1 shows DAG model with five tasks.

Computing resource model consists of 's' number of cloud servers, which is represented by $CS = \{CS_1, CS_2, ..., CS_s\}$ and each cloud server may consist of one or more virtual machines. Virtual machine $V = \{R_1, R_2, ..., R_m\}$ consists of m number of resources which are interconnected through very high-speed network. Here resources are considered as virtual machines. The communication time (CT) between two virtual machines is considered negligible if they belong to same cloud server otherwise it will be included. Here also considered that no preemption and no any interrupt during execution of the tasks on virtual machine. Figure 2 [6] shows mapping of tasks and resources. *A major objective of the task scheduling algorithm is to minimize the makespan.*

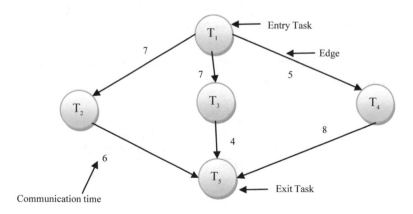

Fig. 1 DAG model with five

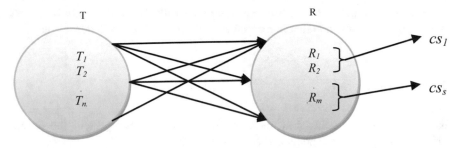

Fig. 2 Mapping T to R

4 Scheduling Attributes

This section discusses some attributes which are required in our proposed algorithm and they are as follows:

- **Estimated Computation Time (ECT)** [7]

$$
ECT_{ij} = \begin{bmatrix} ECT_{11} & ECT_{12} & \cdots & ECT_{1n} \\ ECT_{21} & ECT_{22} & \cdots & ECT_{2n} \\ ECT_{m1} & ECT_{m2} & \cdots & ECT_{mn} \end{bmatrix} \tag{1}
$$

where ECT_{ij} is estimated computation time of task t_i on resource R_j.

- *Average ECT (AECT)* [8] of a task T_i is defined as

$$
AECTi = \frac{\sum_{j=1}^{m} ECTi, j}{m}. \tag{2}
$$

- *Critical Path (CP)* [9, 10] of given DAG is the largest path from entry task to exit task. It is defined as follows:

$$
CP = \max_{path \in DAG} \{length(path)\} \tag{3}
$$

$$
length(\ path) = \sum_{Ti \in T} AECT(T_i) + \sum_{e \in E} Ct(T_i, T_j) \tag{4}
$$

- *Earliest Start TimeST* [6] is defined as follows:

$$
EST(T_i, R_j) = \begin{cases} 0 & \text{if } T_i \in T \text{entry} \\ \max_{T_j \in pred(T_i)} \{EFT(T_j, R_j) + MET(T_i) + Ct(T_i, T_j)\} & \text{otherwise} \end{cases} \tag{5}
$$

Table 1 Performance metrics

S. No.	Metrics
1	Makespan [14]: Execution time of last task of a given DAG, i.e., $\mathbf{Ms} = \min\{\mathbf{EFT}(\mathbf{T_{exit}}, \mathbf{R_j})\}$ Where T_{exit} is the last task of given DAG execute on Resource R_j
2	Speedup [15]: It is ratio of minimum sum of ECT of all tasks on same resource and makespan, i.e., $\mathbf{Sp} = \dfrac{\mathbf{Min.}\{\sum_{j=1}^{m} \mathbf{ECT_{i,j}}\}}{\mathbf{Ms}}$
3	Efficiency [15]: It can be defined as $\mathbf{Eff} = \dfrac{\mathbf{Sp}}{\mathbf{m}}$ Where m is number of computing resources
4	Scheduling Length Ratio (SLR) [15]: It can be calculated as $\mathbf{SLR} = \dfrac{\mathbf{Ms}}{\mathbf{CP}}$

- *Minimum Execution Time MET* [11] is defined as follows:

$$\mathrm{MET}(T_i) = \min.\{\mathrm{ECT}(T_i, R_m)\} \tag{6}$$

- *Earliest Finished Time EFT* [11] is defined as follows:

$$\mathrm{EFT}(T_i, R_j) = \mathrm{ECT}_{ij} + \mathrm{EST}(T_i, R_j) \tag{7}$$

- *Static Level SL* [10, 11]: It is similar to B-level attribute but excludes the communication time between the tasks. It can be defined as follows:

$$\mathrm{SL}(Ti) = \mathrm{AECT}_i + \max_{Tj \,\epsilon\, \mathrm{Succ}(T_i)}\{SL(T_j)\}$$

5 Performance Metrics

Performance metrics are used to compare the proposed algorithm and heuristic algorithms. Following metrics are briefly explained in Table 1.

6 Proposed Strategy

The proposed strategy is based on two well-known attributes critical *path* (*CP*) [9] and static level (SL) [9]. *CP* is the longest path from entry task to exit task of given DAG whereas *SL* is the longest path from the task T_i to exit task but it excludes the communication time between the tasks. We have developed a new priority attribute by using *CP* and *SL* which is called as critical path static (CPS) attribute. CPS of the task Ti can be calculated by using

Table 2 Details of proposed algorithm

Step 1	Input: DAG with nth tasks of given graph
Step 2	Compute CP of given DAG using following: $$CP = \max_{\text{path} \in \text{DAG}} \{\text{length (path)}\} \text{ and length (path)} = \sum_{T_i \in T} \text{AEC}T(T_i) + \sum_{e \in E} Ct(T_i, T_j)$$
Step 3	Compute SL of the task Ti where $i = 1, 2, 3, \ldots,$ nth using $$SL(Ti) = \text{AEC}T_i + \max_{T_j \in \text{Succ}(T_i)} \{SL(T_j)\}$$
Step 4	Compute CPS Attribute using $CPS(T_i) = CP + SL(T_i)$
Step 5	Sort the tasks in decreasing order by CPS values and Allocate to Queue (Q)
Step 6	*while* Q is not empty *do*
Step 7	Remove the first task T_i from Q
Step 8	*If* T_i is satisfied Precedence Constraint *then* *Compute ECT of* T_i *on Each* R_j *as following* *Compute EST* (T_i, R_j) *and EFT* (T_i, R_j) *Allocate T* $_i$ *on* R_j *which takes Minimum EST and EFT*
Step 9	*Else* *Removed Ti will be inserted at end of Q.*
Step 10	*End while*
Step 11	*Makespan* M_s *=Min. {EFT(Texit, VMj}*
Step 12	*Stop*

$$CPS(T_i) = CP + SL(T_i)$$

$$CP = \max_{\text{path} \in \text{DAG}} \{\text{length(path)}\}$$

$$\text{length (path)} = \sum_{Ti \in T} \text{AEC}T(T_i) + \sum_{e \in E} Ct(T_i, T_j)$$

$$SL(T_i) = \text{AEC}T_i + \max_{Tj \in \text{Succ}(Ti)} \{SL(T_j)\}$$

This method operates in two phases such as task priority and computing resource selection. *Phase-I: Task Priority:* This phase is used to find the priority of the tasks using CPS Attributes, sort the tasks in decreasing order and allocate the sorted tasks into a Queue. *Phase-II: Computing Resource Selection:* It is a very important phase for selecting the computing resource. The selection of resource-based *EST and EFT* is in Eqs. 5 and 7. Details of proposed method are in Table 2.

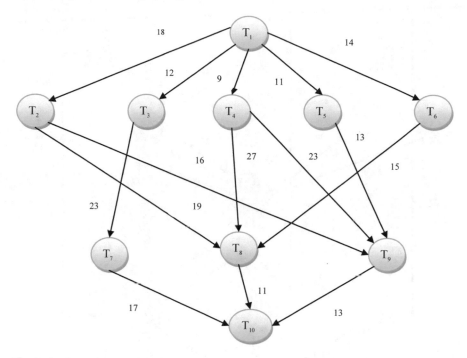

Fig. 3 DAG$_1$ model with ten tasks

7 Performance Analysis

Performance of the proposed strategy is evaluated by using two sets of DAGs with ten and fifteen tasks. *DAG with ten tasks*. This DAG [6] having ten tasks, single entry and considered two cloud servers CS$_1$ and CS$_2$. CS$_1$ consists of two resources R_1 and R_2, and CS$_2$ consists of one resource R_3. Figure 3 shows DAG with ten tasks and Table 3 shows ECT of given DAG (Table 4).

This DAG$_2$ [11] *having fifteen tasks*, multiple entry as shown in Fig. 4 and considered two cloud servers CS$_1$ and CS$_2$. CS$_1$ consists of two resources R_1 and R_2, and CS$_2$ consists of two resources R_3 *and* R_4. It consists of fifteen tasks and the ECT matrix [11] of this DAG shown in the Table 5.

In the above DAG$_1$, its CP value is 108, which is calculated using Eq. (3). The value of both *SL and CPS attributes* computed of all the tasks which is shown in Table 4.

The value of CPS attribute of the tasks is sorted in decreasing order and allocated to a Queue (Q). These are known as priority of the tasks which will be allocated in computing resources as per their EST and EFT. The proposed strategy gives 63 units of makespan which is shown in the Gantt chart in Table 7.

Similarly, for DAG$_2$ has CP value is 184.5. The value of CPS attribute of the tasks is shown in Table 6. By using DAG$_2$, the proposed strategy gives 142 units of

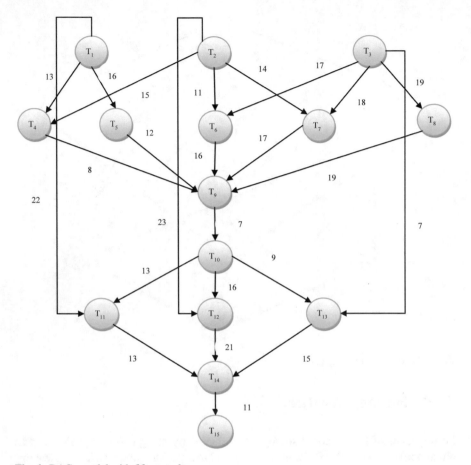

Fig. 4 DAG$_2$ model with fifteen tasks

makespan which is shown in the Gantt chart in Table 8. Proposed method giving 63 and 142 units of makespan which is less as compared to HEFT and also other metrics giving good results as shown in Table 9.

The Scheduling length (Makespan) of calculated based on both DAGs is shown in Fig. 5. The speedup and efficiency are presented in Figs. 6 and 7. SLR is obtained from makespan and critical path for both DAGs is shown in Fig. 8.

8 Conclusion

This paper presented a novel contribution in scheduling strategy in cloud computing platform, and it is based on two well-known priority attributes. The performance analysis has been done based on two DAG models with ten and fifteen tasks. Also,

Table 3 ECT [6] matrix for DAG_1 model

	T_1	T_2	T_3	T_4	T_5	T_6	T_7	T_8	T_9	T_{10}
$CS_1\ R_1$	14	13	11	13	12	13	7	5	18	21
$CS_1\ R_2$	16	19	13	8	13	16	15	11	12	7
$CS_2\ R_3$	9	18	19	17	10	9	11	14	20	16

Table 4 Computing CPS attribute of the tasks

T_i	T_1	T_2	T_3	T_4	T_5	T_6	T_7	T_8	T_9	T_{10}
$SL(T_i)$	61.01	48.01	40.00	44.01	41.01	37.34	25.67	24.67	31.34	14.67
$CPS(T_i)$	169.01	156.01	148.00	152.01	149.01	145.34	133.67	132.67	139.34	122.67
Sorted order of T_i in Q	T_1	T_2	T_4	T_5	T_3	T_6	T_9	T_7	T_8	T_{10}

Table 5 ECT matrix for DAG$_2$ model

	T_1	T_2	T_3	T_4	T_5	T_6	T_7	T_8	T_9	T_{10}	T_{11}	T_{12}	T_{13}	T_{14}	T_{15}
CS$_1$ R_1	17	14	19	13	19	13	15	19	13	19	13	15	18	20	11
CS$_1$ R_2	14	17	17	20	20	18	15	20	17	15	22	21	17	18	18
CS$_2$ R_3	13	14	16	13	21	13	13	13	13	16	14	22	16	13	21
CS$_2$ R_4	22	16	12	14	15	18	14	18	19	13	12	14	14	16	17

Table 6 Computing CPS attribute of the tasks

Ti	T_1	T_2	T_3	T_4	T_5	T_6	T_7	T_8	T_9	T_{10}	T_{11}	T_{12}	T_{13}	T_{14}	T_{15}
$SL(T_i)$	116.25	111.75	114.5	96	99.75	96.5	95.25	98.5	81	65.5	48.75	51.5	49.75	33.5	16.75
$CPS(T_i)$	300.75	296.25	299	280.5	284.25	281	279.75	283	265.5	250	233.25	236	234.45	218	201.5
Sorted order of T_i in Q	T_1	T_3	T_3	T_5	T_8	T_6	T_4	T_7	T_9	T_{10}	T_{12}	T_{13}	T_{11}	T_{14}	T_{15}

Table 7 Gantt chart for proposed strategy for DAG_1 with 10 tasks

CS1	R1			20~32 T_5	32~39*	39~46 T_7	46~51*	51~56 T_8	55~56*	56~63 T_{10}
	R2			18~26 T_4	26~39 T_3		43~55 T_9			
CS2	R3	0~9 T_1	9~27 T_2	27~36 T_6						

Table 8 Gantt chart for proposed strategy for DAG$_2$ with 15 tasks

CS$_1$	R_1	$0{\sim}14$ T_2	$14{\sim}26^*$	$26{\sim}39$ T_4	$39{\sim}86^*$				$86{\sim}99$ T_{11}	
	R_2									
CS2	R_3	$0{\sim}13$ T_1	$13{\sim}26$ T_8	$26{\sim}39$ T_6	$39{\sim}47^*$	$47{\sim}60$ T_9	$60{\sim}73^*$	$73{\sim}89$ T_{13}	$89{\sim}112^*$	$112{\sim}125$ T_{14}
	R_4	$0{\sim}12$ T_3	$12{\sim}13^*$	$13{\sim}28$ T_5	$28{\sim}42$ T_7	$42{\sim}60^*$	$60{\sim}73$ T_{10}	$73{\sim}87$ T_{12}	$87{\sim}125^*$	$125{\sim}142$ $\boldsymbol{T_{15}}$

Table 9 Comparison of Scheduling Algorithms based on Metrics

Metrics	DAG$_1$ scheduling algorithms		DAG$_2$ scheduling algorithms	
	Proposed	HEFT	Proposed	HEFT
SL	63	73	142	152
Sp	2.015	1.716	1.626	1.5
Eff	0.671	0.572	0.406	0.375
SLR	0.583	0.685	0.769	0.834

Fig. 5 Makespan

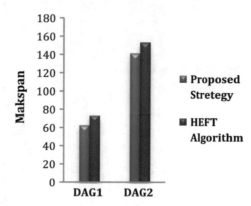

Fig. 6 Speedup

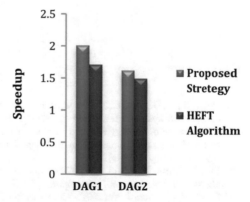

we have considered two cloud servers. Makespan of proposed strategy is evaluated on DAGs with 10 and 15 tasks which give 63 and 142 units which are less as compared to heuristic algorithm HEFT. To minimize the makespan, it is a major objective of task scheduling algorithm. Also, other performance metrics such as speedup, efficiency, and scheduling length ratio give better results in proposed strategy as compared to HEFT. This method may extend using some soft computing methods such as GA and PSO to reduce the makespan and other metrics, and it can be tested on more DAGs with varying size.

Fig. 7 Efficiency

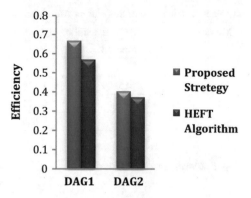

Fig. 8 Scheduling Length Ratio (SLR)

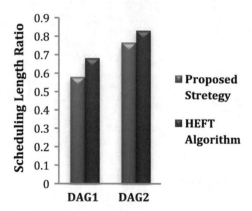

References

1. Deelman, E., D. Gannon, M. Shields, and I. Taylor. 2009. Workflows and e-science: an overview of workflow system features and capabilities. *Future Gener. Comput. Syst.* 25: 528–540.
2. Xue, Shengjun, Wenling Shi, and Xiaoong Xu. 2016. A heuristic scheduling algorithm based on PSO in the cloud computing environment. *International Journal of u-and e-Service, Science and Technology* 9 (1): 349–362.
3. Papadimitriou, C., et al. 1990. Towards an architecture independent analysis of parallel algorithms. *SIAM Journal of Computing* 19: 322–328.
4. Cao Y., C. Ro, and J. Yin. 2013. Comparison of job scheduling policies in cloud computing. In *Future information communication technology and applications*, vol 235, ed. H.K. Jung, J. Kim, T. Sahama, C.H. Yang. Lecture Notes in Electrical Engineering. Springer, Dordrecht (2013).
5. Topcuoglu, H., Hariri, S., and M.-Y. Wu. 2002. Performance-effective and low-complexity task scheduling for heterogeneous computing. *IEEE Transactions on Parallel and Distributed Systems* 13 (3), 260–274.
6. Kumar, M.S., I. Gupta, and P.K. Jana. 2017. Delay-based workflow scheduling for cost optimization in heterogeneous cloud system. In *2017 Tenth International Conference on Contemporary Computing (IC3)*, 1–6, Noida.
7. Frederic, NZanywayingoma, and Yang Yang. 2017. Effective task scheduling and dynamic resource optimization based on heuristic algorithms in cloud computing environment. *KSII Transactions on Internet and Information Systems* 11 (12), 5780–5802

8. Haidri, R.A., C.P. Katti, and P.C. Saxena. 2017. Cost effective deadline aware scheduling strategy for workflow applications on virtual machines in cloud computing. *Journal of King Saud University—Computer* and Information Sciences. (In Press)
9. Kwok, Y.K., and I. Ahmad. 1999. Static scheduling algorithms for allocating directed task graphs to multiprocessors. *ACM Computing Surveys* 31 (4): 406–471.
10. Sinnen, O. 2007. Task scheduling for parallel systems, Wiley-Interscience Publication.
11. Gupta, I., M.S. Kumar, P.K. Jana. 2018. Efficient workflow scheduling algorithm for cloud computing system: A dynamic priority-based approach. *Arabian Journal for Science and Engineering*.
12. llavarasan, E., P. Thambidurai, and R. Mahilmannan. 2005. Performance effective task scheduling algorithm for heterogeneous computing system. In Proceedings of ISPDC, IEEE Computer Society, 28–38.
13. Pandey, S., L. Wu, S.M. Guru, and R. Buyya. 2010. A particle swarm optimization-based heuristic for scheduling workflow applications in cloud computing environments. In *24th IEEE international conference on Advanced Information Networking and Applications (AINA)*, 400–407, IEEE.
14. Muhammad Fasil Akbar, Ehsan Ullah Munir et al. 2016. List-based task scheduling for cloud computing, 2016, IEEE International Conference on Internet of Things and IEEE Green Computing and Communication (GreenCom) and IEEE Cyber, Physical and Social Computing (CPSCom) and IEEE Samrt Data (SmartData).
15. M. F. Akbar, E. U. Munir, M. M. Rafique, Z. Malik, S. U. Khan and L. T. Yang. 2016. List-based task scheduling for cloud computing, 2016, IEEE International Conference on Internet of Things (iThings) and IEEE Green Computing and Communications (GreenCom) and IEEE.

Web-Based Movie Recommender System

Mala Saraswat, Anil Dubey, Satyam Naidu, Rohit Vashisht
and Abhishek Singh

Abstract Recommender systems guide users to find and cull items such as restaurants, books, and movies from the huge collection of available options on the Web. Based on the user's taste and preferences, recommender system recommends a set of items from a large set of available items. Recommendation systems include intelligent aides for filtering and selecting Web sites, news stories, TV listings, and alternative info. The users of such systems typically have various conflicting desires, variations in personal preferences, social and academic backgrounds, and personal or skilled interests. As a result, it looks fascinating to possess personalized intelligent systems that suit individual preferences. The need for personalization has proliferated the event of systems that adapt themselves by ever-changing their behavior supported by the inferred characteristics of the user interacting with them. In this paper, we develop a Web based movie recommender system that recommends movies to users based on their profile using different recommendation algorithms. We also compare various recommendation algorithms such as singular value decomposition, alternating least squares and restricted Boltzmann machines.

Keywords Recommender systems · Collaborative filtering · KNN · Singular value decomposition · Alternating least squares and restricted Boltzmann machines

1 Introduction

"Enhancing the performance of recommender systems" deals with improving performance of recommender systems applicable to various domains [1]. Netflix recommends TV serials and movies based on what you have watched and what other Netflix users with the same interest have watched. Amazon also recommends a product item to a user based on what other customers who purchased that item in which

M. Saraswat (✉) · A. Dubey · S. Naidu · R. Vashisht · A. Singh
Department of Computer Science and Engineering, ABES Engineering College, Ghaziabad, India

A. Dubey
e-mail: anil.dubey@abes.ac.in

© Springer Nature Singapore Pte Ltd. 2020
Y.-C. Hu et al. (eds.), *Ambient Communications and Computer Systems*, Advances in
Intelligent Systems and Computing 1097,
https://doi.org/10.1007/978-981-15-1518-7_24

291

a user might be interested. Broadly recommender systems are categorized into collaborative filtering (CF) and content-based filtering (CB). CF identifies similarity among users based on their past behavior and then recommends items to the user which are liked, bought, or rated highly by similar users [2]. This recommender system can predict items to the user might have an interest, even though the user has never expressed explicit interest. This is generally called user-user collaborative filtering. The opposite of user-user collaborative filtering is to find items similar to a given item and recommend items to users who have also liked, bought, or rated other similar items highly. This goes by the name item-item collaborative filtering.

Content-Based Filtering relies on features of the items based on their content [3]. Based on users' ratings on existing items, a user profile is created and the ranks provided by the users are given to those items. Based on user liking for items, content-based filtering learns the distinct properties of those items and recommend additional items with similar properties. In content-based filtering, the user is recommended items based on their preferences. This does not involve how other users have rated the items.

Recommender systems deal with certain limitations such as *cold start, data scalability* and *sparsity*. Cold-start problem is concerned with the issue that the recommender system cannot draw any inferences for users or items who are new to the system and have not rated the items. There are basically two types of problem. The first one is a new user cold-start problem where a new user without having any ratings to show his/her taste enters the system. Second is new item cold-start problem when a new item which has not been sufficiently been rated enters the system. The next problem is scalable data. With advent of Web 2.0, there is proliferation of e-commerce sites. People sell hundreds of millions of products to millions of users online. So new algorithms and scalable systems should be designed to produce high-quality recommendations even for very large-scale problems. The third problem is data sparsity. Users who are very active contribute to the rating for a few number of items available in the database of movies or books items. Besides some popular items are only rated. Because of rating sparsity, the similarity between two users or items cannot be defined, rendering collaborative filtering useless.

Recommendation system is a software tool that provides suggestions for items to a user [4]. In this work, we design a Web interface and compare various recommendation algorithms for movie domain. Our Web-based recommendation system is a component of a larger system which is a music streaming and downloading Web application. The application integrates our system to main Web application as a Web service. The service provides the opportunity of displaying latest movies and user lists along with accurate information. Moreover, there are two types of lists, professional lists provided by application and user lists created by individual user. Users can display both the types of lists. All these actions generate dataset to our system for accurate recommendations based on users' and items' similarities. The dataset is the most important part of our system; therefore, main system's existence and wide usage are important for generating recommendations on our system.

2 Different Approaches for Recommendation

2.1 K-Nearest Neighbor (KNN)

KNN algorithm for collaborative filtering finds k most similar neighbors of users or items in the feature space based on the similarity of ratings. Based on the similarity of ratings users' are recommended items similar to those as those liked by similar users. K-nearest neighbor finds the k most similar items to a particular instance based on a given distance metric like Euclidean, Jaccard similarity, Minkowsky, etc. [5, 6].

2.2 Singular Value Decomposition (SVD)

Besides CF and CF-based recommender systems, latent factor-based filtering recommendation methods attempt to discover latent features to represent user and item profiles by decomposing the ratings [7]. Unlike the content-based filtering features, these latent features are not interpretable and can represent complicated features. For instance, in a movie recommendation system, one of the latent features might represent a linear combination of humor, suspense, and romance in a specific proportion.

Generally, for already rated items, the rating r_{ij} given by a user i to an item j can be represented as:

$$r_{ij} = u_i^T v_j \tag{1}$$

where u_i is the user profile vector based on the latent factors and v_i is the item vector based on the same latent factors

$$R = USV^T = US \tag{2}$$

One of the ways these user and item profiles can be created is by performing singular value decomposition (SVD) on the rating matrix after filling in the missing values by some form of mean values across the users and items as appropriate. According to SVD, the rating matrix R can be decomposed as follows:

$$R = USV^T = US^{\frac{1}{2}}S^{\frac{1}{2}}V^T \tag{3}$$

We can take the user profile matrix as $US^{\frac{1}{2}}$ and then transpose of the item profile matrix as $S^{\frac{1}{2}}V^T$ to form the latent factor model.

2.3 *Alternating Least Squares (ALS)*

ALS is an iterative optimization process that builds a matrix factorization CF-based model. For every iteration, we try to arrive closer and closer to a factorized representation of our original data [8].

In ALS, we have a rating matrix R of size $u \times i$ where u is the number of users, i is the number of item. The rating matrix is factored into one matrix with users and hidden features U of size $u \times f$ and one V with items and hidden features of size $f \times i$. U and V matrices have weights corresponding to how each user/item relates to each feature. We find values of U and V so that their product approximates R as closely as possible as shown in Eq. 4:

$$R \approx U \times V. \tag{4}$$

2.4 *Restricted Boltzmann Machines (RBM)*

RBMs have two layers: input layer or **visible** layer and outer or the **hidden** layer. The neurons in each layer communicate with neurons in the other layer but not with neurons in the same layer. Thus, there is no intra-layer communication among the neurons [9] as shown in Fig. 1.

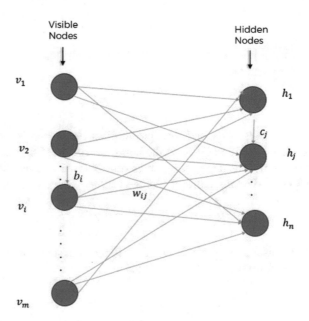

Fig. 1 Restricted Boltzmann machines

Restricted Boltzmann machines can be used to build a recommender system for items ratings. The RBM recommender system can learn the probability distribution of ratings of items for users given their previous ratings and the ratings of users to which they were most similar to. Then RBM recommender system used the learned probability distribution to predict ratings on never-before-seen items [10].

- The hidden layer is used to learn features from the information fed through the input layer.
- The input is going to contain X neurons, where X is the amount of movies in our dataset.
- Each of these neurons will possess a normalized rating value varying from 0 to 1: 0 meaning that a user has not watched that movie and the closer the value is to 1, the more the user likes the movie that neurons representing.
- These normalized values will be extracted and normalized from the ratings dataset.
- After passing in the input, we train the RBM on it and have the hidden layer learn its features.
- These features are used to reconstruct the input, which will predict the ratings for movies that the input has not watched, which is what we can use to recommend movies.

3 Experiments and Results

The Web application has a user-friendly graphical user interface (GUI) to provide ease of use and effectiveness to the users with different roles. Our Web service will be integrated to Web application and the recommendation result will be displayed through Web application GUI. In the Web service interface, there are four pages Login, Lists, Recommendations and Profile. This Login page contains a login text fields and a few buttons. The users of the system who want to get recommendation fill the text field with proper user id. After he fills the text field and click the submit button, new page is opened and recommendations are shown in this page. In Lists page, there will be different sections according to the number of lists created by the user. The Profile page will show user information like name, user id, movie ratings, etc. Figure 2 shows the use case diagram. Figures 3, 4 and 5 depict data flow diagram level 0–2. Recommendation System is composed of the following fundamental features:

(1) Users:
     (a) Generate Data (b) Get Recommendation
(2) Inter-agent:
     (a) Get Recommendation (b) Provide Dataset (c) Update Dataset (d) Integrate Web Service Experiments are conducted on the data that user fill based on their liking in the web interface (Fig. 6).

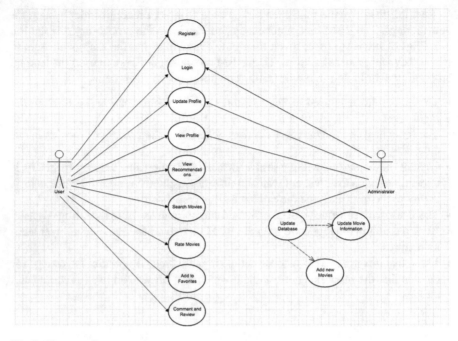

Fig. 2 Use case diagram

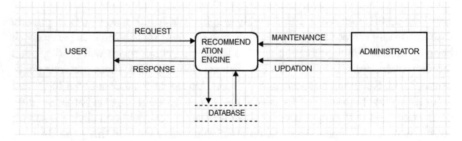

Fig. 3 Data flow diagram level 0

3.1 Evaluation Metrics

The goal of recommender system is not to recommend the same product that the user used/liked before. So to evaluate your recommendation model some metrics are designed such as precision, recall, F-measure, Root Mean Square Error (RMSE) and Mean Absolute Error (MAE) [1]. For our models, we have used RMSE and MAE. They are used to evaluate accuracy of a filtering technique by comparing the predicted ratings directly with the actual user rating:

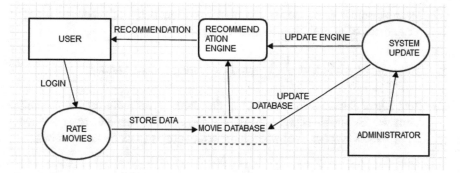

Fig. 4 Data flow diagram level 1

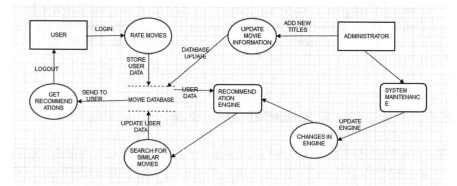

Fig. 5 Data flow diagram level 2

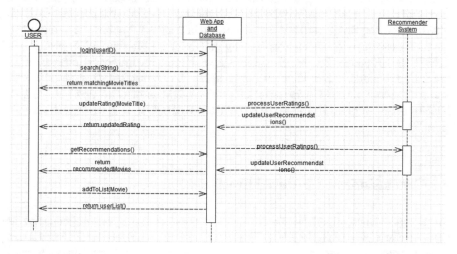

Fig. 6 Sequence diagram

(1) Mean Absolute Error: MAE is a measure of deviation of predicted rating or recommendation from user's actual rating

$$MAE = \frac{1}{N} \sum |\text{predicted rating} - \text{actual rating}|$$ (5)

(2) Root Mean Square Error: RMSE is the standard deviation of the prediction errors. RMSE provides higher weights to large errors compared to trivial errors. Thus, RMSE is more used when large errors are particularly undesirable.

$$RMSE = \sqrt{\sum \text{predicted} - \text{actual})^2}$$ (6)

Recommendation engine predicts more accurately user ratings with lower MAE and RMSE values. RMSE and MAE evaluate how accurate our prediction ratings are compared to actual rating and thus enable us to find out the quality of our recommendations.

3.2 Comparison of Various Recommender System Algorithms

We have developed a Web-based recommender system that recommends movies using various algorithms. We then compare these algorithms using performance metrics. Figure 7 shows the snapshot of Web interface of our Web-based recommender system through which user login to get recommendations.

Figure 8 shows the snapshot of various movies recommended to the users based on his interest. Table 1 summarizes the performance of various recommender system algorithms using RMSE and MAE as discussed in Sect. 3.1. For both user- and item-based K-nearest neighbor approach using collaborative filtering, KNN produces a 0.9375 RMSE and 0.7263 MAE which is good but the recommendations produced are not up to mark. Single Value Decomposition (SVD) has RMSE score of 0.9002 and MAE score of 0.6925 whereas ALS produces RMSE of 1.069 and MAE 0.803. RBM produces highest RMSE of 1.1897 and MAE of 0.9935 with worst performance (Fig. 9).

4 Conclusions and Future Directions

From experiments, algorithm based on SVD provides better recommendation as compared to KNN, ALS and RBM. The restricted Boltzmann model (RBM) of Deep Learning produces better recommendations when compared to several content-based and Collaborative-Based algorithms but when compared on the basis of RMSE (Root Mean Square Error), MAE (Mean Absolute Error), we see that the performance is

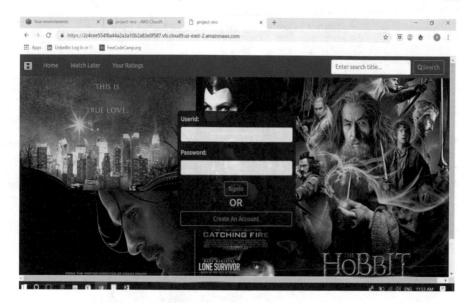

Fig. 7 Snapshot showing user login Web interface

- Pitch Perfect (2012)
- Misérables, Les (2012)
- Batman: The Dark Knight Returns, Part 2 (2013)
- Grandmaster, The (Yi dai zong shi) (2013)
- Now You See Me (2013)
- Man of Tai Chi (2013)
- Inside Llewyn Davis (2013)
- American Heist (2015)
- Tracers (2015)
- 13 Hours (2016)
- Room (2015)
- Ip Man 3 (2015)
- The Crew (2016)
- Jason Bourne (2016)
- Marauders (2016)
- Vigilante Diaries (2016)
- The Magnificent Seven (2016)
- Geostorm (2017)

Fig. 8 Snapshot showing various movie recommendations to the user

Table 1 Performance of various recommendation algorithm		RMSE	MAE
	SVD	0.9002	0.6925
	ALS	1.0687	0.803
	KNN	0.9375	0.7263
	RBM	1.1897	0.9935

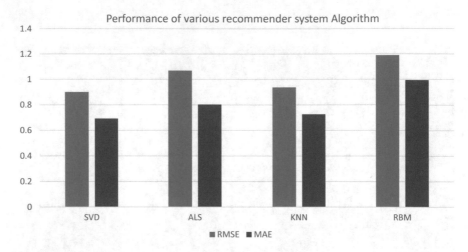

Fig. 9 Chart showing performance of various recommender system algorithm

not good. This is because user needs to enter sufficient ratings to learn the model. Users need to input their ratings to produce good quality recommendations.

Future direction is to build deep factorization machines and is to combine the power of factorization machines with the power of deep neural networks to create even more powerful and improved recommendations. If we somehow manage to mix these two powerful features together, we could unlock the ways to achieve commendable recommendations which could fulfill everyone's requirements.

References

1. Adomavicius, Gediminas, and Alexander Tuzhilin. 2005. Toward the next generation of recommender systems: A survey of the state-of-the-art and possible extensions. *IEEE Transactions on Knowledge and Data Engineering* 6: 734–749.
2. Pazzani, Michael J., and Daniel Billsus. 2007. Content-based recommendation systems. *The adaptive web*, 325–341. Berlin, Heidelberg: Springer.
3. Schafer, J.B., D. Frankowski, J. Herlocker, and S. Sen. 2007. Collaborative filtering recommender systems. *The adaptive web*, 291–324. Berlin, Heidelberg: Springer.
4. Resnick, P., and H.R. Varian. 1997. Recommender systems. *Communications of the ACM* 40 (3): 56–58.
5. Linden, G., B. Smith, and J. York. 2003. Amazon.com recommendations: Item-to-item collaborative filtering. *IEEE Internet Computing* 7 (1): 76–80.
6. Konstan, J., B. Miller, D. Maltz, J. Herlocker, L. Gordon, and J. Riedl. 1997. GroupLens: Applying collaborative filtering to Usenet news. *Communications of the ACM* 40 (3): 77–87.
7. Sarwar, B.M., G. Karypis, J.A. Konstan, and J. Riedl. 2000. Application of dimensionality reduction in recommender system—A case study. In *ACM WebKDD'00 (Web-mining for ECommerce Workshop)*.
8. https://datasciencemadesimpler.wordpress.com/tag/alternating-least-squares/.

9. Hinton, G.E., and R.R. Salakhutdinov. 2006. Reducing the dimensionality of data with neural networks. *Science* 313 (5786): 504–507.
10. Salakhutdinov, Ruslan, Andriy Mnih, and Geoffrey Hinton. 2007. Restricted Boltzmann machines for collaborative filtering. In *Proceedings of the 24th international conference on machine learning*. ACM.

A Two-Step Dimensionality Reduction Scheme for Dark Web Text Classification

Mohd Faizan and Raees Ahmad Khan

Abstract Dark web is infamous for the presence of unethical and illegal content on it. The intelligence agencies are increasingly using an automated approach to detect such content. Machine learning classification techniques can be used to detect such content in textual data from dark Web sites. However, their performance suffers due to the presence of irrelevant features in the dataset. In this paper, a two-step dimensionality reduction scheme based on mutual information and linear discriminant analysis for classifying dark web textual content is proposed. This scheme filters out the irrelevant features using mutual information scheme in the first step. The remaining features are then transformed into a new space for a reduction in the number of features using linear discriminant analysis. The proposed scheme is tested on the dark web dataset collected explicitly from dark Web sites using a web crawler and on the Reuters-21,578 dataset for benchmarking purpose. Three different classifiers were used for classification. The results obtained on the two datasets indicate that the proposed two-step technique can positively improve the classification performance along with a significant decrease in the number of features.

Keywords Dark web · Text classification · Feature selection · Dimensionality reduction

1 Introduction

Dark web is a part of the Internet that requires special tools for accessing it. It is a collection of Web sites called hidden services similar to that accessed on the regular Internet but they use onion routing [1] technique for establishing a connection to the client. Tor, a freely available browser, can be used to surf the dark web. The covert nature of the dark web has attracted several researchers to identify the characteristics of content being hosted there. A study has reported that 45% of the dark web content was unethical of which 18% was of child abuse [2]. This was further confirmed by other studies where it was found that majority of

M. Faizan (✉) · R. A. Khan
Department of Information Technology, Babasaheb Bhimrao Ambedkar University,
Lucknow, India

© Springer Nature Singapore Pte Ltd. 2020   303
Y.-C. Hu et al. (eds.), *Ambient Communications and Computer Systems*, Advances in
Intelligent Systems and Computing 1097,
https://doi.org/10.1007/978-981-15-1518-7_25

dark web content was illegal like drugs, arms and ammunition, child abuse content, forged documents, and counterfeits [3–6].

Due to the hidden nature of the dark web, there is no exclusive search engine that indexes such Web sites. To monitor the activities that are carried out in dark web, there is a need to identify the Web sites that are hosting unacceptable content. The machine learning classification technique can be used to identify such content that can categorize the textual data into predefined classes. The idea is to define several categories of unethical dark web content and to use machine learning classification techniques to detect the content of hidden services and put them into its specific category.

Classification of the text document is a supervised machine learning technique where each text document is mathematically represented as an n-dimensional vector (n is the length of the vector). The elements of this vector are called terms or features where each feature is a word or a group of words in a document. A large number of documents in a dataset usually results in a high-dimensional feature space. A feature space or a feature set is a collection of all the unique features present in the entire dataset under study. The feature space contains a high number of inessential and unwanted features that increases the computational complexity of machine learning algorithms. Moreover, the presence of noisy features may also result in low classification accuracy. Dimensionality reduction (DR) techniques are employed to get rid of useless features and their associated problems. Dimensionality reduction can be performed by either feature selection (FS) or feature extraction (FE) or a combination of both. Feature selection is the method of selecting only the most suitable and relevant terms from the original feature set based on some criteria. There are three different approaches to achieve this task: filter, wrapper, and embedded approach. The filter approach makes use of statistical techniques for selecting good features. Wrapper approach uses a learning algorithm to evaluate the performance of the selected features. The embedded approach generally uses machine learning algorithms for classification and the searching criteria for the optimal feature subset are built into the classifier construction. Some of the commonly used filter approaches are: mutual information (MI) [7], reliefF [8], chi-square (CHI) [9], document frequency (DF) [10], information gain (IG) [11], etc. Feature extraction reduces the feature space by projecting it into a new low-dimensional feature space using mathematical transformations. The aim is to substitute redundant features with another small number of new features that can properly represent the information exhibited by the original feature space. Principal component analysis (PCA) [12], latent semantic indexing (LSI) [13], and linear discriminant analysis (LDA) [14] are some of the methods of feature extraction. An effective DR process should be capable of removing unsuitable and redundant features while ensuring the presence of informative features. This should also positively improve the performance of the learning algorithm.

The objective of the paper is to propose a two-step DR technique for classifying dark web textual content. The first step implements feature selection on the original feature space to obtain a reduced feature set. Mutual information method is used to rank the features from high to low based on their scores in the original space. High-ranking features are then selected for the second step. Feature extraction is carried out on the reduced feature set obtained in the first step. Linear discriminant analysis is used to obtain the new feature space with a significant decrease in its dimension. The performances of the learning algorithms are evaluated on the new feature space. The results show a significant increase in the classification performance on our dataset of dark web text documents in terms of *f-score*. Moreover, to confirm the effectiveness of the proposed technique, the same experiment is conducted on the Reuters-21,578 collection. Rest of the paper is organized as follows: Sect. 2 states the related work already undertaken and Sect. 3 describes the proposed technique. Section 4 describes the experimental setup required to test the effectiveness of the proposed scheme. Section 5 states the results obtained and their interpretation. Finally, Section 6 draws the conclusion about the work.

2 Related Work

A number of research works have been done to remove redundant features without significantly affecting the performance of the algorithm [15–21]. As the single-stage DR method could not reduce feature space to a large extent, the combination of two-stage DR or hybrid models using filter–filter approach [17], filter–wrapper approach [22] was proposed in studies. They may be a combination of feature selection and feature extraction methods. The two-stage dimensionality reduction was also used in microarray data classification wherein the author used k-means clustering for filtering the feature space followed by the application of PSO for further processing [23]. In yet another hybrid approach, IG–GA and IG-PCA were, respectively, used to carry out dimensionality reduction on Classic-3 and Reuters-21,578 datasets [24]. The first step used the IG method as an FS tool to filter out discriminative features. In the second step, GA and PCA were used separately as an FS and FE tool, respectively, on the selected features obtained from the first step. The proposed technique gave the promising results on K-nearest neighbor (KNN) and decision tree C4.5 classifiers. However, the IG-GA method being a filter–wrapper approach requires constant interaction with the classifier making it computationally expensive task. A recent study has proposed a new DR method called improved canonical correlation analysis (ICCA) [25]. This new method not only used conventional target data for training but also the Universum data that holds additional information related to the domain of the target data. Experimental results obtained for several synthetic and real-world datasets show similar or improved performance levels on the proposed scheme as compared with other existing techniques.

In another study, an ensemble feature selection technique was proposed that is capable of stable feature selection without compromising with the performance as claimed by the authors [26]. An unsupervised decision graph-based feature selection was performed in another study where the decision graph for the features was created and their decision score is calculated [27]. The highest scoring features are then selected to form an optimal subset such that the correlation between selected features is small. The performance of the proposed scheme was compared with other unsupervised feature selection methods on sixteen different datasets. The statistical verification of the experimental results shows that the proposed feature selection method outperforms other methods irrespective of the sample size and distribution. A two-stage feature selection for text categorization was proposed using a novel category correlation degree (CCD) filter technique [28]. CCD was formulated based on the frequency of feature occurrence and the corresponding category. In the first stage, CCD score is used to select the features while in next stage LSI was used to obtain the lower-dimensional feature space. The reduced feature space was supplied to SVM to assess the performance in terms of accuracy and *f-score*. The results show the improvement in accuracy when using CCD with LSI as compared to DF, expected cross-entropy (ECE), and CHI.

3 Proposed Technique

The proposed two-stage DR method is represented in Fig. 1. In the first step of DR, MI of each of the feature in the vocabulary is computed. Mutual information (MI) is a concept from information theory that can be used to measure the mutual dependence between the feature and its class.

The MI between two variables X and Y is given by Eq. 1.

$$\text{MI}(X:Y) = \sum_{i=1}^{n} \sum_{j=1}^{n} p(X(i), Y(j)).\log\left(\frac{p(X(i), Y(j))}{p(X(i)).p(Y(j))}\right) \tag{1}$$

The features and their computed scores are then ranked from high to low with the highest scoring feature getting top rank. Once the features have been ranked, k best features from the feature space of size m are selected to reduce its dimension (where $k \ll m$). In the second stage, LDA is applied to the reduced feature space from stage one. LDA [14] is a statistical technique that uses class labels to find a new reduced number of axes from original dataset such that the distance among different classes is maximized. Given a dataset having d classes with feature space of length k obtained from the first stage, LDA can find new dimensions of size $\min(d - 1, k)$. To achieve this, eigenvectors of the training data are computed and

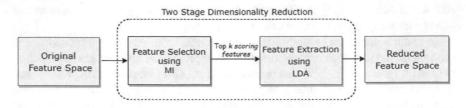

Fig. 1 Proposed DR technique

stored in inter- and intra-class scatter matrices. The eigenvalues associated with each of the eigenvectors are arranged in the decreasing order of their length. Only eigenvectors with the highest eigenvalues are kept as they are more informative than the smaller length eigenvectors. The selected eigenvectors are used to transform the dataset into a new space. At the end of this step, the dataset is left with the feature space of size $d - 1$.

4 Experimental Setup

Dark web dataset and the standard Reuters-21,578 dataset used in this study are described as follows. For dark web, a customized Python web crawler was developed to collect data from the Tor dark web. The crawler was fed with a list of initial URLs from the surface web. The crawler connects to each of the URL and downloaded the HTML of each of the hidden service in a text file. No multimedia content was downloaded. At the same time, the crawler also collects the fresh URL from the visited Web site. In the end, the crawler was able to capture the content of 4000 hidden services. The content of a total of 1006 hidden services was selected for further analysis. The selected content was parsed with all the HTML tags removed and only plain text was left. These 1006 documents were then manually classified into six categories. Table 1 shows the distribution of documents into five categories of our dataset. The dataset is unbalanced. Reuters-21,578 collection is a benchmark dataset for text classification problems. The corpus is a collection of documents from the Reuters news agency. The original version contains 135 categories and 21,578 documents [29]. We are taking only the top ten categories consisting of 9980 documents for our experiment as shown in Table 2.

The documents in the dataset were preprocessed by converting complete text to lowercase, removal of stopwords, symbols (like punctuation marks, numeric, and special characters), and finally splitting the document into individual words. After preprocessing, the documents are represented in the bag of words (BoW) model and term frequency–inverse document frequency (tf-idf) was used to assign weights to the terms in document vector. For testing purpose, three different classifiers were

Table 1 Distribution of categories in the dark web dataset

Category	Document count
CC dumps	271
Counterfeits	37
Drugs	179
Forged documents	40
Violence	97
Others	382
Total	1006

Table 2 Distribution of categories in the Reuters-21,578 dataset

Category	Document count
Acquisition	2369
Corn	237
Crude	578
Earn	3964
Grain	582
Interest	478
Money-fx	717
Ship	286
Trade	486
Wheat	283
Total	9980

used: logistic regression (LR), support vector machine (SVM), and Naïve Bayes (NB). LR is a linear classification algorithm that uses a sigmoid function to obtain a probability value for mapping data points to a particular class. SVM computes a decision hyperplane that separates examples belonging to different classes. NB is a probabilistic classification technique and is one of the most simple and fast classifiers. The ease of implementation and low cost of training the NB has made it a popular algorithm for text classification.

The effectiveness of text classification is measured using precision, recall, and accuracy. While accuracy can be used when the dataset is almost balanced, it does not provide a good measure when there is an unequal distribution among categories (as in our case). To get a balanced measurement, *f-score* is used which is harmonic mean of precision and recall. For the problem of multiclass classification, micro-averaged and macro-averaged *f-score* is used [30]. In this paper, micro-averaged *f-score* is computed for evaluation.

5 Results

In order to test the effectiveness of the proposed scheme, an experiment was conducted on the dark web and Reuters-21,578 dataset. For both the datasets, the following method was adopted. After preprocessing of the datasets, the feature space was obtained of size 1886 and 15,665 for dark web and Reuters-21,578, respectively. The MI score of each feature was calculated and are ranked in decreasing order of their MI score. Top $k\%$ $(k = 10, 11 \ldots 16)$ of features are then selected separately. The selected subset of features is then further transformed into a new space using LDA. The final dimension of feature space is five and nine for dark web and Reuters-21,578, respectively (one less than the number of classes). The reduced feature space is then fed to the three different classifiers, i.e., NB, SVM, and LR. A

10-*fold cross-validation* method is applied to all the classifiers for optimum result. Python scikit-learn package [31] was used for DR and classification. The final result in terms of *f-score* for dark web is summarized in Table 3 and for Reuters-21,578 in Table 4. It can be seen in Table 3 that NB classifier gives the highest classification performance when 16% of features are selected. SVM and LR give their best performance when 15% and 16% of features are selected, respectively. All three classifiers produce best results when features are selected in the range of 15–16%. At the same percentage of selected features with MI without the application of LDA, the NB, SVM, and LR give 73.14, 82.81 and 87.56 of *f-score,* respectively.

In Table 4, the highest *f-score* is achieved by NB and LR when 16% of features were selected. Although SVM produce its best result when features were selected in the range of 11–13%, it could not surpass the performance of NB and LR. LR emerges as the best performer with the highest score among all the three. At the same percentage of selected features with MI without the application of LDA, the NB, SVM, and LR give 70.65, 85.68, and 82.11 of *f-score,* respectively.

Table 5 shows the comparison of three different FE methods when used in combination with MI on the dark web and Reuters-21,578 datasets. For both datasets, results are compared when 16% of features are selected, i.e., when the best result is

Table 3 Performance of NB, SVM, and LR in terms of *f-score* for dark web dataset

Percentage of feature	Number of features	Classifier performance (*f-score*)		
		NB	SVM	LR
10	189	93.63	93.63	93.33
11	207	93.53	93.83	93.93
12	226	94.73	94.73	94.63
13	245	95.72	95.42	95.61
14	264	95.32	95.12	95.92
15	283	95.52	95.92	96.22
16	302	95.93	95.62	96.42

Table 4 Performance of NB, SVM, and LR in terms of *f-score* for Reuters-21,578 dataset

Percentage of feature	Number of features	Classifier performance (*f-score*)		
		NB	SVM	LR
10	1566	88.84	89.01	89.62
11	1723	88.98	89.08	89.73
12	1880	89.06	89.02	89.67
13	2036	89.26	89.08	89.83
14	2193	89.42	88.82	89.96
15	2350	89.50	88.86	89.95
16	2506	89.65	88.91	90.04

Table 5 Comparison of the proposed scheme with the other FE techniques

Dataset	Dimensionality reduction scheme	Classifier		
		NB	SVM	LR
Dark web	MI-PCA	55.96	65.51	69.00
	MI-LSI	61.22	65.31	69.50
	MI-LDA	95.93	95.92	96.42
Reuters-21,578	MI-PCA	69.21	73.76	73.67
	MI-LSI	68.06	74.07	74.58
	MI-LDA	89.65	88.91	90.04

shown by MI-LDA approach. The number of components for both PCA and LSI is set to the same values as used for LDA, i.e., one less than the number of classes.

It can be seen that MI-LDA combination outperforms the other two FE techniques with tremendous improvement in classification in terms of f-$score$. Moreover, the feature space also gets a drastic reduction in its size thus decreasing the computational cost of the classifiers. The results obtained above show that the best performance is achieved when around 15–16% of features are selected based on MI scores followed by the application of LDA. Moreover, results also show that MI alone is not capable of delivering an optimal feature subset that can produce good classification performance. Thus, the transformation of feature subset into a new space with the simultaneous reduction in its size by LDA positively improves the classification performance. However, the same results could not be achieved when LDA was replaced by other FE techniques like PCA and LSI. Thus, the proposed two-step MI-LDA technique gives promising classification performance for classifying the dark web textual content.

6   Conclusions

This paper has introduced a two-stage dimensionality reduction method for feature space of dark web text documents to detect unethical content present in it. Using mutual information followed by the application of LDA, a reduced feature subset is obtained which is free from irrelevant terms. The dark web dataset and the Reuters-21,578 dataset were used for testing and evaluating the effectiveness using three different text classifiers. The results obtained for both the datasets indicate that good classification accuracy is possible with the proposed scheme. In addition, the significant reduction in feature space also reduces the computational overheads. The future work shall focus on increasing the number of classes in dark web dataset so as to evaluate the performance of the proposed scheme. In addition, the evolutionary algorithm in combination with FS techniques shall be used for effective DR of the feature space.

References

1. Reed, M.G., P.F. Syverson, and D.M. Goldschlag. 1998. Anonymous connections and onion routing. *IEEE Journal on Selected Areas in Communications* 16 (4): 482–494.
2. Guitton, C. 2013. A review of the available content on Tor hidden services: The case against further development. *Computers in Human Behavior* 29 (1): 2805–2815. https://doi.org/10.1016/j.chb.2013.07.031.
3. Biryukov, A., et al. 2014. Content and popularity analysis of Tor hidden services. In *Proceedings of the IEEE 34th international conference on distributed computing systems workshops*, 188–193. Washington: IEEE Computer Society.
4. Faizan, M., and R.A. Khan. 2019. Exploring and analyzing the dark web: A new alchemy. *First Monday* 24(5). https://doi.org/10.5210/fm.v24i5.9473.
5. Owen, G., and N. Savage. 2016. Empirical analysis of Tor hidden services. *IET Information Security* 10 (3): 113–118. https://doi.org/10.1049/iet-ifs.2015.0121.
6. Al Nabki, M.W., et al. 2017. Classifying illegal activities on tor network based on web textual contents. In *Proceedings of the 15th conference of the European chapter of the association for computational linguistics*, 35–43. Stroudsburg: ACL.
7. Battiti, R. 1994. Using mutual information for selecting features in supervised neural net learning. *IEEE Transactions on Neural Networks and Learning Systems* 5: 537–550.
8. Kononenko, I. 1994. Estimating attributes: Analysis and extensions of relief. In *Proceedings of the European conference on machine Learning*, 171–182.
9. Li, Y., C. Luo, and S. Chung. 2008. Text clustering with feature selection by using statistical data. *IEEE Transactions on Knowledge and Data Engineering* 20: 641–652.
10. Liu, L., et al. 2005. A comparative study on unsupervised feature selection methods for text clustering. In *Proceedings of the IEEE international conference on natural language processing and knowledge engineering*, 597–601. China: IEEE.
11. Mitchel, T. 1997. *Machine learning*. New York: McGraw-Hill.
12. Jolliffe, T. 2002. *Principal component analysis*. New York: Springer-Verlag.
13. Song, W., and S. Park. 2009. Genetic algorithm for text clustering based on latent semantic indexing. *Computers and Mathematics with Applications* 57: 1901–1907.
14. Fisher, R.A. 1938. The statistical utilization of multiple measurements. *Annals of Human Genetics* 8 (4): 376–386.
15. Labani, M., et al. 2018. A novel multivariate filter method for feature selection in text classification problems. *Engineering Applications of Artificial Intelligence* 70: 25–37. https://doi.org/10.1016/j.engappai.2017.12.014.
16. Wang, Y., and L. Feng. 2018. Hybrid feature selection using component co-occurrence based feature relevance measurement. *Expert System with Applications* 102: 83–99. https://doi.org/10.1016/j.eswa.2018.01.041.
17. Zhang, Y., C. Ding, and T. Li. 2008. Gene selection algorithm by combining ReliefF and MRMR. In *Proceedings of the IEEE 7th international conference on bioinformatics and bio engineering*, 127–132. Boston: IEEE.
18. Jadhav, S., H. He, and K. Jenkins. 2018. Information gain directed genetic algorithm wrapper feature selection for credit rating. *Applied Soft Computing* 69: 541–553. https://doi.org/10.1016/j.asoc.2018.04.033.
19. Khammassi, C., and S. Krichen. 2017. A GA-LR wrapper approach for feature selection in network intrusion detection. *Computers & Security* 70: 255–277. https://doi.org/10.1016/j.cose.2017.06.005.
20. Zheng, Y., Y. Li, G. Wang, et al. 2018. A novel hybrid algorithm for feature selection. *Personal and Ubiquitous Computing* 22 (5–6): 971–985. https://doi.org/10.1007/s00779-018-1156-z.
21. Xue, X., M. Yao, and Z. Wu. 2018. A novel ensemble-based wrapper method for feature selection using extreme learning machine and genetic algorithm. *Knowledge and Information Systems* 57 (2): 389–412. https://doi.org/10.1007/s10115-017-1131-4.

22. Solorio-Fernández, S., J. ArielCarrasco-Ochoa, and J. Fco. Martínez-Trinidad. 2016. A new hybrid filter–wrapper feature selection method for clustering based on ranking. *Neurocomputing* 214, 866–880. https://doi.org/10.1016/j.neucom.2016.07.026.
23. Sahu, B., and D. Mishra. 2012. A novel feature selection algorithm using particle swarm optimization for cancer microarray data. *Procedia Engineering* 38: 27–31.
24. Uguz, H. 2011. A two-stage feature selection method for text categorization by using information gain, principal component analysis and genetic algorithm. *Knowledge-Based Systems* 24 (7): 1024–1032.
25. Chen, X., and L. Wang. 2018. A new dimensionality reduction method with correlation analysis and universum learning. *Pattern Recognition and Image Analysis* 28 (2): 174–184. https://doi.org/10.1134/S1054661818020189.
26. Ben Brahim, A., and M. Limam. 2018. Ensemble feature selection for high dimensional data: A new method and a comparative study. *Advances in Data Analysis and Classification* 12 (4): 937–952. https://doi.org/10.1007/s11634-017-0285-y.
27. He, J., et al. 2017. Unsupervised feature selection based on decision graph. *Neural Computing and Applications* 28 (10): 3047–3059.
28. Wang, F., et al. 2015. A two-stage feature selection method for text categorization by using category correlation degree and latent semantic indexing. *Journal of Shanghai Jiaotong University (Science)* 20 (1): 44–50.
29. Reuters-21578 text categorization collection, distribution 1.0. http://kdd.ics.uci.edu/databases/reuters21578/reuters21578.html.
30. Sebastiani, F. 2002. Machine learning in automated text categorization. *ACM Computing Surveys* 34 (1): 1–47.
31. Pedregosa, F., et al. 2011. Scikit-learn: Machine learning in python. *Journal of Machine Learning Research.* 12: 2825–2830.

TLC Algorithm in IoT Network for Visually Challenged Persons

Nitesh Kumar Gaurav, Rahul Johari, Samridhi Seth, Sapna Chaudhary, Riya Bhatia, Shubhi Bansal and Kalpana Gupta

Abstract SIoT (Social Internet of things), as the name suggests, is the social network of things. SIoT could be useful in providing social help in healthcare industry. This paper describes how it can be used for the social well-being of the people with disabilities (PwD). A particular scenario for the blind persons has been taken into consideration, that is, how to ease the travelling for blinds, especially in crossing a traffic light. The main focus is to make them walk and travel independently without the help of others. A traffic light crossing (TLC) algorithm has been proposed to help the visually challenged persons to cross the traffic light in IoT Network. In this chapter, different computing types for IoT and tools supporting have also been discussed.

Keywords Social Internet of things · Walking stick · GPS · Physically handicapped · Health care · Traffic light · Homo sapien

N. K. Gaurav (✉) · R. Johari · S. Seth · S. Chaudhary · R. Bhatia · S. Bansal
SWINGER: Security Wireless IoT Network Group of Engineering and Research Lab, University School of Information, Communication Technology (USICT), Guru Gobind Singh Indraprastha University, Sector-16C, Dwarka Delhi, India
e-mail: swinger@ipu.ac.in

R. Johari
e-mail: swinger@ipu.ac.in

S. Seth
e-mail: swinger@ipu.ac.in

S. Chaudhary
e-mail: swinger@ipu.ac.in

R. Bhatia
e-mail: swinger@ipu.ac.in

S. Bansal
e-mail: swinger@ipu.ac.in

K. Gupta
Centre for Development of Advanced Computing (C-DAC), Noida, India

© Springer Nature Singapore Pte Ltd. 2020
Y.-C. Hu et al. (eds.), *Ambient Communications and Computer Systems*, Advances in Intelligent Systems and Computing 1097,
https://doi.org/10.1007/978-981-15-1518-7_26

313

1 Introduction

Internet of things is automated interaction of non-living things, that is, electronic devices, home appliances and other devices connected to each other, each called a network of devices. Internet of things is aimed at connecting non-electrical appliances too to each other. Kevin Ashton coined the term Internet of things. IoT applications are spread over various sectors: commercial, consumer, infrastructural and industrial spaces. IoT can be applied in any sector, be it agriculture, industry, health care, smart homes, transportation, automation and energy conservation. IoT-enabled devices are characterized as ubiquitous, intelligent and a little complex [1]. The most important part in IoT, as the name suggests, is Internet. Everything is made connected to Internet. The name is self-explanatory: connecting each and everything through Internet. The devices connected in the network are considered to be smart because of the capabilities of the devices to send information or receive information or both [2].

IoT is transforming the world totally. Today, machines are being made intelligent so that that the machines can do every human task on its own without human interaction. This is what is IoT. Making machines capable enough to interact with each other on their own and further take smarter decisions when there is a need for communication [3]. Devices connected in the network contain sensors which are used for integration, interaction and analysis [4, 5].

Multiple services are provided by the IoT platform for different features. Respective services are used by industry to execute their tasks. The three types of IoT analytics platforms are [6]:

1. Descriptive analytics.
2. Predictive analytics.
3. Prescriptive analytics.

IoT is connections of things. Different things which are used for networking can be classified into six categories [7]: products; fleets; assets; infrastructure; people.

Market. Real-world applications of Internet of things are [8]: smart home; wearables; connected cars; industrial Internet; smart cities; agriculture; smart retail; energy engagement; health care; poultry and farming.

2 Literature Survey

Chaudhary et al. [9] discuss the new and innovative application of IoT such as usage of IoT in healthcare, IoT in agriculture, IoT in smart home, IoT in natural calamities along with a detailed and systematic literature survey. Gupta et al. [10] discuss approaches which would be followed depending on the nature of IoT application. For small areas or regions in city containing street lamp pole where a number of appliances (read poles) grow only in one particular direction, wired configuration as well as wireless configuration can be used to help person with disability (Pwd) to seamlessly traverse through the streets.

Johari et al. [11] propose a POSOP algorithm to effectively utilize the DTN contact which would help the villagers to communicate with doctors in the hospitals and discuss their ailments and diseases. Farahani et al. [12] focus on IoT eHealth ecosystem holistic architecture which is used to discuss the applicability of IoT in medicine and health care. The proposed system discuss about the health care which is patient-centric and not clinic-centric. The patient-centric agents like patient, hospital and services are flawlessly connected to each other and need multi-layer architecture for complex data handling. The challenges like scalability, regulation, interoperability, security and privacy and the challenges of IoT eHealth are also addressed in this paper. Yeh et al. [13] focus on demonstration of IoT-based healthcare system which is secure and is operated through body sensor networks architecture. Robust crypto-primitives are utilized to construct efficiency and robustness simultaneously for ensuring entity authentication and confidentiality among smart objects. Ge et al. [15] focus on the implementation and design of an IoT healthcare system. In order to reduce the loss of data and to enhance the interoperability between the devices and also while transmitting the measured information, CoAP and IEEE/ISO PHD standards are used. Ukil et al. [16] focus on healthcare analytics which is a challenging task in which anomaly detection plays a major role. Reducing FNR is a challenging task of anomaly detection. Healthcare analytics adds new dimensions to maintenance and retrieval of electronic health records, medical image analysis, biomedical signal analysis, affordable and remote healthcare system and big data mining in health care. The primary goal of this is to minimize the diagnosis error and to maximize the disease early detection.

Catarinucci et al. [14] focus on personnel, patient and biomedical devices used for automatic tracking and monitoring-based smart architecture. A smart hospital system is proposed which relies on complementary technologies and is able to collect physiological parameters and environmental conditions in real time. The proposed system would provide handling of emergency immediately and tracking of nursing staff, patients within nursing institute and hospitals. Ugrenovic et al. [17] focus on the IoT remote healthcare monitoring system. The system through Web browser provides the patient condition. For remote data representation and data access, CoAP is selected as application-level protocol. For connecting 6LoWLAN network with the Internet and the outside world, the required steps are described, and CoAP protocol implementation in Mozilla Firefox is the main focus of the work done in this paper. Also, the use of CoAP protocol overcomes some resource-constrained issues. Maksimovic et al. [18] focus on providing medical assistance directly and appropriate means for healthcare delivery. For healthcare system infrastructure development, the WSNs enable the global approach and provide set of information applicable to all stakeholders like patients, medical staff and health insurance. In this paper, Do-It-Yourself (DIY) solution is presented for patient-oriented infrastructure development where patients are adaptable and sustainable.

Ahuja et al. [19] discuss about how today many vendors and distributors are logging into field of IoT and are designing new, creative and innovative hardware and software applications; The author discusses and analyses a case study of healthcare management information system (HMIS). Turcu et al. [20] focus on the people's access to healthcare services improvement by using radio frequency identification and IoT technologies. The objective of this paper is to show the weaknesses and strength of multi-model approach for IoT development applied in health care. Rohokale et al. [21] focus on better health control and monitoring of rural and poor human being's health parameters like haemoglobin, blood pressure, blood sugar and so on and propose an IoT approach like safe motherhood program for rural healthcare applications.

3   Methodology Adopted

A usual problem with physically disabled, particularly blinds, while going alone on the road is that they are not sure about the routes. The main issue is that people cannot be trusted whether they will lead them to the right direction or not. Further, at some places where there is no one to guide, physically challenged people can hurt themselves while crossing the roads. The difficult situation for them is to cross a traffic light. So to resolve this issue, a proposal has been made. The walking stick or similar devices they use can be made smart. By smart, it is meant that devices can warn and alert the physically challenged when to cross and when to not. It has been proposed that latitudinal and longitudinal directions are tracked which can help them in guiding further. Secondly, a unique ID can be created using location as the base. The physically challenged people can store the destination in the device. The smart device using navigation can lead them to the destination. Further, using unique ID created, the family members or relatives of physically challenged people can track them easily. The Homo sapien ID is dynamic in nature so that it can hold the current location of individual person. As an example, it is considered that the unique ID created is of 31 digits. The first 19 digits are static and based on the place of birth and family the person belongs to. Last 12 digits are dynamic and depends on the current location of the person. The first 19 digits are distributed into 7 fields. The 7 fields are continent code (1 digit), country code (2 digits), state code (2 digits), district code (3 digits), Tehsil code (5 digits), family code (caste, creed, et al) (3 digits) and person code (3 digits). The dynamic 12 digits are divided into 6 digits each of latitude and longitude code. Figure 1 describes the unique ID. For example, consider a person named "PC (321)" is born in family "FC (876)" in tehsil "TC (25347)", district "DC (232)", state code "SC (29)", country code "CC (54)" and continent code "CC (9)", then, static code for the person will be
    "9542923225347876321"

CC (Continent code) 1 digit	CC (Country code) 2 digits	SC (State code) 2 digits	DC (District code) 3 digits	TC (Tehsil code) 5 digits	FC (Family code) 3 digits	PC (Person code) 3 digits	Latitude 6 digits	Longitude 6 digits

← ———————————— Static part ———————————— →|← Dynamic part →

Fig. 1 Block diagram for unique ID

Let the latitude code be "25.4527" and longitude code be "78.8839", then, the code for person "A" will be

"954292322534787632125.452778.8839"

Further, the current research work also describes how in real life the unique Homo sapien ID is created.

The prerequisites of the system are that the user needs to enter the destination in the walking stick. For entering the destination, voice recognition system can be embedded. The walking stick or any other device using the help of navigation system and the destination point can have a contact with traffic signal. The traffic signal also needs to be made smart; that is, they should work in a unidirectional way with these devices. The traffic signal will guide the physically challenged whether to cross the road now or not. It will prompt the user when the signal is green for them to cross and warn them to stop in the other case.

The traffic light case has been elaborated in this work. The case is of a blind person standing on a square junction to cross the road. There could be many scenarios for crossing the road on a square junction such as moving straight or moving left or moving right. In this work, the major scenarios which cover the other cases have been taken into consideration. First, if the blind person wants to go in straight direction, the smart stick with the help of pre-stored routes and direction will sense the correct red light and guide the person for appropriate action. The second case is similar to the first that is the case for going in left direction. If the person wants to turn right, then he needs to cross two roads. The smart stick will sense the nearby traffic lights and guide him accordingly to move straight or turn right when appropriate. For example, in Fig. 2, the physically challenged person needs to go straight so as reach his destination. His smart stick would sense the nearby traffic lights and guide him to wait or move as per the signals. The man would wait as his nearest traffic light as shown in Fig. 3.

3.1 Concept of Homo sapien ID

Suppose a person named "Amit" born in continent "Asia", country "India", state "Uttar Pradesh", district "Ghaziabad" and tehsil "Modinagar". Let the family code for the person be "026" and the person code be "009". The family has been created

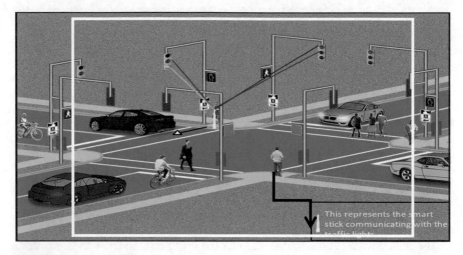

Fig. 2 Traffic light scenario

Fig. 3 Traffic light scenario depicting a blind crossing a road

based on factors such as caste, creed and other factors. As an illustration, here caste is taken, so, 026 is the caste and 009 is the id of the particular person.

The details about the Continent Codes, Number of Countries in Continent, Country Code and State Code are described in Table 1, 2, 3 and 4 respectively.

Table 1 Continent codes [22]

S. No.	Continent	Code
1.	Africa	1
2.	Asia	2
3.	Europe	3
4.	Australia	4
5.	Antarctica	5
6.	North America	6
7.	South America	7

Table 2 Number of countries in each continent [23]

S. No.	Continent	Number of countries
1.	Africa	54
2.	Asia	50
3.	Europe	51
4.	Australia	14
5.	Antarctica	00
6.	North America	23
7.	South America	12

Table 3 Sample chart for country code for Asia [23]

S. No.	Country	Code
1.	Afghanistan	01
2.	Armenia	02
3.	Azerbaijan	03
4.	Bahrain	04
5.	Bangladesh	05
6.	Bhutan	06
… … … … …	… … … …	… … … … …
12	India	12
… … … … …	… … … …	… … … … …
49.	Vietnam	49
50.	Yemen	50

States are further divided into districts. There are 75 districts and 349 tehsils in Uttar Pradesh. Modinagar comes under Ghaziabad district [13]. The code for "Modinagar" is "00738" and for Ghaziabad is "140" [27, 28].

Table 4 Sample chart of state code for India [24]

S. No.	State/UT	Code
1.	Andhra Pradesh	01
2.	Arunachal Pradesh	02
3.	Assam	03
4.	Bihar	04
5.	Chhattisgarh	05
6.	Goa	06
7.	Gujarat	07
… … … …	… … … …	… … … …
28.	Uttar Pradesh	28
… … … …	… … … …	… … … …
35.	Lakshadweep	35
36.	Puducherry	36

Latitude and Longitude coordinates for Modinagar are 28.8344°N, 77.56990°E [29].

Therefore the UNIQUE code for Amit is:

212281400073802600928.8344°N77.5699°E

4   Traffic Light Signal Crossing (TLC) Problem Algorithm

Algorithm : Traffic Light Signal Crossing Problem Algorithm

Notation :

BP_i<- i[th] blind person, where, i ranges from 1 to 100.
Sen_G<- sensor detects Green light on traffic signal.
Sen_R<- sensor detects Red light on traffic signal.

Prerequisites:
BP_i needs to enter the destination via voice recognition.
GPS enabled sending device with internet access.
do
{
    If (GPS is working)
    {
        Stick senses the traffic signal

An alarm is raised to stop the BP_i.
User is waiting to cross the traffic light signal.
With the help of destination added, stick senses the appropriate traffic light.
do
{
    if (Sen_G)
    {
        Stick using GPS senses correct traffic light.
        Stick guides to cross the road.
    }
    else if (Sen_R)
    {
        Stick raises an alarm to stop.
        BP_istops.
        BP_i is waiting.
        Until the signal is green, $BP_i$is waiting.
    }
    } while (BP_i crosses the traffic light signal)
}
else If (GPS is not working)
{
    Stick senses the traffic signal.
    An alarm is raised to stop.
    User is waiting to cross the traffic light signal.
    If (BP_i is walking and Sen_R)
    {
        Alarm is raised that one of the traffic signal is red
        BP_i needs to be aware before taking the right path
    }
    else if (Sen_G)
    {
        BPi keeps walking according to the known route.
    }
}
} while (BP_i reaches destination)

The proposed algorithm considers two scenarios: one, when the GPS is working and other, when GPS is not working properly. The second case is the situation where network is an issue. Generally, in villages and outskirt regions, GPS connectivity is an issue. In this situation, the GPS tracking cannot work. Therefore, the blind person needs to confirm the route himself. While travelling on the road, if a traffic signal is sensed by the stick, sensor identifies whether it is green or red for the person to cross the road. If the signal is red, stick will raise an alarm to stop and again raise an alarm when the signal is green for him to cross. In the other case, when GPS is enabled, the appropriate traffic signal according to the added destination will be sensed, and the blind person will be guided when to cross the road depending upon whether the signal is green or red. There may be a situation when there are multiple traffic signals on a single point for different directions. In this situation, stick with the help of destination and GPS tracking will sense the appropriate traffic signal and guide the blind person. In a special case, if the person needs to go in diagonally opposite direction, then with the help of GPS, he will need to cross the traffic signal two times to reach the diagonally opposite road. GPS will keep on rerouting accordingly and guide the blind person.

5 Conclusion and Future Work

In this paper, an algorithm has been discussed to help the persons with disability (PwD), for example blind persons in crossing the road. An algorithm uses the GPS-enabled system for guiding the person to reach to its destination. Also, an IoT-based healthcare system is developed for blind persons for sensing the traffic light signal. An alarm system for guiding the blind based on the sensed information is used inside the stick.

In future, it is proposed to work on "Intelligent Helmet" project for visually disabled persons. It is proposed that helmet would have interfaced accelerometer module, LEDs, micro-vibration motor and ultrasonic sensor with Arduino module. Accelerometer module helps LEDs to glow according to the respective change in axis of head movement of blind person. LEDs would be used as indicator for left and right turn. When ultrasonic sensor detects the obstacle around the blind person, micro-vibration motor would start vibrating for alert when there is any risk for blind person.

References

1. https://en.wikipedia.org/wiki/Internet_of_things.
2. https://www.iotforall.com/what-is-iot-simple-explanation/.
3. https://www.happiestminds.com/Insights/internet-of-things/.
4. https://www.ibm.com/blogs/internet-of-things/what-is-the-iot/.
5. https://www.iotforall.com/4-computing-types-for-iot/amp/.
6. https://www.networkworld.com/article/3324238/internet-of-things/3-types-of-iot-platform-analytics.html.
7. https://www.digitalistmag.com/iot/2017/05/02/iot-6-categories-of-connected-things-05060901/amp.
8. https://www.worldometers.info/geography/continents/.
9. Chaudhary, S., R. Johari, R. Bhatia, K. Gupta, and A. Bhatnagar. 2019. CRAIoT: Concept, review and application (s) of IoT. In *2019 4th international conference on internet of things: Smart innovation and usages (IoT-SIU)*, 1–4. IEEE.
10. Gupta, A.K. and R. Johari. 2019. IOT based electrical device surveillance and control system. In *2019 4th international conference on internet of things: Smart innovation and usages (IoT-SIU)*. 1–5. IEEE.
11. Johari, R., N. Gupta, and S. Aneja. 2015. POSOP routing algorithm: A DTN routing scheme for information connectivity of health centres in Hilly State of North India. *International Journal of Distributed Sensor Networks* 11 (6): 376861.
12. Farahani, B., F. Firouzi, V. Chang, M. Badaroglu, N. Constant, and K. Mankodiya. 2018. Towards fog-driven IoTeHealth: Promises and challenges of IoT in medicine and healthcare. *Future Generation Computer Systems* 78: 659–676.

13. Yeh, K.H. 2016. A secure IoT-based healthcare system with body sensor networks. *IEEE Access* 4: 10288–10299.
14. Catarinucci, L., D. De Donno, L. Mainetti, L. Palano, L. Patrono, M.L. Stefanizzi, and L. Tarricone. 2015. An IoT-aware architecture for smart healthcare systems. *IEEE Internet of Things Journal* 2 (6): 515–526.
15. Ge, S.Y., S.M. Chun, H.S. Kim, and J.T. Park. 2016. Design and implementation of interoperable IoT healthcare system based on international standards. In *2016 13th IEEE annual consumer communications & networking conference (CCNC)*, 119–124. IEEE.
16. Ukil, A., S. Bandyoapdhyay, C. Puri, and A. Pal. 2016. Iot healthcare analytics: The importance of anomaly detection. In *2016 IEEE 30th international conference on advanced information networking and applications (AINA)*, 994–997. IEEE.
17. Ugrenovic, D., and G. Gardasevic. 2015. CoAP protocol for web-based monitoring in IoT healthcare applications. In *2015 23rd telecommunications forum telfor (TELFOR)*, 79–82. IEEE.
18. Maksimović, M., V. Vujović, and B. Perišić. 2015. A custom Internet of Things healthcare system. In *2015 10th iberian conference on information systems and technologies (CISTI)*, 1–6. IEEE.
19. Ahuja S., R. Johari, and C. Khokhar 2016. *IoTA: Internet of things application. Second international conference on computer and communication technologies. Advances in intelligent systems and computing*, vol. 381. Springer.
20. Turcu, C.E., and C.O. Turcu. 2013. Internet of things as key enabler for sustainable healthcare delivery. *Procedia-Social and Behavioral Sciences* 73: 251–256.
21. Rohokale, V.M., N.R. Prasad, and R. Prasad. 2011. A cooperative internet of things (IoT) for rural healthcare monitoring and control. In *2011 2nd international conference on wireless communication, vehicular technology, information theory and aerospace & electronic systems technology (Wireless VITAE)*. 1–6. IEEE.
22. https://www.countries-ofthe-world.com/continents-of-the-world.html.
23. https://simple.wikipedia.org/wiki/List_of_countries_by_continents.
24. https://www.mapsofindia.com/maps/schoolchildrens/statesandcapitals.htm.
25. https://www.analyticsvidhya.com/blog/2016/08/10-youtube-videos-explaining-the-real-world-applications-of-internet-of-things-iot.
26. http://mospi.nic.in/sites/default/files/6ec_dirEst/ec6_district_code_&_names.html.
27. https://www.mapsofindia.com/villages/uttar-pradesh/.
28. http://www.tips4holiday.com/villages-in-muzaffarnagar-tehsil/.
29. https://www.google.com/search?q=latitude+and+longitude+of+atali+up&oq=latitude.

Educational App for Android-Specific Users—EA-ASU

S. Siji Rani and S. Krishnanunni

Abstract M-learning or mobile learning is the fastest-evolving learning technology, and it has ample opportunities in global learning technical industry. If the app is designed very well with respect to its implementation, then it will definitely fulfill the purpose of learning and discovery. In accordance with this context, we have aimed at designing of an educational app EA-ASU. The proposed app aims at teaching and self-learning, even by any person who does not even have any professional knowledge regarding this. EA-ASU models any real-world learning scenarios where any user can access the various digital collection of study materials. At the same time, user can modify or upload any content to the center repository. It has also been concluded that EA-ASU, when deployed among students, was found to be very beneficial to them and it was able to improve their learning capability at ease.

Keywords Mobile learning · EA-ASU · Educational application system · E-learning

1 Introduction

Nowadays, mobile learning has become widespread among students-user community [1]. The present statistics confirms that it is growing rapidly among students of all ages. One of the main reasons for this growing popularity is that mobile phone is easily available and carried among students. At the same time with the increase of smartphones, many apps have been developed for different contexts. One basic app which is highly recommended among students community is 'mobile learning' apps. From the findings of a survey conducted by Lynda.com in 2015 [2], it is reported that 30% of smartphone users are using their phones to consume organizational training content for different educational purpose. Keeping this idea in mind, the proposed work discussed here deals with one such educational app EA-ASU.

At present, hundreds of educational applications are available in the market but only few of them are up to the mark and helping the students for better learning.

S. Siji Rani (✉) · S. Krishnanunni
Department of Computer Science and Engineering, Amrita Vishwa Vidyapeetham Amritapuri, Kollam, India

© Springer Nature Singapore Pte Ltd. 2020   325
Y.-C. Hu et al. (eds.), *Ambient Communications and Computer Systems*, Advances in Intelligent Systems and Computing 1097,
https://doi.org/10.1007/978-981-15-1518-7_27

Most of the applications are not satisfying the students and teachers up to their expectations. A well-organized educational application can result in effective control reduced training time and improved productivity in contrast with the traditional teaching/training scenarios. An Android application is a software application that runs on the Android platform [3]. Here, we develop an Android app for an educational system.

The proposed system focuses on the design and development of an innovative application for E-learning built on a web-mobile platform. By this app, any student can search for concerned topics or courses without any difficulties involved. In addition, he/she may upload or download materials on desired subjects by doing so. Again, it helps to share the concerned topics among their friends' circle.

In the current century, we can see technology everywhere. Due to upgrade in technologies, smartphones have been produced in large quantity and been used by people of all age groups. Most recent study says that self-learning helps students to grasp things very easily. Students use mobile learning platform to enhance their studies. Also, it is needed to understand a particular thing, how it works and what is the reason behind it [2]. It can be explained by using mobile learning. Mobile learning helps students to study any topic in detail clearly. They can study anytime they want, they can pause it, and they can continue it whenever they wish to continue, and this process can be repeated any number of times without a problem. This will help student to learn any topic without obstacles. In mobile learning approach, contents will be uploaded to application which can be explained using videos, audios recordings, PDFs, etc. Mobile learning will help be useful to students who travel a lot while going to school, college, etc. The course content in mobile app is downloadable content, so that students who cannot go to educational universities can use this to improve their studies. Students can take the content offline by downloading it. Mobile learning and its additional features are described in paper [4].

The rest of this paper is organized as follows: Sect. 2 covers the related works in this domain. Where Sect. 3 discusses the various methodologies used for implementing EA-ASU. Section 4 based on experimentation and testing results follows it. We conclude the proposed app in Sect. 5.

2 Related Works

Knowledge gaining with the help of mobile devices has become a common standard. Since the mobile is portable, flexible, and economical, the mobile devices are fascinated by all kinds of human. Currently, students prefer more mobiles than their computers. With the recent advancement in social networking systems, the amount of time people spend on these social systems is also constantly increasing. And it is also seen that there are many peoples, mostly students, in the social networking area like Facebook, Instagram, etc., who are almost always active in this area. Studies show that 60% of social media time spent is facilitated by mobile devices. This paves the way for further growth of online education. Today, there are many online education systems available. Upon evaluating the existing mobile learning systems, we have selected few applications, which are similar to the proposed one. These applications were selected by searching for keywords like e-learning, educational app, digital library, etc., and also considering its functional benefits in the process of learning. The selected applications were downloaded and tested for their important features, and the outcomes of this work help as the basis for additional assessment. Additional useful features for the proposed system were obtained as a result. Certain insights obtained as a result of evaluation on these apps are as follows.

Considering the Seesaw App [5] which is an Android-based educational application with a user-friendly interface, although this app can strengthen the connection between school and home, it cannot be used by all and the learning contents are not course-specific. Instructors can boost students to create, reflect, share, and collaborate. Also students who are added by the instructors to the group, only such students have the privileges to access the contents. The paper [6] describes LMS App. Mainly, it provides the elementary requirements for mobile learning-based applications. This study does not include a common syllabus. Not anyone with LMS App can access its contents; it is bounded for few foreign universities and these universities who provide the credentials for users. WhatsApp has more than one billion monthly active users, which make WhatsApp groups as a major file sharing platform. It can be particularly used as an educational tool for sharing knowledge and clearing doubts like a student-teacher problem-solving platform. However, the shared materials will not be organized, and these materials will not remain accessible all the time.

A detailed review of these existing methodologies demonstrates that even though there has been lot of work done on educational systems, they were not standardized. There is a need for a re-conceptualization of the precise nature of mobile learning. So this paper approaches a student-focused digital group that allows students to freely document and share what they are learning [7]. In an Android-based application [8], user interfaces are developed and it provides users to manage several activities like user management and user rating feature. The main objective of the paper [9] is to propose a primary outline for theorizing about mobile learning and to complement theories of infant, classroom, office, and informal learning.

Basically, educational applications play a very important role in the improvement of education. Moreover, it is considered as a large impact on educational quality [10]. Kim explains the significant part of communications among a teacher and students in humanities classes [11]. A professor could not exchange opinions with students especially in large-scale classes. They developed a smartphone app, in order to overcome the limit of interaction between a professor and students in large-scale classes. The app, they have developed, exploited smartphone's data processing and wireless communication ability, and it showed that smartphone could be a useful tool for increasing the education effect of large-scale classes since most of the students use smartphones as a medium for communication. The journal of M. Sharples [12] discusses the history of mobile learning. Here, it is discussed that the evolution of hardware software and communications is briefly discussed. Also, construction, conversation, and control of effective learning are discussed. According to Sharples, book reading can be considered to enable individual, casual, exceedingly interactive, and collaborative learning at anytime, anyplace. Even though its revolutions is not ever flourished to auxiliary learning through collaboration with subjective, mobile devices: The paper [13] is a similar mobile application that has been designed for the blind that routes them to a destination without others help. It also detects the obstacles and nature of road through which the blind is walking. The referred papers are not mentioning about the problems faced by the current generation users. The problems like

- The user may be busy with their regular busy schedule and might not have time to read the materials.
- Students not usually get the chance to gather around and meet for group discussions.
- Some students would not have the proper idea to fetch the materials online.

With our findings of the modern problems, we have tried to overcome these limitations by proposing our new application.

3 Proposed Approach

Traditional educational systems concentrate on providing abundant resources mostly in the form of electronic documents. Users spend a considerable time searching for the required materials. For students who intend to learn by using the right learning resources, at any time and any place they desire, the services that the existing systems provide are inadequate. In this paper, a new context of one mobile learning system is designed and implemented where any student or teacher with the application can access its contents and contribute to it at the same time. This approach on educational system presents a common engineering syllabus, which is not restricted for any universities and holds no information regarding any events happening at universities. However, the system can be considered as a model while developing similar systems for other syllabi in future. The instructors can also utilize in so many ways from this study, the proposed system provides them with an efficient model to exhibit their lessons-related formulas even outside classrooms.

The main contributions include

- Students enrolled in courses can start chat rooms for group discussions for better clarification of the topic.
- The application provides podcast facility for each topic for the busy users, as they can use this facility while traveling or doing small chores.
- The application provides a platform to access and download certain research papers that are not accessible for the regular pupil.

Figure 1 shows the functional architecture of educational app system EA-ASU. The functional architecture shows the main function modules of the proposed system. The key contribution of this innovative model is to give mobile learner (i.e., student) with virtual learning contents anytime and anywhere. Upload module allows users

Fig. 1 Architecture of EU-ASU app system

to upload the learning materials in an organized manner. The learning materials are built by combining data from different users using the upload option. Since the teachers may not be always free to involve, the uploading privilege is also extended for students. This lays the base of knowledge sharing functionality.

Students will be able to have access to courses in a well-ordered manner. This helps students and instructors to save their time by choosing the exact content at required time. Chat room module provides a platform for students to put queries and for teachers to solve doubts. Students can filter required topic from bundles of topics using the search topic module. This module also helps the students to save their time. Students can save the important points of the studies for using last-minute notes, before going to examination.

Mobile learning is all about the process of effective learning, and for that it is essential that learning contents should be easily accessible at the exact time when it is required. The system should be knowledge sharing and course-specific at the same time. Students visit the application because they are looking for something and the main goal should be to help them find it. While the contents of application are useful, effective, and well written, the developers should make sure that it is clearly organized. It is important to define a clear information structure for any educational system, so that the visitors will be able to quickly scan and find the content they are looking for. After developing a clear structure for contents, developers should look into more detail and decide on where to display that content inside the application. Starting with displaying critical content is always the best method. The developed educational application presents each course as a hierarchy of topics and sub-topics, which helps students for better and fast learning. Listing all services on one page, instead of individual landing pages, is a wrong way to organize content for a well-optimized system. Here, the contents are organized into logical categories with each service having its own page. Main pages serve as the main categories. The system funnels traffic through multiple entry points or landing pages for better optimization.

Search term data is an important functionality, which provides valuable insight into what the users are interested in. It is very likely that students are having the same queries regarding the topic. Including the search box in the system improves the user experience and solves this problem. Users could boost the searching process to find their desired topic materials or to practice-related services. The proposed system uses a search bar module which is powered by Google search engine. Its algorithm works in such a way that it will search for the keyword in the native site and then search on the web. The results will be arranged accordingly. Therefore, this module not just helps a student to search for hidden required topic, but it also opens a scope of research while staying within the environment for the student.

In addition, this design includes a chat room in the application, a subject-wise chat space for doubt clearance and discussions. As a result, mobile learner will get responses in the form of answers that correspond to his queries. The instructor's interaction is highly needed to engage mobile learner into an interactive learning environment. A major benefit of this study is shown by offering learners who are continually mobile with an alternative approach of understanding courses-related tasks at their time convenience, rather than attending the traditional classroom-based instructor-led session.

This design also includes links to the video of various important topics related to various engineering branches. For example, when computer science students are searching for topics related to sorting, they have access to video topics so that they will understand the topic very clearly. This also helps teachers to reduce classroom lectures so that student will be able to access the video at any time they want. Also, teachers will be able to note list of the students who went through the corresponding topics. This app includes a summary section of all the topics chapter wise, which will help them to revise for the test. Student can access the previous year question paper from the app. These papers are stored in the database, and it is very easy to download.

The system uses Android open-source code to develop a service system for the application. Instead of using native Android packages, here the system uses the concept of mobile-web application as a practical alternative to native app development for developing a mobile-web presence. Developing a mobile website for the system has many advantages over developing a native Android application. This can be accessed on web also. The data flow in mobile web system is shown in Fig. 2. It can be installed in tablet, PC, or smartphones or any Android wearable. Some of the inherent advantages of mobile websites are broader accessibility, compatibility, upgradeability, and findability. Native applications can be useful when achieving a specific purpose that cannot be effectively accomplished via a web browser. But for an educational system to offer mobile-friendly content to the widest possible audience, a mobile website becomes more suitable. These mobile-websites can be later developed as database-driven web applications that act very much like native apps using Android studio tool.

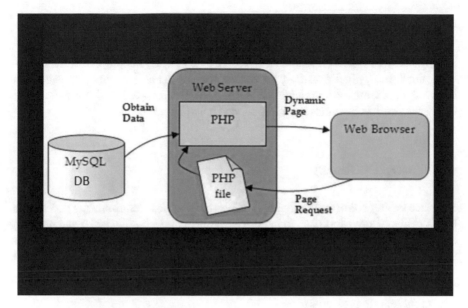

Fig. 2 Data flow in mobile web system

4 Results and Performance Analysis

This study was designed to develop an educational system using mobile learning technology for better learning and teaching at the university level. A web/mobile application is developed that allows students to access course materials and participate in discussions. The application was first analyzed for errors on emulators and afterward on different real mobile devices. The results of these tests are reported in the following sections of this chapter: (1) using emulators, (2) on real mobile devices, (3) testing for responsiveness, (4) functionality and compatibility testing, (5) summary of results.

4.1 Using Emulators

The software emulator under Android SDK was used to test the application during initial stages of development. Even though they are designed for testing native apps, they also include the default web browsers for each device which will show you a very good approximation of how pages will be rendered. Upon testing the application using this feature, different cross-browser issues were spotted and solved.

4.2 Analysis on Real Mobile Devices

The system was tested on real devices for user experience factors like variable network conditions, pixel densities, the relative size of tap targets, and real page load times. The contents were readable with varying pixel densities. The download feature of study materials becomes useful under bad network conditions.

4.3 Interface Testing

Interface testing confirms that our app meets its functional requirements and attains high quality. Simple approach to interface testing is that a human tester can perform a set of user operation on target app and validate that it is behaving correctly. For testing Android apps, we can apply either UI tests that span a single app method or UI tests that span multiple apps method. Testing of menu options, buttons, chat, settings, and navigation flow of the application is done here. Figure 3 shows interface design for learning materials.

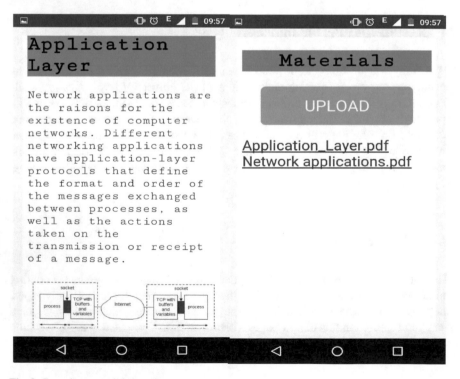

Fig. 3 Learning materials interface

4.4 Testing for Responsiveness

Today, smartphones come with different screen sizes and hardware configurations, so it is important for the developed system to be responsive in all variations. This makes testing for responsiveness an essential part of testing and analysis. Responsive design is no longer a luxury, but a necessity. Responsive design is a perfect all-in-one solution so that the size of the device no longer matters.

4.5 Functionality and Compatibility Testing

The application is tested on phones which are flagship phones to low budget phones like Iphone, Samsung, Oneplus, Asus, and Xiaomi. The application was tested on both Android and iOS. Since the system is a mobile-web application, frequent OS updates will not have much affects.

4.6 Summary of Results

Apart from the above-mentioned testing methods, the application was made to undergo usability testing to make sure the application is user friendly and provide a satisfactory experience. All the services and functionalities such as memory usage, low-level resource, and auto-deletion of temperature files, local database growing issue, backup and recovery, upgrading the application, privacy and security were tested to give a satisfactory result and experience. Operational testing—Testing of backups and recovery plan if battery goes down, or data loss while upgrading the application from store, Installation testing—Validation of the application by installing/uninstalling it on the devices and Security Testing Testing an application to validate if the information system protects data or not. Some of the main functionalities which have been included in EA-ASU are as follows.

- Course-specific uploading and downloading of materials
- A chat system
- Comprehensive and valuable content searching facility
- Easy navigation
- Previous years question paper accessing.

5 Conclusions

This section contains summary based on the outcomes from the study. Throughout this paper, we highlighted the design and implementation of a mobile-web educational application system for students and used information system success factors to validate their satisfaction with the system. With the developed system, the university students have got a platform for sharing learning materials and discussions. The study has shown that the use of mobile learning has done big impact on inside and outside classroom learning. The results of the study may help faculty understand if and how to best incorporate mobile learning strategies into teaching and learning. With our proposed system, we have solved the problem of lack of educational system for learning purpose. And as a result, students become a lot more receptive and even more willing to learn. One of the highlights of mobile learning apps is that it encourages self-learning. The handy little downloads in the developed system provide a collection of opportunities that permit for learning outside the classroom. Even though mobile learning is in the development stage, there is so many works needed to be done. And there is always a scope for improvement. Future research may want to include multiple institutions and examine differences based on region, available resources, and faculty technology training. Additional research could also be done to include students while developing the application.

References

1. Klaen, Andre, Marcus Eibrink-Lunzenauer, and Till Gloggler. 2017. *Requirements for mobile learning applications in higher education 2017 [Online]*. Available https://ieeexplore.ieee.org/document/6746846.
2. Lynda Report. 2009. *The benefits of mobile learning 2009 [Online]*. Available http://cdn.lynda.com/lyndacom-whitepaper-mobile-learningbenefits.pdf.
3. Arya, G.S., O.R. Varma, S. Sooryalakshmi, V. Hariharan, and S. Siji Rani. 2015. Home automation using android application. In *6th international conference on advances in computing, control, and telecommunication technologies, ACT 2015*, 245–253.
4. Antony, D.Asir, Gnana Singh, E.Jebamalar Leavline, Janani Selvam. 2013. *Mobile application for m-learning 2013 [Online]*. Available https://www.researchgate.net/publication/316432736.
5. Android Application. *Seesaw: The Learning Journal [Online]*. Available https://play.google.com/store/apps/id=seesaw.shadowpuppet.co.classroom.
6. Klassen, Andre, Till Gloggler. 2013. *Requirements for mobile learning applications in higher education—LMS App*. Anaheim, USA [Online]. Available http://ieeexplore.ieee.org/document/6746846/.
7. Sharples, M. 2007. *A short history of mobile learning and some issues to consider*. Online presentation mLearn, Doctoral Consortium.
8. Binu, P.K., V.S. Viswaraj. 2017. Android based application for efficient carpooling with user tracking facility. In *2016 IEEE international conference on computational intelligence and computing research, ICCIC 2016, art. no. 7919536*.
9. Shih, J.L., G.J. Hwang, Y.C. Chu, and C.W. Chuang. 2011. An investigation-based learning model for using digital libraries to support mobile learning activities. *The Electronic Library* 29 (4): 20.

10. Lee, J., and J. Choi. 2012. Implementation of application for vocabulary learning through analysis of users needs using smart phone. *The Journal of Korean Association of Computer Education* 15 (1): 4353.
11. Kim, I.-M. 2011. Android phone app. Development for large scale classes. *The Journal of Digital Policy Management* 9(6): 343–354.
12. Sharples, M. Professor of Educational Technology, *A short history of mobile learning and some issues to consider*.
13. Akhil, R., M.S. Gokul, S. Sanal, V.K. Sruthi Menon, and L.S. Nair. 2018. Enhanced navigation cane for visually impaired. In *Ambient communications and computer systems. Advances in intelligent systems and computing*, vol. 696, eds. Perez, G., S. Tiwari, M. Trivedi, K. Mishra. Singapore: Springer.

Personality Prediction System Through CV Analysis

Alakh Arora and N. K. Arora

Abstract Personality prediction is to scientifically finding out, appraising and comprehending an individual's personality based on CV. In this paper, an automated personality prediction system through CV is presented, which classifies the personality traits as well as profession of an individual naturally with the assistance of a PC framework, without the human intercession in which sets of characters to identify the personality traits related to a person, which helps in choosing the profession of that individual.

Keywords Artificial intelligence · Data mining · Fuzzy logic · Classification · Association · Curriculum vitae

1 Introduction

Enrollment, or the way toward choosing the correct competitors from a huge pool of applicants, has dependably been an extremely key issue the extent that businesses are concerned. Conventional methods [1–3] included directing identity and other specialized qualification assessment tests, meetings and gathering exchanges. Generally, with the appearance of innovation around us, we have seen a move in the way enrollments are being led. Online select frameworks [3–5] are information administration frameworks to utilize work searchers, by using the extent of Web 2.0 and other long-range informal communication locales. The utilization of Web-based social networking as an enrollment instrument presents more up to date openings and difficulties for managers. Online networking clearly offers speed, effectiveness and the capacity to target and draw in particular, especially correlated occupation searchers from the tremendous pool of competitors [6]. For example, receiving on the Web enlistment brought about 44% cost advancement and a critical diminishment in time required filling an opening in SAT telecoms [7].

A. Arora (✉)
Amity University, Noida, Uttar Pradesh, India

N. K. Arora
AICTE, New Delhi, India

© Springer Nature Singapore Pte Ltd. 2020
Y.-C. Hu et al. (eds.), *Ambient Communications and Computer Systems*, Advances in Intelligent Systems and Computing 1097,
https://doi.org/10.1007/978-981-15-1518-7_28

It is not just our gender that can shape what we do online. In the last few years, a growing body of research has found that many of our behaviors—including social media interactions, emotional responses to adverts and susceptibility to persuasion techniques—can be profoundly influenced by **personality** [8, 9].

Often it requires a notable designation of time and assets for filling an employment opportunity selecting supervisors which involves regular observation with selecting resumes from several hopefuls with abundant capabilities. The most ideal model or approach to work through the enlisting procedure is to have a meticulous arrangement. It is advised that the arrangement should start with building a successful activity post. Composing a workable activity post will differentiate and separate from the competitors that we are occupied with; in same manner, it will help in getting rid of those applicants that are not suitable for the position from the very starting point. Thus, building an arrangement for ability recognition proof will be the basic requirement each framework ought to be work particular. In the event that we take mind in building up a particular framework, which in turn should possess capacity to peruse resumes with relation to our set objectives and prerequisites for the activity and once when we are prepared to start perusing resumes, which would be done after an underlying screening process post an examination and IQ test. Through these steps, we can save our time and energy dedicated to candidates who are not worthy for the position.CV analysis can be sometimes one of the most time-consuming tasks of a recruiter looking for considerable skill to perform accurately yet quickly. As always number of applicants grow as compared to recruiting resources which remains same, or sometimes even reduce, thus most recruiters are opting resume analysis software.

2   Background Details and Related Work

For this purpose of meeting the various objectives under consideration, secondary data is collected and analyzed. Secondary data is collected through various Web sites, institutional publications, journal and government publications of repute and prominence. The available literature online is analyzed to finalize the project.

A.   *MAIN OBJECTIVES*

To empower a more compelling approach to short rundown submitted hopeful CVs from a substantial number of candidates.

To rank the CV depends on the experience and other key aptitudes which are required for specific employment profile.

To push the HR division to effortlessly waitlist the hopeful in view of the CV positioning approach.

To assist the human asset office with selecting right contender for specific occupation profile, which thusly give master workforce to the association.

3 Proposed Approach

Personality prediction is done for various purposes; however, it is done widely for identifying a right candidature for jobs.

The personality prediction empowers a more compelling approach to short run-down submitted hopeful CVs from an extensive number of candidates giving a predictable and reasonable CV positioning arrangement, which can be legitimately supported.

In spite of the fact that we might be satisfied to get an expansive number of reactions, it can be troublesome and tedious to deal with every one of the CVs adequately.

Here is the thing that to consider while evaluating whether a candidate is reasonable for the part. By utilizing the ones most important to opportunity one ought to have the capacity to quickly extricate the imperative data you require from each of the candidates' CVs.

Key skills and achievements

Do their key abilities associate with the activity and how might their aptitudes advantage you?
Have they demonstrated that they can set objectives and accomplish them?
What makes them special? Have they been engaged with different examinations, exercises?
Have they voyaged or increased common experience??

Qualifications and education

Do they have the scholastic foundation expected to finish the activity?
What have they done to build up their abilities further?
Have they exemplified an expert standard in their investigations?

Previous employment

What work encounter did they pick up?
Have they given an explanation behind leaving their present position?
Who have they worked for and to what extent have they been with every business?
Consider association size, area and nature of the business.
What were their obligations in their position and who did they answer to?
What customers have they worked with?
Have they set forward any activities?
What victories have they accomplished?
Have they been jobless for a long span?
Do they have any establishing or regulatory aptitudes that would help them in the position?
Is it true that they were equipped for working and learning in the meantime?

Looking into above requirements framework would rank the experience and major aptitudes needed for specific occupation post. After this, framework would rank the CV depends upon the experience and other major aptitudes those are necessary for specific employment post. The framework would push the HR office to effectively waitlist the hopeful in light of the CV positioning approach. This framework will center in capability and experience as well as spotlights on other essential angles those are necessary for specific employment post. This framework will assist the human asset division with selecting right possibility for specific employment profile which thus give master workforce to the association. Applicant here will enroll him/herself with every one of its points of interest and would transfer the own particular CV into the framework which will be additionally utilized by the framework to waitlist their CV. Applicant can likewise give an online test which will be directed on identity inquiries and also inclination questions. In the wake of finishing the online test, applicant can see their own test brings about graphical portrayal with marks.

This paper describes two frameworks in which personality dimensions relevant to health, such as Conscientiousness, can be used to inform interventions designed to promote health aging [10]. In this paper, the author catalogs and discusses the findings of several hundred scientific titles in the field of personality description and measurement [11]. The immediate stimulus for this book came from unresolved problems associated with assessment criteria and design [12].

Methods Involved in CV Analysis for Personality Prediction

Identifying the personality and its prediction through CV is a tedious task which becomes easy with the help of software. Some of the personality traits which are to be predicted are diagrammatically shown below:

Steps involved in personality prediction involved are mentioned below:

Step 1: Review the Candidate's Career Path. Special attention should be made not only to sort the occupation of the applicant but also to the work environment in which has developed. In the event that our position has constrained portability, it might not identify hopeful candidates who are looking to rapidly work to achieve higher positions and are genuinely after a professional Additionally, on the off chance that we would want a competitor who will make an effort to accomplish advancements, thus requiring to stay away from applicants who appear to be occupied with repetitive work environment and duty assigned

Step 2: Look for Accomplishments. We should survey and identify that the aspirant not just meet the criteria of the nature of employment what he is looking for but also should consider their growth in the working environment On the off chance that our position has limits regarding job versatility, which would likely not engage a hopeful candidate who has already worked rapidly up the positions and is searching for genuine career progress. Most importantly, on the off chance that we want an applicant who will strive to attain advancements, our aim should be to keep away from candidates who appear occupied but are staying at a similar level of responsibility. Consideration should be made to look for industry honors or working environment achievements mentioned in the resume. A rundown of obligations will just express what a hopeful was required to achieve once a day as indicated by their activity obligations. Be that as it may, a rundown of achievements will address the sort of worker the applicant is. Watch out for competitors who comprehend the objectives of the business in general and in addition their area of expertise.

Step 3: Check for Consistency. A decent resume is reliable and does not have any huge holes between occupations and instruction. At the point when a competitor's educational programs vitae displays indications of long stretches of joblessness, this might be the aftereffect of a few things. The first is that they discovered work between employments, yet they are either humiliated about the idea of the activity or they consider it to be superfluous. A moment alternative, more typical among more youthful candidates, is that they required significant investment off for some reason. In the event that the competitor has a solid resume with the exception of a couple of holes, we may consider enquiring further for a clarification.

Step 4: Look for Detail. Check for authentication of subtle elements in resumes. Applicants should mention their begin and end date of their past work alongside the insights about their activity position. In the event that they list instructive points of interest, they should list the date of graduation and also the capability accomplished.

Step 5: Review Education. While surveying a competitor's instruction, it is important to remember our base training prerequisites. Look consciously through word overpowering instruction areas to recognize what was picked up in the midst of these periods. Observation should be made apart of whether the competitor would school low maintenance while working all day. This shows dedication to accomplishing an instructive objective while possessing the potential to be flexible with different obligations. We ought to likewise know about an applicant who neglects an extensive rundown of informative encounters. They might cover an absence of down-to-earth involvement with a considerable rundown of end of the week courses.

Step 6: Look for Clarity. Competitors should utilize clear, compact and important dialect in their resumes. They ought to maintain a strategic distance from the utilization of nearby language and rather utilize general industry catchphrases that are effortlessly perceivable.

Application of personality prediction

1. Job Applications

   Getting information about personality of a livelihood applicant without making him careful that he is being attempted. Discovering attributes of away candidates without influencing them to travel long separations. Discovering quality and shortcoming of a worker before talking process. Aides in making right inquiries amid meet. Selecting specialist, sitter, bookkeepers and so on.

2. Employee and Teams

   Finding qualities of individual along these lines enhancing cooperation. Enhancing representative execution through direct counsel with HR. Guide for execution evaluation. Making achievement display by profiling great and poor staff. Distinguish best workers amid cost-cutting. Help while giving profession guiding

3. Personal

   Enhancing connections, well-being, accomplishing objectives. Vocation directing. Enhancing correspondence with relatives. Picking present for birthday festivity.

4. Investigation

   Knowing character of business accomplices and clients.

   Comprehend the kind of individual who composed mysterious letter.

   Comprehension of the creator of debilitating letter.

   Comprehend perspective of individual who submitted suicide.

   See candidly unsecure individual.

   Remove unacceptable jury competitors.

4 Results

This framework will naturally decide the key expertise trademark by characterizing every master's inclinations and positioning choices. The exhibited framework computerizes the procedures of necessities particular and candidate's positioning. The proposed framework produces positioning choices that were moderately exceptionally reliable with those of the human specialists. This framework will empower a more powerful approach to short rundown submitted hopeful CVs from an extensive number of candidates giving a steady and reasonable CV positioning strategy.

Limitations

This framework requires vast storage space because of information that is to be stored identified with CV. Should have a dynamic Web association. May give off base outcomes if information not entered legitimately.

5 Conclusions

This framework can be utilized as a part of numerous business divisions that may require master competitor. This framework will decrease workload of the human asset office. This framework will assist the human asset division with selecting right contender for specific employment profile which thusly gives master workforce to the association. Administrator or the worry individuals can without much of a stretch waitlist a hopeful in light of their online test checks and can choose a fitting contender for wanted occupation profile.

References

1. Broughton, J.Andrea, Beth Foley, Stefanie Ledermaier, and Annette Cox. 2013. *The use of social media in the recruitment process*. The Institute for Employment Studies.
2. Amdouni, S., and W. Ben Abdessalem Karaa. 2010. Web-based recruiting. In *Proceedings of international conference on computer systems and applications (AICCSA), 2010*. 1–7.
3. De Meo, P., G. Quattrone, G. Terracina, and D. Ursino. 2007. An XML-based multi-agent system for supporting online recruitment services. *Systems Man and Cybernetics Part A: Systems and Humans* 37: 464–480.
4. Kessler, R., J.M. Torres-Moreno, and M. El-Beze. 2007. E-Gen: Automatic job offers processing system for human resources. In *Proceedings of MICAI'07, 2007*, 985–995. Springer-Verlag.
5. Radevski, V., and F. Trichet. 2006. Ontology-based systems dedicated to human resources management: An application in e-recruitment. *On the Move to Meaningful Internet Systems* 4278: 1068–1077.
6. Liu, T. 2009. Learning to rank for information retrieval. *Foundations and Trends in Information Retrieval* 3: 225–331.
7. Pande, S. 2011. E-recruitment creates order out of chaos at SAT telecom: System cuts costs and improves efficiency. *Human Resource Management International Digest* 19: 21–23.
8. Cattell, R.B. 1957. *Personality and motivation: Structure and measurement*. New York: World, Book.
9. Mairesse, F., M.A. Walker, M.R. Mehl, and R.K. Moore. 2007. Using linguistic cues for the automatic recognition of personality in conversation and text. *Journal of Artificial Intelligence Research* 30: 457–500.
10. Chapman, Benjamin P., Sarah, Hampson, and John Clarkin. 2014. *Personality-informed interventions for healthy aging: Conclusions from a National Institute on aging work group*.
11. Cattell, R.B. 1946. *Description and measurement of personality*. England.
12. Stern, G.G., and M.I. Stein, Bloom. 1956. *Methods in personality assessment*. New York.

Intelligent Image Processing

Image Enhancement: A Review

Prem Kumari Verma, Nagendra Pratap Singh and Divakar Yadav

Abstract Image enhancement is the main function in image processing. Different image enhancement techniques exist in the literature. The goal of image enhancement technique is to improve the quality and characteristics of image in such a way that the important information of image is easily extracted. The contrast enhancement techniques are useful in various medical image modality, such as X-ray, MRI, ultrasound, PET, SPECT, etc. Enhancement process is performed on original image to improve the quality of visibility and it is applied in various domains such as spatial, frequency and fuzzy domain. By the process of enhancement of the image become more convenient than original image. The main objective of this process is to improve the quality of image in different medical imaging modality in different domain. Here, we show the importance of enhancement technique which is used in various field.

Keywords Image enhancement techniques · CLAHE · Histogram equalization

1 Introduction

Image enhancement is a process of improving the visual appearance and enables to identify the desired and required area of an image. Today, medical image is playing the lead role in modern diagnosis. Contrast enhancement process is very helpful for the surgeon to detect an identified, the affected or abnormal region. Image enhancement is a superior process to the comparison of the plane image. The quality of an image

P. K. Verma (✉) · D. Yadav
M.M.M. University of Technology, Gorakhpur, India

D. Yadav
e-mail: dsycs@mmmut.ac.in

N. P. Singh
National Institute of Technology, Hamirpur, India
e-mail: nps@nith.ac.in

© Springer Nature Singapore Pte Ltd. 2020   347
Y.-C. Hu et al. (eds.), *Ambient Communications and Computer Systems*, Advances in
Intelligent Systems and Computing 1097,
https://doi.org/10.1007/978-981-15-1518-7_29

is affected by various factor like illumination and equipment of image. Contrast enhancement process in image processing improves the quality of the image by increasing the dynamic range of gray levels so that the resulting image is better than the original image. The principal objective of image enhancement is to get a more suitable image comparison to the original image. In digital image processing, various techniques are used in image enhancement as like median filter, discrete Fourier transform, discrete Wavelet transform and so on. Image enhancement also improves the quality of the image by removing the noises present in the image. Various processes are used to remove the noises, such as salt and pepper noise and Gaussian noise. Gaussian noise is more effective but it makes image blurry and little effect to remove other noises. Image enhancement is used to emphasize the edge details of the image. Enhancement technique involves the mapping of image intensity data into the given transform technique. Image enhancement is applicable and useful in every field like medical image analysis, satellite images analysis, etc.

Image extraction is also a part of medical image processing which is use to detect the abnormal part of effected region [14, 15]. This is simple, effective and able process to detect the retinal blood vessel by applying various types of filters and thresholding process.

2 Literature Survey

Image enhancement is an important factor of image processing because it can help or relieve in correct identification and treatment planning [7]. There are various medical image enhancement techniques exist in literature. The details of the existing techniques, used dataset and their performance are mention in Table 1. Image processing has various technology for enhancement process.

2.1 Histogram Equalization

Histogram equalization is simple and easy to improve the image quality. HE is used to adjust image intensities to enhance contrast. An histogram is a graphic representation of image. Graphics scale shows the frequency intensities [18] of pixel. Histogram equalization is a point process that is used to redistribute the image intensity. Histogram equalization is following some steps:

– Compute the histogram of the image.
– Calculate the normalize sum of histogram.
– Transform the input image to an output image.

Figure 1(a) baby original image (b) internal brain original image and (c) retinal blood vessels original image and his histogram image shown, respectively.

After applying CLAHE shown in Fig. 2, respectively, and his histogram also after applying CLAHE . Here, we see the difference between both histograms.

Table 1 Various image enhancement techniques used in medical images with their performance

Authors	Enhancement techniques	Dataset	Result
Badra and Giragama [1]	Adaptive contrast enhancement (SAUCE)	Retinal fundus image	0.9411
Knika Kapoor et al., Kapoor and Arora [5]	Histogram	i-Knee	HE=0.4681
		ii-Shoe	HE=0.2830
		iii-Tree	HE=.3641
		iv-Sun	HE=0.0424
		v-Forest	HE=0.2590
		vi-CCTV	HE=0.7048
		vii-Plane	HE=0.0958
		viii-Building	HE=0.0636
Ahmed and Mohamed Ben et al., Miri and Mahloojifar [9]	Wavelet transform, K-mean	cerebral MRI images	PSNR = 21.6272 MSAD=17.776
Mohamed ab del-nasser	Deep-learning	Ultrasound images	PSNR=35.267
Kamil Dimiler et al.	Image processing with neural network	MRI brain images	83% correct identification
Remya K.R., Aiswarya Raj	Discrete Wavelet transform, PCA	Satellite image	Better preserved
Kale vaishnaw G	i-Spatial filtering	Lung X-ray image	PSNR=11.16
	ii- Gaussian filter		PSNR=31.74
	iii- median filter		PSNR=20.54
	iv- Laplacian filter		PSNR=10.14
	v- Derivative filter		PSNR=12.33
	vi- High boosting filter		PSNR=11.19
Er. Nancy et al., Nancy and Kaur [10]	i-Image negation	Moter wheel	PSNR=2.4073
	ii-Log transform		PSNR=9.0046
	iii-Exponential transform		PSNR= 40.3298
	iv-Power low transform		PSNR= 51.9616
	v-Gray level slicing		PSNR= 16.1414
	vi-Contrast enhancement		PSNR= 42.7688
	vii-Mean filter		PSNR=22.5375
	viii-Median filter		PSNR=22.6221
Mookiah et al., Jordan et al. [4]	Local configuration pattern and statistical ranking	STARE area of retina	Acc=97.6%, Acc=90.7%

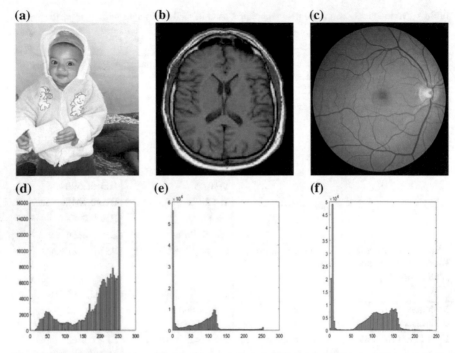

Fig. 1 Histogram **a**, **b**, and **c** of Original images given in this figure (a), (b), and (c), respectively

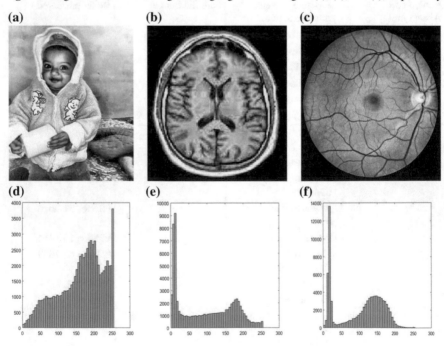

Fig. 2 Histogram **a**, **b**, and **c** of enhanced images given in this figure (a), (b), and (c), respectively

2.2   Contrast Limited Adoptive Histogram Equalization

Contrast limited adoptive histogram equalization (CLAHE) is advanced form of adoptive histogram equalization (AHE). CLAHE is used to enhance the visibility of image by increasing the local pixel region. CLAHE is normally used to control the image [13, 14, 16, 17, 19] quality. CLAHE has basically been developed for medical image. CLAHE is applied to each pixel from which the change function occurs. Which helps in making the image.
  Algorithm:

- Read an input image.
- Pre-process the input image (Removing the noise, if present in the image).
- Process each contextual region thus producing gray level.
- Illustrate gray-level mapping in order to assemble CLAHE image.

2.3   Adoptive Histogram Equalization

Adaptive histogram equalization (AHE) is based on histogram equalization (HE). This method creates each histogram of sub-image to redistribute the original chavi of the brightness of the images. Histogram equalization (HE) works on the full image but it docs not provide complete sense. Unlike HE, adaptive histogram equalization (AHE) improves the local image of the retina in the blood vessel. Therefore, AHE has more pressurized effect than HE.

2.4   Frequency Domain Filtering

Frequency domain filtering is the filtering device that applies to change the signal. Fourier transforms are used in the frequency domain in image processing when the input images are in the spacial domain [3]. The frequency of image enhancement is used for the domain method prominence 2D image, which is mainly composed of unitary changes. Steps in frequency domain filtering:
  1. Image must be transformed by spatial domain using the fast Fourier transform.
  2. The resulting complex image must be multiplied by a filter.
  3. The filtered image must be transformed to the spatial domain.

2.5   PCA Based Image Enhancement

Principal component analysis (PCA) is a data analysis tool, i.e. used to reduce the dimensionality of the large number of pixel. It is also used to identify the pattern of data and to express the data dimension [12]. The PCA D-correlated genuine dataset focuses on the small subset of transformer datasets, we use the PCA to eliminate its dementia from the original image, using it also shows the position of the pixel in the image, for image processing, it is very important.

Steps of PCA:

– Let X1,X2,…,Xn denote the random variable for size n. The mean of the dataset is a random variable defined by:

$$Mean(X) = \left(\frac{1}{n}\right) \sum_{i=1}^{n} X^i \tag{1}$$

– Now we calculate the standard deviation(SD):-

$$SD = \sqrt{\left(\frac{1}{n}\right) \sum_{i=1}^{n} (Xi - \bar{X})^2} \tag{2}$$

– The formula for covariance is very similar to the formula for variance. The formula-

$$con(X, Y) = \sum_{i=1}^{n} \frac{(X_i - \bar{X})(Y_i - \bar{Y})}{n} \tag{3}$$

2.6 Local Pixel Grouping

Local pixel grouping is used to remove the noise to the original image. Pixels are identified based on the spatial coordinate and their gray scale value (intensity value) [20]. The selection of samples of training for some blocks from the training block in LPG is obtained using block -matching technique. LPG is used to select the pixel that is similar to the image of LPG. The LPG is the parent component of PCA.

2.7 Median Filtering

Median filtering is used to remove the noise from image which is non-linear technique used in image processing. Median filters work as an edge detection from the image, which comes under the image preprocessing [11]. This process is widely used in digital image processing. Mean filtering is used to remove the noise of salt and pepper, this process is also used to remove the noises. A median filter works on windows scaling technique. Median filter is used in 2D images. A median filter is a more effective process when the goal simultaneously reduces noises.

2.8 Noise Removal Using a Weiner Filtering

Weiner filtering is used for signal processing. Weiner filters are used to estimate the images and the noises found in the liner-time invariant. Basically, the Weiner filter is used to eliminate noise from the original image. Weiner filter is mainly two types:

– Weiner filter in space domain.
– Weiner filter in frequency domain.

Weiner filter in space domain is designed to minimize the mean-square-error between desired and assuming signal. Weiner filter in frequency domain is used to minimize the impact of noise at frequencies which have poor signal-to-noise ratio. Weiner filter works on auto-regressive model. Weiner filter builds the relation between AR parameter of clean speech and noise speech signal. The Weiner filter is used in various fields, such as signal processing, image processing, control system and digital communication.

2.9 Unsharpe Mask Image

The unsharpe mask filter is used to correct blurring introduced during scanning, re sampling or printing. It is used to see the print of the image on-line [2]. USM is used to create a negative image of the original image. The first original negative is copied in the USM and then it was turned into positive. This filter increases the frequency of the image.

2.10 Filtering with Morphological Operators

Morphological image processing is a collection of non-linear operation related to the shape or morphology features like boundaries, skeletons, etc. Morphological operator is based on set theory [8]. The fundamental operation of morphology operator is erosion and dilation. The dilation operation is used to grow the size of object. The erosion process is a compliment of dilation process. It is use to shrink the size of image.

3 Conclusion

In this paper, we analyze the various method of image enhancement. Enhancement technique is used to improve the quality of image [6]. Image enhancement techniques are usually applied in remote sensing data to improve the visual analysis for human. Enhancement is used to find the good result and it is also helpful to identify the effected region of human body. Image enhancement technology has become an important part of preprocessing in digital image processing; it is an important tool for the vision processing application. In this review, the various process used to improve the quality of image have been discussed.

References

1. Bandara, A., and P. Giragama. 2017. A retinal image enhancement technique for blood vessel segmentation algorithm. In *2017 IEEE international conference on industrial and information systems (ICIIS)*, 1–5, IEEE.
2. Bedi, S., and R. Khandelwal. 2013. Various image enhancement techniques—A critical review. *International Journal of Advanced Research in Computer and Communication Engineering* 2 (3).
3. Choudhary, R., and S. Gawade. 2016. Survey on image contrast enhancement techniques. *International Journal of Innovative Studies in Sciences and Engineering Technology* 2 (3): 21–25.
4. Jordan, K.C., M. Menolotto, N.M. Bolster, I.A. Livingstone, and M.E. Giardini. 2017. A review of feature-based retinal image analysis. *Expert Review of Ophthalmology* 12 (3): 207–220.
5. Kapoor, K., and S. Arora. 2015. Colour image enhancement based on histogram equalization. *Electrical & Computer Engineering: An International Journal* 4 (3): 73–82.
6. Kaur, S., and P. Kaur. 2015. Review and analysis of various image enhancement techniques. *International Journal of Computer Applications Technology and Research* 4 (5): 414–418.
7. Khidse, S., and M. Nagori. 2014. Implementation and comparison of image enhancement techniques. *International Journal of Computer Applications* 96 (4).
8. Maragos, P. 2005. Morphological filtering for image enhancement and feature detection. *Analysis* 19: 18
9. Miri, M.S., and A. Mahloojifar. 2009. A comparison study to evaluate retinal image enhancement techniques. In *2009 IEEE international conference on signal and image processing applications (ICSIPA)*, 90–94, IEEE.
10. Nancy, E., and E.S. Kaur. 2013. Comparative analysis and implementation of image enhancement techniques using matlab. *International Journal of Computer Science and Mobile Computing* 2 (4).
11. Nirmala, D. 2015. Medical image contrast enhancement techniques. *Research Journal of Pharmaceutical Biologycal and Chemical Sciences* 6 (3): 321–329.
12. Patil, V.D., and S.D. Ruikar. 2012. PCA based image enhancement in wavelet domain. *International Journal of Engineering Trends and Technology* 3 (1).
13. Singh, N.P., and R. Srivastava. 2016. Retinal blood vessels segmentation by using gumbel probability distribution function based matched filter. *Computer Methods and Programs in Biomedicine* 129: 40–50.
14. Singh, N.P., and R. Srivastava. 2017. Weibull probability distribution function-based matched filter approach for retinal blood vessels segmentation. In *Advances in computational intelligence*, 427–437. Heidelberg: Springer.

15. Singh, N.P., and R. Srivastava. 2018. Extraction of retinal blood vessels by using an extended matched filter based on second derivative of gaussian. In *Proceedings of the National Academy of Sciences, India Section A: Physical Sciences*, 1–9.
16. Singh, N.P., T. Nagahma, P. Yadav, and D. Yadav. 2018. Feature based leaf identification. *2018 5th IEEE Uttar Pradesh section international conference on electrical, electronics and computer engineering (UPCON)*, 1–7, IEEE.
17. Singh, S., N.P. Singh. 2019. Machine learning-based classification of good and rotten apple. In *Recent trends in communication, computing, and electronics*, 377–386. Heidelberg: Springer.
18. Singla, N., and N. Singh. 2017. Blood vessel contrast enhancement techniques for retinal images. *International Journal of Advanced Research in Computer Science* 8 (5).
19. Yadav, P., and N.P. Singh. 2019. Classification of normal and abnormal retinal images by using feature-based machine learning approach. In *Recent trends in communication, computing, and electronics*, 387–396. Heidelberg: Springer.
20. Zhang, L., W. Dong, D. Zhang, and G. Shi. 2010. Two-stage image denoising by principal component analysis with local pixel grouping. *Pattern Recognition* 43 (4): 1531–1549.

Diagnosis of Cough and Cancer Using Image Compression and Decompression Techniques

Ashish Tripathi, Ratnesh Prasad Srivastava, Arun Kumar Singh,
Pushpa Choudhary and Prem Chand Vashist

Abstract This paper is dedicated to provide a technique with an innovative approach which can efficiently compress and recognize medical images. Since medical images are huge in size, therefore, compression of medical images is needed. Then, recognition capability is tested with the compressed and the uncompressed images. Basically, in this paper, two steps have been used to identify the disease. In the first step, the physical size of the medical image is reduced, and in the second step, differentiation of the image of particularly lung part of the human body at different disease states is performed. The objective of this paper is to detect and analyze the lung part of the human body based on cough state before entering into the cancer state so that the disease can be cured.

Keywords Medical images compression · Discrete cosine transformation (DCT) · Principle component analysis (PCA) · Medical images recognize

1 Introduction

Coughing is generally caused by due to weather changes, daily lifestyle, respiratory infection and many other conditions. It takes a week or two to become illness. Occasional cough is normal and it happens and cured with general medications, but a persistent cough may be symptoms of severe health conditions and cause of lung cancer. In the early stage, the symptoms of lung cancer may not be noticeable, and people may diagnose at the advanced stage. Therefore, the objective of this paper is to detect and analyze the lung part of the human body based on cough state before entering into the cancer state so that the disease can be cured. Cough is a very common symptom for general cancer patients approximately 24–38% patients and approximately 46–87% of lung cancer patients [1]. In this, the author took 100 cancer patients data, using the Memorial Symptom Assessment Scale from the beginning

A. Tripathi (✉) · A. K. Singh · P. Choudhary · P. C. Vashist
Department of IT, G. L. Bajaj Institute of Technology and Management, Greater Noida, India

R. P. Srivastava
Department of Information Technology, College of Technology, GBPUAT, Pantnagar, India

© Springer Nature Singapore Pte Ltd. 2020
Y.-C. Hu et al. (eds.), *Ambient Communications and Computer Systems*, Advances in
Intelligent Systems and Computing 1097,
https://doi.org/10.1007/978-981-15-1518-7_30

state of cancer treatment to 3, 6 and 12 months state representation, a prevalence of 42.9%, 39.2%, 35.1% and 36.1%, respectively, similarly to the experience of breathlessness, although less distressing than breathlessness [2]. These numbers almost doubled in the lung cancer subgroup analysis.

2 Related Work

The effect of image compression is evaluated by Blackburn et al. on image recognition. They used two sets, the gallery set (images of the person known to the system) and the probe set (unknown images). Then, they compressed the set of unknown images. They concluded that recognizing images with the compressed images as probe set does not drop the recognition accuracy to a significant extent even the highest performance drop they analyzed was below than 0.2 bpp. They claimed that the recognition accuracy even becomes better at some compression ratio [1].

For the recognition of the image, Kim et al. [2] used the wavelet-based compression technique. They compressed both the probe set images and the gallery set images for their experiment. The compression is done at 0.5 bpp. They used PCA method for image recognition. Wat et al. [3] analyzed the results by applying both important techniques of image recognition, i.e., PCA and LDA [4]. McAuliffe et al. [5] tested the effect of JPEG compression during the project development entitled "Image Recognition Format for Data Interchange." They used the rate of compression, i.e., 10:1 for the experiment. Wijaya et al. [6] performed image recognition on compressed images which were compressed up to 0.5 bpp. To do the experiment, CMU pie database and MACE filter-based classifier had been used by them. They concluded that the effect of the compression does not affect the recognition in adverse manner.

The comparative study of the JPEG compression effect on image recognition was done by Delac et al. [7]. Wahba et al. [8] [9–12] measured the performance of different data compression algorithms based on different parameters on binary, gray level and RGB images. The recognition system used in the experiment was twofold; i.e., in the first part, only the probe set was compressed and the gallery set was uncompressed. Images used for experiment were taken by a surveillance camera [13, 14]. In the second part, both the probe set and the gallery set were compressed. By the joint opinion of the authors, the compression does not deteriorate the recognition of enhancement however it increases the proficiency of image recognition slightly in some cases.

3 Methodology

3.1 JPEG Compression

The JPEG compression method proceeds as follows (the input image may be gray level values in a matrix or RGB color model image mathematically shown by three-dimensional matrix):

- RGB standard transformation values to YCbCr color system.
- Split the image in 8 × 8 pixel blocks.
- Apply DCT to individual blocks.
- Quantize each individual block.
- Apply zigzag scan.
- Apply Huffman encoding on the resulting bit data.

3.2 Discrete Cosine Transformation

Figure 1 shows the basic operation technique of DCT. This is the main step of image compression. Attentions of DCT are more on lower frequencies due to better transforming of linear sequences. Lower frequencies are better than the higher frequencies, and block will be transformed column-wise followed by row-wise [15].

A transformation function is represented by $F(y, x)$, original function is represented by $F(x, y)$ and coefficients of the frequency domain transform is given by $F(u, v)$ as shown in Eq. 1.

$$F(u, v) = 1/4 Cu Cv \ 7x = 0 \ 7y = 0 \ F(x, y) \cos[(2x + 1)u/16] \cos[(2y + 1)v/16]$$
Where, $Cu, Cv = 1/2$ for $u, v = 0$; otherwise $Cu, Cv = 1$ \hfill (1)

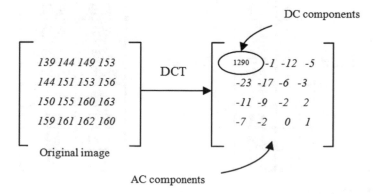

Fig. 1 DCT calculation [9]

3.3 Principal Component Analysis

It identifies patterns in data and expresses the data in such a way as to highlight their similarities and differences. The method of PCA includes the following steps:

Step 1: Get some data: For example, let us use the following data set of two dimensions as shown in Fig. 2. Further, subtract the mean $xi - x_{avg}$ and $yi - y_{avg}$ to adjust the data.

Step 2: Calculate the covariance matrix: Since the data is two dimensional, the covariance matrix will be 2×2 shown in Eq. 2.

$$\text{cov} = \begin{pmatrix} 0.6165555556 & 0.6154444444 \\ 0.6154444444 & 0.7165555556 \end{pmatrix} \tag{2}$$

Step 3: Calculating the eigenvectors and eigenvalues from the covariance matrix as shown in Eq. 2, and the values of eigenvectors and eigenvalues are represented in Eqs. 3 and 4:

$$\text{eigenvalues} = \begin{pmatrix} 0.0490833989 \\ 1.28402771 \end{pmatrix} \tag{3}$$

$$\text{eigenvectors} = \begin{pmatrix} -0.735178656 & -0.677873399 \\ 0.677873399 & -0.735178656 \end{pmatrix} \tag{4}$$

Step 4: Choosing components and forming a feature vector: After calculating the eigenvectors of the covariance matrix, find out the highest to lowest eigenvalue. It is

$$\text{Data} = \begin{array}{|c|c|} \hline x & Y \\ \hline 2.5 & 2.4 \\ \hline 0.5 & 0.7 \\ \hline 2.2 & 2.9 \\ \hline 1.9 & 2.2 \\ \hline 3.1 & 3.0 \\ \hline 2.3 & 2.7 \\ \hline 2 & 1.6 \\ \hline 1 & 1.1 \\ \hline 1.5 & 1.6 \\ \hline 1.1 & 0.9 \\ \hline \end{array}$$

x	Y
2.5	2.4
0.5	0.7
2.2	2.9
1.9	2.2
3.1	3.0
2.3	2.7
2	1.6
1	1.1
1.5	1.6
1.1	0.9

DataAdjust =

x	Y
0.69	0.49
-1.31	-1.21
0.39	0.99
0.09	0.29
1.29	1.09
0.49	0.79
0.19	-0.31
-0.81	-0.81
-0.31	-0.31
-0.71	-.101

Fig. 2 Data gathering

essential to calculate number of n eigenvalue and eigenvector in multi-dimensional data.

The formula for a matrix of vector is shown in Eq. 5 for calculating a feature vector:

$$\text{feature vector} = \left(\text{eig}_1, \text{eig}_2, \text{eig}_3, \ldots, \text{eig}_n\right) \tag{5}$$

In the example set of data, there are two eigenvectors, so the feature vector will be as

$$\text{feature vector} = \begin{pmatrix} 0.677873399 \ 0.735178656 \\ 0.735178656 \ 0.677873399 \end{pmatrix} \tag{6}$$

Step 5: Deriving the new data set: The concluding step in PCS contains the parameter of eigenvectors which required in the data to select and generate a feature vector after that a transpose of that vector to multiply on the left of the original data set, transposed. The final result is shown in Eq. 7.

In the last step of algorithm, once the components (eigenvectors) that needs to be kept in our data are chosen and a feature vector is formed then a transpose of the vector is taken and it is multiplied on the left of the original data set, then transposed. The final calculation is shown in Eq. 7.

$$\text{final data} = \text{Row Feature Vector} \times \text{Row Data Adjust} \tag{7}$$

Column transposed of the matrix with eigenvectors is "row feature vector"; after transposition, most significant eigenvector now appear at the top row of the matrix, and row data adjust is the mean-adjusted data transposed. After computation, final data items are appeared in column of the matrix and row display dimensions [16].

3.4 Steps for Image Compression

3.4.1 Tiling

The image is divided into a number of rectangular parts called tiles. This is mainly done for the purpose if one wants to access only particular areas of the image. It has no effects on compression efficiency. One another reason of tiling may be that different levels of compression can be performed on the different tiles.

3.4.2 Transform Standard RGB Values to the YCbCr Color System

First, the image should be converted from RGB into a different color space called YCbCr. In YCbCr notation, Y represents the brightness, and the Cb and

Cr components represent the chrominance of the pixel for blue and red colors, respectively.

When the image has been separated into three categories, then the unwanted informative noise without the uses of human eye interaction detecting the difference can be easily removed in the chrominance components. These mechanisms also increase the compression ratio.

3.4.3   DCT Calculation

Specification of image uses an orthogonal discrete cosine transform (DCT) which provided better result at transforming linear sequence due to focus on lower frequencies. Transformation of the block reads in such a way that left to right for column-wise and after that top to bottom for row-wise.

Uses a direct formula for the function written in with respect to the pixel values, $f(y, x)$, and the coefficients of frequency domain transform are $F(v, u)$, shown in Eq. (8).

$$F(u, v) = 1/4 Cu\, Cv\, 7x = 0\; 7y = 0\; F(x, y) \cos[(2x+1)u/16] \cos[(2y+1)/16]$$
Where $Cu, Cv = 1/2$ for $u, v = 0$; otherwise $Cu, Cv = 1$ (8)

3.4.4   Quantizing

A way to decide when to keep or not the frequency data is quantizing. High-frequency data items may be discarded without losing too much important information, because human eye cannot perceive high-frequency brightness and has even more trouble with high-frequency color changes [17]. Quantizing compresses certain frequency ranges to a single value; of course, these single values cannot be decoded back to their original values, but just this can be decided that in which range the original value would have been.

The equation for quantization is as follows:

$$(v, u) = [F(,)(Q(,)/2)]/Q(u, v) \qquad (9)$$

We are trying to find out the value of the quantized coefficient $C(v, u)$; in that, sequence value of frequency coefficient for DCT is $F(v, u)$, and for the pixel (v, u) in the block, the value of quantizing step size is $Q(v, u)$.

3.4.5 Zigzag Ordering

This process is done after quantization. Zigzag encoding stores the matrix elements in a 64×1 vector (from 8×8 matrix). In the vector, the lower frequency data is in the lower ranges of the vector, and the higher frequency data is in the higher ranges of the vector. The vector is built up by moving the matrix in the zigzag order. The first value of the vector is called the DC value, and the rest all are known as AC values. The DC value has the lowest frequency and in large perceptive finds the main value of the entire block. The AC values are small, restrained variation of DC value, reflected in the main block. Role of high frequencies AC value is responsible for the encoded JPEG images.

Each block's DC value of the first block is explicitly stored in the form of the differential encoding. Then, the difference between the first DC and DC of the second block is stored. Actual predication is that the differences are usually small, and small numbers can be stored in fewer bits.

3.4.6 Run-Length Encoding/Huffman Encoding

The quantized AC coefficient of image contains sequence of consecutive runs of zeros and these values of RLE use to encode that. Huffman algorithm is used to transform nonzero AC coefficient along with the DC coefficient to compress the data.

Steps for image compression

Get the data: Collect images to create a trainee set.
Subtract the mean: Then calculate the average image as shown below in Fig. 3.
The mean image is calculated as follows (Fig. 4).
Then subtract it from the training images as depicted in Fig. 5
Now, we build the covariance matrix which is $N2$ by M shown in Eq. 10:

$$A = \left[\vec{a}_m \vec{b}_m \vec{c}_m \vec{d}_m \vec{e}_m \vec{f}_m \vec{g}_m \vec{f}_m \right] \tag{10}$$

The covariance matrix which is $N2$ by $N2$ given in Eq. 11:

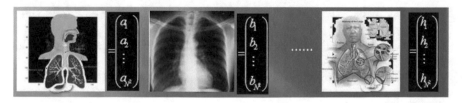

Fig. 3 Example showing different input images and their corresponding vectors

Mean Image

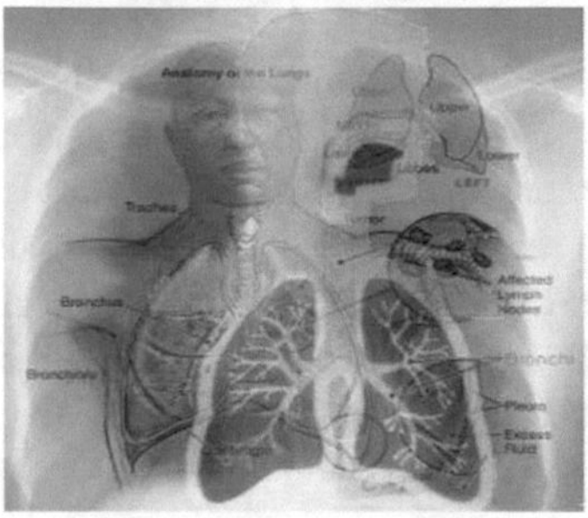

Mean Calculation

$$\vec{m} = \frac{1}{M} \begin{pmatrix} a_1 + b_1 + \cdots + h_1 \\ a_2 + b_2 + \cdots + h_2 \\ \vdots \quad \vdots \quad \quad \vdots \\ a_{N^2} + b_{N^2} + \cdots + h_{N^2} \end{pmatrix}, \quad where\ M = 8$$

Fig. 4 Mean image (left and calculation of mean image (right))

$$\vec{a}_m = \begin{pmatrix} a_1 - m_1 \\ a_2 - m_2 \\ \vdots \quad \vdots \\ a_{N^2} - m_{N^2} \end{pmatrix}, \quad \vec{b}_m = \begin{pmatrix} b_1 - m_1 \\ b_2 - m_2 \\ \vdots \quad \vdots \\ b_{N^2} - m_{N^2} \end{pmatrix}, \quad \vec{c}_m = \begin{pmatrix} c_1 - m_1 \\ c_2 - m_2 \\ \vdots \quad \vdots \\ c_{N^2} - m_{N^2} \end{pmatrix}, \quad \vec{d}_m = \begin{pmatrix} d_1 - m_1 \\ d_2 - m_2 \\ \vdots \quad \vdots \\ d_{N^2} - m_{N^2} \end{pmatrix},$$

$$\vec{e}_m = \begin{pmatrix} e_1 - m_1 \\ e_2 - m_2 \\ \vdots \quad \vdots \\ e_{N^2} - m_{N^2} \end{pmatrix}, \quad \vec{f}_m = \begin{pmatrix} f_1 - m_1 \\ f_2 - m_2 \\ \vdots \quad \vdots \\ f_{N^2} - m_{N^2} \end{pmatrix}, \quad \vec{g}_m = \begin{pmatrix} g_1 - m_1 \\ g_2 - m_2 \\ \vdots \quad \vdots \\ g_{N^2} - m_{N^2} \end{pmatrix}, \quad \vec{h}_m = \begin{pmatrix} h_1 - m_1 \\ h_2 - m_2 \\ \vdots \quad \vdots \\ h_{N^2} - m_{N^2} \end{pmatrix}$$

Fig. 5 Subtraction of mean images with training images

$$\mathrm{cov} = AA^{\mathrm{T}} \tag{11}$$

Find eigenvalues and eigenvectors of the covariance matrix. Then choose components and form a featured vector.

4 Implementation

4.1 Images Compression

Step1. The raw data (image) will be divided into 8×8 blocks, i.e., 8 rows and 8 columns.

Step2. Then, the image will be converted from RGB color space to YCbCr color space.

Step3. Each component Y, Cb, Cr of each 8×8 block is converted to frequency domain representation using discrete cosine transform (DCT).

Step4. Quantization.

Step5. Zigzag ordering.

Step6. Apply Huffman encoding.

4.2 Images Reorganization

The following step involved:

1. 1 Database preparation
2. Training
3. Testing

Database preparation: The database is created with 64 images of human lung in four distinct states. These states are normal state of lung, images of lung in cough state, images of lung in pneumonia state and images of lung in the cancer state. Database is also prepared for testing phase by taking 4–5 images of four distinct states [18, 19].

Training: Open the (.jpg) file using open file dialog box using train database.

- By using patient's train folder, read all the medical image data.
- Regulate and monitor all the medical images.
- Find the suitable eigenvectors of reduced covariance matrix (RCM).
- After that, display the result in the eigenvectors of covariance atrix.
- Compute the identifying pattern vectors (RPV) for individual medical image.
- Calculate each patient's maximum out of the distances

Testing: Testing is carried out by the following steps:

- Choose the medical image which is tested using open file dialog box.
- Read the image and normalize.
- Calculate the RPV of image using eigenvector of covariance matrix.
- Calculate the distance of medical image RPV from average RPVs from patient's database medical images.
- After that, get the minimum distance value of the patient's medical images.
- If the minimum distance of testing image is less than the maximum distance of training image of that patient medical images then identified the patient medical images.

5 Results

5.1 Graphical User Interface

Developed Software:

Performing image recognition by the software
Main window of image recognition is shown in Fig. 6.
On the basis of above data, the image compression module of the software has a capability of compressing an image by 86% shown in Table 1.
On the basis of above data, the image recognition module of the software has a capability of recognizing an image by 80% depicted in Table 2.

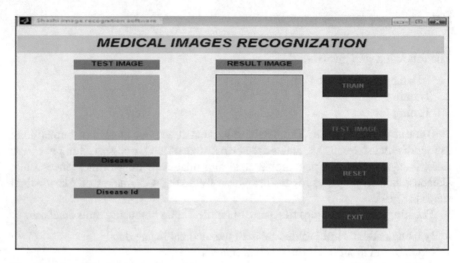

Fig. 6 Main window of image recognition

Table 1 Compression capability of software

S. No.	Size of uncompressed image(in KB)	Size of compressed image (in KB)	Compressed (%)
1.	858	118	86.25
2.	826	121	85.35
3.	757	62.9	91.69
4.	762	143	81.23
5.	759	105	86.16

Table 2 Recognition capability of software

S. No.	Test image	Correctly matched	Incorrectly matched	Accuracy (%)
1.	Human lung in normal state	5	0	100
2.	Human lung in cough state	3	2	60
3.	Human lung in pneumonia state	3	2	60
4.	Human lung in cancer state	5	0	100

A. **Some snapshots of important results**
   See Fig. 7.

A. Some snapshots of important results

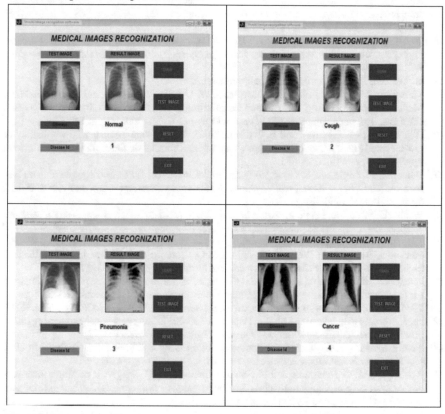

Fig. 7 Snapshots of image recognition

6   Conclusion

In this paper, the technique is being developed for the detection of cancer through analyzing cough with the help of chest image. Images are compressed within a certain limit so that recognition performance should not be dropped. Comparison is done between test images and trained images of database. In the applied method, images sizes have been reduced approximately 86% and found 80% average accuracy in the prediction of lung cancer as shown in Table 2.

References

1. Delac, K., G. Mislav, and G. Sonja. 2007. Image compression effects in face recognition systems. In *Face Recognition*. IntechOpen.

2. Kim, Byung S., Sun Kook Yoo, and Moon H. Lee. 2006. Wavelet-based low-delay ECG compression algorithm for continuous ECG transmission. *IEEE Transactions on Information Technology in Biomedicine* 10 (1): 77–83.
3. Wat, Kshitij, and S.H. Srinivasan. 2004. Effect of compression on face recognition. In *Proceedings of the 5th international workshop on image analysis for multimedia interactive services, WIAMIS 2004*.
4. http://en.wikipedia.org/wiki/Image_compression.
5. McAuliffe, M.J., F.M. Lalonde, D. McGarry, W. Gandler, K. Csaky, and B.L. Trus. 2001. Medical image processing, analysis and visualization in clinical research. In *Proceedings 14th IEEE symposium on computer-based medical systems, CBMS 2001, IEEE*. 381–386.
6. Wijaya, S.L., M. Savvides, and B.V. Kumar. 2005. Illumination-tolerant face verification of low-bit-rate JPEG2000 wavelet images with advanced correlation filters for handheld devices. *Applied Optics* 44 (5): 655–665.
7. Delac, K., M. Grgic, and S. Grgic. 2005. Effects of JPEG and JPEG2000 compression on face recognition. In *International conference on pattern recognition and image analysis*, 136–145. Springer, Berlin, Heidelberg.
8. Wahba, Z. Walaa, and Ashraf, Y.A. Maghari. 2016. *Lossless image compression techniques comparative study*. In *Lossless Image Compression Techniques Comparative Study 3.2*
9. Jain, A.K. 1981. Image data compression: A review. *Proceedings of the IEEE* 69 (3): 349–389.
10. Panpaliya, Neha, et al. 2015. A survey on early detection and prediction of lung cancer. *International Journal of Computer Science and Mobile Computing* 4 (1): 175–184.
11. Avinash, S., K. Manjunath, and S. Senthil Kumar. 2016. An improved image processing analysis for the detection of lung cancer using Gabor filters and watershed segmentation technique. In *2016 International Conference on Inventive Computation Technologies (ICICT)*. Vol. 3, IEEE.
12. Joon, Preeti, Aman Jatain, and Shalini Bajaj Bhaskar. 2017. Lung cancer detection using image processing techniques. *International Journal of Engineering Science* 6497.
13. Assefa, Mickias, et al. 2013. Lung nodule detection using multi-resolution analysis. In *2013 ICME international conference on complex medical engineering, IEEE*.
14. Krishnaiah, V., G. Narsimha, and N. Subhash Chandra. 2013. Diagnosis of lung cancer prediction system using data mining classification techniques. *International Journal of Computer Science and Information Technologies* 4 (1): 39–45.
15. Elveren, Erhan, and Nejat Yumuşak. 2011. Tuberculosis disease diagnosis using artificial neural network trained with genetic algorithm. *Journal of Medical Systems* 35 (3): 329–332.
16. Hamad, Aqeel Mohsin. 2016. Lung cancer diagnosis by using fuzzy logic. *IJCSMC* 5 (3): 32–41.
17. Xu, Feng, et al. 2008. Detection of intrafractional tumour position error in radiotherapy utilizing cone beam computed tomography. *Radiotherapy and Oncology* 89 (3): 311–319.
18. Seppenwoolde, Yvette, et al. 2007. Accuracy of tumor motion compensation algorithm from a robotic respiratory tracking system: A simulation study. *Medical Physics* 34 (7): 2774–2784.
19. Genant, Justin W., et al. 2002. Interventional musculoskeletal procedures performed by using MR imaging guidance with a vertically open MR unit: Assessment of techniques and applicability. *Radiology* 223 (1): 127–136.

3D Lung Segmentation Using Thresholding and Active Contour Method

Satya Prakash Sahu, Bhawana Kamble and Rajesh Doriya

Abstract Lung segmentation is the first step to identify any lung-related disease. It is an image processing-based process to obtain the boundary of the lung area from thoracic on CT images. To challenge this scenario, advanced diagnosis methods are needed that requires CT scan images of patient. Radiologists need huge amount of time to detect if any person is having lung cancer or not. To help radiologists, several researchers had proposed many computer-aided diagnosis systems to detect lung-related disease at early stages. In the present work, lung segmentation is done in three dimensions. Image processing techniques are applied named thresholding, morphological operation and active contour to achieve this. At first, preprocessing is done to normalize the value and then Otsu thresholding to divide image into two regions. Morphological operation like erosion is applied to eliminate unwanted region. Active contour is applied at last to segment lung in 3D. Here, 15 subjects are taken from LIDC-IDRI. This method has achieved 0.967 Jaccard index and 0.983 Dice similarity coefficient when compared to ground truth. A 3D view of lung segmentation is also shown.

Keywords CAD system · Lung segmentation · Active contour · Thresholding

1 Introduction

Lung cancer growth emerges among every single other kind of disease for providing one of the highest occurrence rates of mortality. Shockingly, this illness is regularly analyzed late, influencing the treatment result. According to different insights studied by various organizations, the lung disease has been seen as most basic cancer growth

S. P. Sahu (✉) · B. Kamble · R. Doriya
Department of Information Technology, National Institute of Technology, Raipur,
Chhattisgarh, India
e-mail: spsahu.it@nitrr.ac.in

B. Kamble
e-mail: bkamble.mtech2017.it@nitrr.ac.in

R. Doriya
e-mail: rajeshdoriya.it@nitrr.ac.in

© Springer Nature Singapore Pte Ltd. 2020   369
Y.-C. Hu et al. (eds.), *Ambient Communications and Computer Systems*, Advances in
Intelligent Systems and Computing 1097,
https://doi.org/10.1007/978-981-15-1518-7_31

around the world. International Agency for Research on Cancer (IARC) had prepared an approximated statistical figure in 2008 about occurrence and death rate of 27 cancers in GLOBOCAN series for 182 countries [1]. An approximation of 1.27 crore new lung cancer cases and 76 lakhs lung cancer deaths occurred in 2008. In 2012, 18 lakhs new lung malignant growth cases were evaluated [2]. Deaths because of lung malignant growth are more when contrasted with other disease-related demises around the world. In 2012, it has contributed approximately 13% for cases enrolled and in area of death rate because of malignant growth its commitment was 19%, announced by GLOBOCAN (IARC) in the module of cancer surveillance. Lung disease is seen to be the real reason for deaths in mail populaces. Among females, lung malignant growth is the main source of death in growing nations and the second driving reason for death in growing nations [3]. In 2015, European Union encountered an abatement in lung malignancy mortality in men by 6% contrasted with 2009, while malignant growth demise rates expanded in females by 7%, accordingly moving toward male partners [4]. In USA, from the American Cancer Society, cancer certainties and figure 2016 [5], there were around 224,400 new instances of malignant growth representing about 14% and 158,080 related passings that accounts 27% among of a wide range of disease. In India, lung malignant growth establishes 6.9% of all new disease cases and 9.3% of all malignant growth-related passings in both genders, and it is the normal malignancy and reason for malignancy-related mortality in men, with the most elevated events from Mizoram in guys and females (age balanced rate 28.3 and 28.7 per 100,000 populace in guys and females, separately) [6]. Computer-aided diagnosis (CAD) system is nowadays very popular which is taken as second opinion for the various radiologists in the area of lung cancer detection. CAD system is basically the complete package which comprises computer-aided detection (CADe) and computer-aided diagnosis (CADx).

2   Related Work

Nery et al. [7] proposed a method which starts with a 3D Gaussian filter to reduce some noise. The raw lung region is detected using a threshold approach. Lungs are separated by performing sequential erosion. Inferior and superior borders of lungs are obtained, then morphological operation is applied to hold the maximum area of lung, and at last subtraction is done between original image and the border obtained. Degree of similarity achieved is 97.5%. Van Rikxoort et al. [8] proposed a hybrid method. By using automatic 3D algorithm, region growing and morphological operation lung field are segmented and then automatic error detection is performed. At last, multiatlas segmentation is performed. Volumetric overlap = 0.95.

Da Nobrega et al. [9] proposed a method which is detailed in five steps: image addition; location of trachea is located; whole lung segmentation (lungs with respiratory track); trachea is segmented; and then volume is subtracted. This 3D Region Growing (RG) achieved an average result of 98.76%. Yim et al. [10] proposed a method in which at first, by applying an inverse seeded region growing and connected component labeling, the respiratory track and lungs are extracted. Then, trachea and large airways are delineated from the lungs using three-dimensional region growing. At last, to obtain the borders of lung region accurately, the result from the second step is subtracted from the result obtained from first step. When region growing method is compared to the automatic segmentation method, RG has shown better results.

Rebouc et al. [11] proposed a method for segmentation of lung in two dimensions called adaptive crisp active contour model (ACACM) that is able to segment lungs with fibrosis, lungs suffering from emphysema and also healthy lungs. It is possible because the Multilayer Perceptron existing in the calculation of the external energy of the ACACM method that defines the origin of the pulmonary edges can be adjusted. This method works better than 3D region growing. Active crisp contour has shown good results when compared to above-mentioned methods.

The above-mentioned studies have showed the lung segmentation using different image processing techniques with their accuracies mentioned above. These accuracies can be enhanced by using other techniques which are tried in the presented work. The above studies have segmented the lungs in 3D and 2D, and in the presented work we focused on the 3D segmentation by using active contour method. As mentioned above, active crisp contour has shown the better result when compared to other described method. So, here we have tried to apply the active contour method in three dimensions and also achieved a good result.

3 Materials and Method

The objective of lung segmentation here is to segment the lung region from the set of thoracic CT images and show the three-dimensional view of the lung with higher accuracies.

3.1 *Dataset*

To evaluate the proposed method, the data has been taken from LIDC-IDRI dataset [12]. Eight medical imaging and seven academic centers collaborated to create this dataset [13]. Each subject in the dataset contains images of thoracic CT scan with a XML file which contains the results computed by four radiologists. The CT images are in the form of Digital Imaging and Communications in Medicine (DICOM) format which is a standard for storing, transmitting information, handling and printing and in medical imaging. Here, 15 subjects are taken for the segmentation process.

3.2 Proposed Method

A flow process of the presented method is shown in Fig. 1, and the following points describe the flow of the lung segmentation process step by step.

3.2.1 Data Preprocessing

Preprocessing is one of the most crucial processes in lung segmentation process. Initial step of preprocessing is selecting appropriate datasets from LIDC-IDRI database. For this work, CT scan images of 15 patients have been taken which are in DICOM format. All the dimensions of length one are removed. The values of the pixel of the images are transformed to single to normalize the value between [0 and 1] so that its accuracy can be improved. Normalization of value is important because the bigger values will have more impact on the whole data. For normalization, the following equation is used:

$$x_{i_{norm}} = \frac{(x_i - x_{min})}{(x_{max} - x_{min})} \tag{1}$$

where $x_{i_{norm}}$ is the normalized value, x_i is the current value, x_{min} is the minimum value in that column and x_{max} is the maximum value in that column.

3.2.2 Step-by-Step Process

The lung segmentation is implemented by the following steps:

1. To provide input to the active contour mask, two images of the lungs are needed: One is axial view image, and another one is coronal view image. Then, thresholding is applied in both the images. Multilevel image thresholding is used to

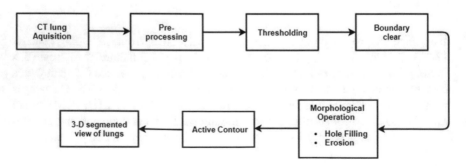

Fig. 1 Flowchart of the lung segmentation process

segment the lungs from the background. It uses Otsu's [14] method for thresholding. This method gives a thresh value to segment lungs. Then, this thresh value is used to binarize image. It computes a threshold for each pixel using the local mean intensity around the neighborhood of the pixel. Then, the image is converted to single image to keep in computable format for further processing.

2. After the first step, the images are complemented to bring the lungs in foreground and remove all the unnecessary background details.
3. Unwanted borders are cleared to lessen the area.
4. Hole-filling algorithm [15] is applied to remove the respiratory ways and other vessels.
5. Morphological operation; erosion [16] is applied to smooth the border of the lungs. It also excludes the touching edges. These above-described steps are applied to both the images of the lungs.
6. Finally to create the 3D mask to give input to the active contour function. Both axial and coronal images are inserted in the proper position. By using this 3D mask, lung is segmented and 3D view of the segmented lung can be seen.

3.3 Otsu Thresholding

Otsu [14] has proposed the automatic method of thresholding. The approximation of threshold procedure is acquired by using straightforward and efficient way to find the gray-level intensity value (t^*) which benefits in maximizing the difference between classes and minimizing the weight within class difference. It has been assumed that histogram is two-mode distribution. In any given computed tomography image, suppose that the pixels are depicted in L gray levels [1, 2..., L], probability distribution (p_k) can be considered for the normalized histogram as:

$$p_k = \frac{n_k}{N}, \; p_k \geq 0, \; \sum_{k=1}^{l} p_k = 1 \tag{2}$$

where the count of pixel having the gray level k is denoted by n_k and $N = n_1 + n_2 + \cdots + n_l$ is the entire pixel count.

$$\sigma_B^2(t) = \frac{[\mu_T s(t) - \mu(t)]^2}{s(t)[1 - s(t)]} \tag{3}$$

Thus, generalized mode of optimal threshold t^* is achieved with the help of histogram and each gray level for which the possibility of the threshold t that maximizes $\sigma_B^2(t)$ has been tested and is given in [17, 18]:

$$\sigma_B^2(t^*) = \max_{1 \leq t \leq l} \sigma_B^2 \tag{4}$$

3.4 *Active Contour*

Active contours also called snakes are the curves that move along the images to find any item's boundary present on the image. Here, active contour without edges method based on Chan–Vese [19] algorithm is applied which is a region-based approach. This method changes the initial curve so that it divides foreground and background into two regions. The method is very strong and gives very good results when there is a difference between the background and foreground regions. The proposed method is a depreciation of an energy-based segmentation. Let us take a image I_o, combination of two zones of having approximately hybrid-constant intensities of definite values I_o^i and I_o^o. I_o^i is the region which is to be detected in the image. I_o^i boundary is denoted by B_o. Now, we have $I_o \approx I_o^i$ inside the object and $I_o \approx I_o^o$; then, the fitting term will be:

$$F_1(E) + F_2(E) = \int_{\text{inside}(E)} |I_0(x, y) - a_1|^2 \mathrm{d}x\mathrm{d}y + \int_{\text{outside}(E)} |I_0(x, y) - a_2|^2 \mathrm{d}x\mathrm{d}y$$

(5)

where E is other variable curve and the constants a_1 and a_2 are the mean of I_o inside E and outside E. The attenuating term minimizer is the boundary of object B_o stated as:

$$\inf_E \{F_1(E) + F_2(E)\} \approx 0 \approx F_1(B_o) + F_2(B_o)$$

(6)

If the contour E is inner side of the object, then $F_1(E) \approx 0$ but $F_2(E) > 0$. If the contour E is outer side of the object, then $F_1(E) > 0$ and $F_2(E) \approx 0$. If the contour S is both inner side and outer side of the object, then $F_1(E) > 0$ and $F_2(E) > 0$. At last, the fitting energy is minimized if $E = B_o$, i.e., if the contour E is on the border of the object. All these above explanations are demonstrated in Fig. 2.

Now, the fitting curve is minimized by the energy function $F(a_1, a_2, E)$ which is defined as:

$$F(a_1, a_2, E) = v.\text{Length}(E) + \mu.\text{Area}(\text{inside}(E))$$

$$+ \omega_1 \int_{\text{inside}(E)} |I_0(x, y) - a_1|^2 \mathrm{d}x\mathrm{d}y + \omega_2 \int_{\text{outside}(E)} |I_0(x, y) - a_2|^2 \mathrm{d}x\mathrm{d}y$$

(7)

$F_1(E)>0, F_2(E)\approx 0$

Fitting >0

$F_1(E) \approx 0, F_2(E)>0$

Fitting >0

$F_1(E)>0, F_2(E)>0$

Fitting >0

$F_1(E) \approx 0, F_2(E) \approx 0$

Fitting ≈ 0

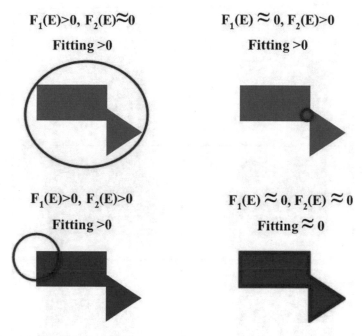

Fig. 2 Curve fitting, Chan–Vese

4 Result and Analysis

The proposed segmented method is performed on CT images taken from the LIDC-IDRI dataset. This method is applied on fifteen subjects of dataset having 3450 CT images. This method is developed in MATLAB with the assistance of volume viewer application which shows the 3D view of segmented area. The performed method is validated through the 3D slicer. Three-dimensional slicer is an open-source accessed for image analysis and visualization. The ground truth is prepared by using this software, and from that volume is calculated and other evaluation measures are also calculated which is described in the next section. Figure 3 shows the sample result of the segmented lungs in 3D by proposed method with the ground truth.

4.1 Evaluation Measures

To evaluate the proposed system, the result of the segmentation has been calculated, based on the described metrics: under-segmentation, over-segmentation, Jaccard index, Dice similarity coefficient, and volume error is also calculated.

The over-segmentation can be described as the area or volumes (voxel count) that are not present in reference standard but are present in the segmented region of

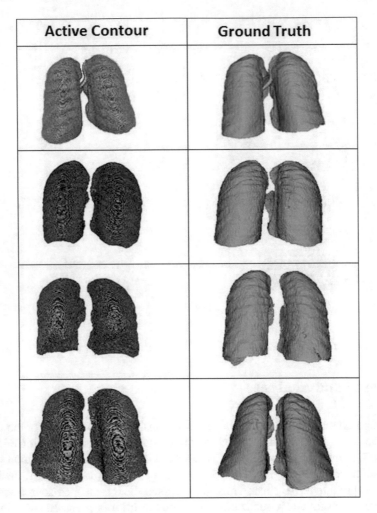

Active Contour	Ground Truth

Fig. 3 Three-dimensional view of lungs

interest. The over-segmentation $O_s(O_p, O_g)$ is calculated as:

$$O_s(O_p, O_g) = \left| \frac{O_p \backslash O_g}{O_g} \right| \quad (8)$$

where $O_p \backslash O_g$ is the relative complement of O_g in O_p, O_p is the number of pixels in the proposed method ROI and O_g is the number of pixels in the ground truth.

The under-segmentation can be described as the area or volumes (voxel count) that are not present in segmented region of interest by presented method but are considered in ground truth. The under-segmentation $U_s(U_p, U_g)$ is calculated as:

$$U_s(U_p, U_g) = \left| \frac{Up \backslash Ug}{U_g} \right| \tag{9}$$

where $O_p \backslash O_g$ is the relative complement of O_p in O_g, O_p is the number of pixels in the proposed method ROI and O_g is the number of pixels in the ground truth.

Dice similarity coefficient (DSC) is another evaluation measure used for observing the segmentation accuracy [20]. To compare the similarity between the reference standard region and the segmented region of interest generated by presented method, Dice similarity is calculated. The DSC is given as the overlap degree of two segments S_p and S_g. The DSC(S_p, S_g) is calculated as:

$$DSC(S_p, S_g) = \frac{2 * (S_p \cap S_g)}{|S_p| + |S_g|} \tag{10}$$

To compare the overlap degree between two segments, Jaccard index JSC(S_p, S_g) is calculated. JSC compares members for two sets to check which members are distinct and which are shared.

$$JSC(S_p, S_g) = \frac{S_p \cap S_g}{|S_p| + |S_g| - |S_p \cup S_g|} \tag{11}$$

Volume error is also calculated by comparing the volumes of both the ground truth and the result of the proposed method. The volume of the 3D lung is calculated by the product of the total number of pixels present in the segmented region and x—spacing, y—spacing, z—spacing. The volume error VE(V_p, V_g) is calculated as:

$$VE(V_p, V_g) = \frac{2 * |V_p - V_g|}{|V_p + V_g|} \tag{12}$$

where V_p is the volume of lungs of proposed method and V_g is the volume of lungs of ground truth.

4.2 Experiment and Results

To calculate the performance of the presented method, it has been experimented on 15 patients' CT images (approximately 3450 CT images) from the database of LIDC-IDRI of National Cancer Imaging Archive. Table 1 shows the result obtained by the proposed work, calculated by the evaluation measures described in Sect. 4.1 when compared to the ground truth prepared by the 3D slicer software. Figure 4 shows the presentation of accuracies in terms of similarity measures which is bar graph of the cumulative probability distribution build on the JSC and DSC for 15 patients' CT scan images.

Table 1 Performance results of the purposed method

S. no.	Over-segmentation	Under-segmentation	Jaccard index	Dice similarity coefficient	Volume error
1	0.029	–	0.971	0.985	0.027
2	–	0.139	0.860	0.925	0.151
3	0.046	–	0.955	0.977	0.045
4	–	0.010	0.989	0.994	0.010
5	0.046	–	0.995	0.997	0.070
6	0.003	–	0.996	0.998	0.289
7	0.034	–	0.967	0.983	0.032
8	0.033	–	0.967	0.983	0.032
9	0.020	–	0.980	0.990	0.190
10	0.001	–	0.998	0.999	0.005
11	0.024	–	0.975	0.987	0.024
12	0.023	–	0.974	0.987	0.024
13	0.041	–	0.959	0.979	0.041
14	–	0.034	0.962	0.980	0.040
15	0.024	–	0.975	0.987	0.024
Mean	**0.024**	**0.061**	**0.967**	**0.983**	**0.069**

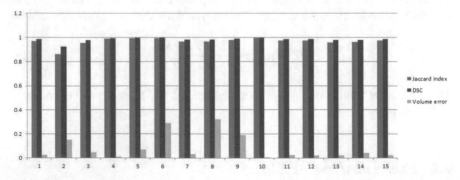

Fig. 4 Bar plots of proposed method of 15 patients: Jaccard index, Dice similarity coefficient, volume error

5 Conclusion and Discussion

The previous work [17] has contributed to the field of lung segmentation using fuzzy clustering and morphological operations and achieved the result of 99.94% of overlap ratio with Jaccard index of 0.9444 and Dice similarity coefficient of 0.9710 obtained in 2D. The proposed work has done the segmentation of lungs in 3D by using another approach with the volumetric analysis.

The efficiency of the presented method is performed on 15 cases of lung thoracic CT scans from LIDC-IDRI datasets that is available publically. The presented method acquired the average under-segmentation of 0.061, over-segmentation of 0.024, average volume error of 0.069, JSC of 0.967 and DSC of 0.983. The experiment is performed on a computer with Windows 8.1 Pro, CPU Intel Core i5-6500, with 4 GB of RAM and a graphic card Intel(R) HD Graphics 530. The presented method can accomplish the necessity of CAD system of lung cancer detection and contribute the precise region of interests for advance processing (nodule detection) toward the diagnosis of lung cancer.

References

1. Ferlay, J., H. Shin, F. Bray, D. Forman, C. Mathers, and D.M. Parkin. 2010. Estimates of worldwide burden of cancer in 2008: GLOBOCAN 2008. *International Journal of Cancer* 127 (12): 2893–2917.
2. Ferlay, J., et al. 2013. *GLOBOCAN 2012 v1. 0, Cancer incidence and mortalityworldwide: IARC Cancer base no. 11 (Online)*. Lyon, France: International Agency for Research on Cancer.
3. Torre, L.A., F. Bray, R.L. Siegel, J. Ferlay, J. Lortet-Tieulent, and A. Jemal. 2012. Global cancer statistics. *CA A Cancer Journal for Clinicians* 65 (2): 87–108.
4. Malvezzi, M., et al. 2015. European cancer mortality predictions for the year 2015: does lung cancer have the highest death rate in EU women. *Annals of Oncology* 26 (4): 779–786.
5. Swensen, S.J., et al. 2002. Screening for lung cancer with low-dose spiral computed tomography. *American Journal of Respiratory and Critical Care Medicine* 165 (4): 508–513.
6. Flehinger, B.J., M.R. Melamed, M.B. Zaman, R.T. Heelan, W.B. Perchick, and N. Martini. 1984. Early lung cancer detection: Results of the initial (prevalence) radiologic and cytologic screening in the Memorial Sloan-Kettering study. *The American Review of Respiratory Disease* 130 (4): 555–560.
7. Nery, F., J.S. Silva, N.C. Ferreira, and F. Caramelo. 2012. 3D automatic lung segmentation in lowdose CT. In *2012 IEEE 2nd Portuguese Meeting in Bioengineering (ENBENG)*, 1–4.
8. Van Rikxoort, E.M., B. de Hoop, M.A. Viergever, M. Prokop, and B. van Ginneken. 2009. Automatic lung segmentation from thoracic computed tomography scans using a hybrid approach with error detection. *Medical Physics* 36 (7): 2934–2947.
9. Da Nobrega, R.V.M., M.B. Rodrigues, and P.P. Reboucas Filho. 2017. Segmentation and visualization of the lungs in three dimensions using 3D Region Growing and Visualization Toolkit in CT examinations of the chest. In *2017 IEEE 30th International Symposium on Computer Based Medical Systems (CBMS)*, 397–402.
10. Yim, Y., H. Hong, and Y.G. Shin. 2005. Hybrid lung segmentation in chest CT images for computer-aided diagnosis. In *Proceedings of 7th International Workshop on Enterprise networking and Computing in Healthcare Industry, HEALTHCOM 2005*, 378–383.
11. Rebouc, P.P., R.M. Sarmento, P.C. Cortez, and V.H.C. De. 2015. Adaptive crisp active contour method for segmentation and reconstruction of 3d lung structures. *International Journal of Computer Applications* 111 (4): 60811–60905.
12. Cha, J. 2018. Segmentation, tracking, and kinematics of lung parenchyma and lung tumors from 4D CT with application to radiation treatment planning. *Electronic Theses and Dissertations. Paper 2938*
13. Clark, K., et al. 2013. The Cancer Imaging Archive (TCIA): maintaining and operating a public information repository. *Journal of Digital Imaging* 26 (6): 1045–1057.
14. Otsu, N. 1979. A threshold selection method from gray-level histograms. *IEEE Transaction on Systems Man and Cybernetics* 9 (1): 62–66.

15. P. Soille. 2013. *Morphological image analysis: principles and applications*, Springer Science & Business Media.
16. Gonzalez, R.C., R.E. Woods, and S.L. Eddins. 2009. *Digital Image Processing Using Matlab*. US: Gatesmark Publishing, LLC.
17. Sahu, S.P., P. Agrawal, N.D. Londhe, and S. Verma. 2017. A new hybrid approach using fuzzy clustering and morphological operations for lung segmentation in thoracic CT images. *Biomedical and Pharmacology Journal* 10 (4): 1949–1961.
18. Sahu, S.P., P. Agrawal, N.D. Londhe, and S. Verma. 2019. Pulmonary nodule detection in CT images using optimal multilevel thresholds and rule-based filtering. *IETE Journal of Research* 1–18.
19. Chan, T.F., and L.A. Vese. 2001. Active contours without edges. *IEEE Transactions on Image Processing* 10 (2): 266–277.
20. Sampat, M.P., Z. Wang, S. Gupta, A.C. Bovik, and M.K. Markey. 2009. Complex wavelet structural similarity: A new image similarity index. *IEEE Transactions on Image Processing* 18 (11): 2385–2401.

DWT-LBP Descriptors for Chest X-Ray View Classification

Rajmohan Pardeshi, Rita Patil, Nirupama Ansingkar, Prapti D. Deshmukh
and Somnath Biradar

Abstract In this paper, we have attempted the problem of chest X ray Image view classification. To do this, we have applied three steps pre-processing, feature computation and classification. Image enhancement based on histogram equalization is carried out in pre-processing. For computing the features hybrid method based on Discrete Wavelet Transform and Uniform Local Binary Pattern is developed and SVM used for classification. We have got encouraging results on NIH dataset with accuracy of 98.00% using 10 fold cross validation.

Keywords Discrete wavelet transform · Local binary patterns · Chest X-ray · View classification · Support vector machines · Cross-validation

1 Introduction

The last decade witnessed tremendous growth of information technology in the medical imaging applications. Easy availability of digital medical imaging devices created mountains of medical imaging data, which is waiting for it processing and analysis. Among the many available modalities Chest x-ray is becoming more popular due to its easy usability and low cost. Automatic Chest X-ray analysis have several applications such as detecting the abnormalities, screening of different diseases such as Pneumonia [1], tuberculosis [2] etc. For screening of disease, inspection of both lateral and frontal views of Chest X-ray is carried out by radiologist. But in automatic system it is first required to identify the view for the further analysis, because

R. Pardeshi
Department of Computer Science, Karnatak Arts, Science and Commerce College, Bidar, Karnataka, India

R. Patil · N. Ansingkar · P. D. Deshmukh
MGM Dr.G.Y.Pathrikar College of Computer Science and IT, Aurangabad (MS), India

S. Biradar (✉)
Department of Electronics, Karnatak Arts, Science and Commerce College,
Bidar, Karnataka, India

© Springer Nature Singapore Pte Ltd. 2020
Y.-C. Hu et al. (eds.), *Ambient Communications and Computer Systems*, Advances in Intelligent Systems and Computing 1097,
https://doi.org/10.1007/978-981-15-1518-7_32

the analysis procedures for the both views are significantly different. Therefore identifying the view of Chest x-ray image is significant pre-cursor to Computer Aided Diagnosis (CAD) system. By motivating from this factor, we have presented an approach, for Chest X-ray Image View Classification.

The remainder of the paper is structured as follows: In Sect. 2 we have presented related work. In Sect. 3 we have briefed our method and Sect. 4 dedicated for experiments and discussion. We concluded in Sect. 5.

2 Related Work

Automatic analysis of chest x-ray images have received the large attention in past years and several algorithms are presented by authors, but very few are focused on view classification problem.

Maximum-Minimum profile length ratio based method is presented in [3] for chest X-Ray Orientation Estimation. They have considered eight directions for estimation of orientation of chest x-ray. In [4] authors developed algorithm for orientation estimation of chest X-ray images using the projection profile based features and neural network. Authors presented a scheme based on template matching for classification of posteroanterior (PA) and lateral views of chest X-rays [5]. Authors in [6] presented a scheme for view classification of chest radiographs, to do this, they have reduced the input image to size of 32×32 and 8×8 pixels and intensity values of reduced images are considered as features. Further, Nearest Neighbor Classifier, Tangent distance and Cross Correlations is considered for classification task. In [7] authors explored the impact of image features size with various distance measures. They have considered the image size of 1×1, 2×2, 4×4, 8×8, 16×16, 32×32, 64×64 for feature computation and K-Nearest Neighbor Classifier was applied with different distance functions for classification. Projection profiles based measures namely body symmetry index and background percentage index is computed from input chest radiographs in [8], these feature fed into linear discriminant analysis classifier for view classification. In [9] authors presented three novel features for classification chest radiographs namely tilt angle of the scapula superior border, the tilt angle of the clavicle and the extent of radiolucence in lung fields which are further fed into linear discriminant analysis for view classification. Authors in [10] developed a method for classification of chest x-ray view, they have employed several features such projection profiles, Body size ratio, Pyramidal Histogram of Oriented Gradients, Contour based shape features and meta classifier with attribute feature selection is applied for the task of classification. In [11] authors presented Force Histogram Descriptors for classification of chest radiograph view with three classifiers namely Support Vector Machines, Random Forest and Multilayer Perceptron.

From the above aforementioned paragraph, we can see that, most of the methods have employed profile based features and local descriptors which are dependent on clean image quality and few of them requires image to be binarized. In this work, we have presented a hybrid method based Discrete Wavelet Transform and Local Binary

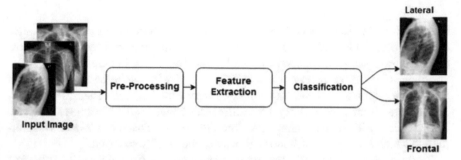

Fig. 1 Schematic diagram of proposed method

Patterns for Feature Extraction for efficient representation of chest radiographs. To exhibit the efficacy of our features we have applied linear support vector machine classifier during the view classification phase.

3 Proposed Method

In this work, our aim is to automatically identify the view of input chest x-ray image. This is one of the important pre-cursor step towards development fully automatic chest x-ray image analysis system for automatic disease screening. To achieve this, we have employed the three steps pre-processing, Feature Computation and classification. During preprocessing we have applied Contrast Limited Adaptive Histogram Equalization (CLAHE) for image enhancement. For feature extraction, we have presented a hybrid scheme by combining Discrete Wavelet Transform (DWT) and Local Binary Patterns (LBP). During Classification stage, we have employed the Support Vector Machine Classifier. Schematic Diagram of our method is shown in Fig. 1.

3.1 Pre-processing:

Pre-processing steps are essential to achieve the better image representation and these steps are different for different applications. In our case, we have to analyze the chest x-ray image which consist of bone structures having dynamic range of intensity values. To achieve the better view, simple image enhancement operation as pre-processing step is considered. To enhance the input chest x-ray image we have employed the basic histogram equalization operation, by which the input grayscale image is enhanced.

3.2 Feature Extraction:

Discrete Wavelet Transform: The concept of multiresolution theory was first introduced by Mallat [14]. Wavelets were consider to be the powerful tool for the signal or image processing and analysis. A scaling function is used in creation of a series approximations of a function or image by factor of 2 in resolution from its nearest neighboring approximations. Its Complementary functions are called as Wavelets. These wavelets are used to encrypt the differences between adjacent approximations. The wavelets with single scaling functions are used by Discrete Wavelet Transformation which serve as an orthonormal basis of the DWT expansion.

Two dimensional scaling function represented as $\varphi(a, b)$, and three 2-D wavelets represented as $\psi^H(a, b)$, $\psi^V(a, b)$, $\psi^D(a, b)$, these are the products of one dimensional scaling and wavelet, which can be represented by:

$$\psi^H(a, b) = \varphi(a)\psi(b) \tag{1}$$

$$\psi^V(a, b) = \psi(a)\varphi(b) \tag{2}$$

$$\psi^D(a, b) = \varphi(a)\varphi(b) \tag{3}$$

These wavelets measure the functional variations—changes in the intensity in images along different directions. Equation (1) measures variations along columns. Equation (2) calculates the variation along rows. Equation (3) corresponds to diagonal variations. 2D DWT can be implemented by using digital filters.

Local Binary Patterns are one the efficient texture descriptors which transforms input image into an image of integer labels describing the small-scale spatial texture of an image [12]. Local binary patterns can be extended with little modifications, while grouping the decimal values in histogram bins. In our process of feature computation we have used uniform local binary patterns, which are summarized as: Input image is converted into binary image, then 3×3 window is considered around each pixel as neighborhood. If input binary string has more than 2 transitions, then it considered as non-uniform code otherwise uniform code. In this way only uniform codes without considering rotation information are considered as feature.

DWT-LBP Descriptors: First Input Image is divided into two parts vertically, then each part of Image is decomposed using Discrete Wavelet Transform and detail sub bands representing Vertical, Horizontal and Diagonal energies. In this way, we have got total 6 sub bands, from each sub band then we have computer Uniform Local Binary Pattern of 10 dimension and hence obtained $6 \times 10 = 60$ dimension feature vector. For better understanding, the process is presented in diagrammatic form in Fig. 2.

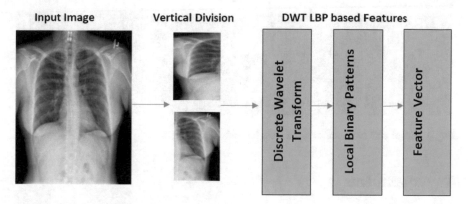

Fig. 2 Schematic diagram of proposed method

3.3 Classification:

Support Vector Machine is one of the popular algorithm in machine learning for the task of classification and regression. It belongs to family of supervised learning algorithms works on statistical learning theory, proposed by Vapnick [13]. The main idea of SVM is it tried to find linear decision surface by transforming the data into hyperplane, where separation between two classes can be achieved efficiently.

Given training data $\vec{X_1}, \vec{X_2}, \ldots, \vec{X_n} \in R_n$, feature vectors of input chest x-ray images, with labels $y_1, y_2, \ldots, y_n \in +1, -1$, where $+1$ is frontal view and -1 is lateral view of chest x-ray image. The linear classifier tries to separate instances by maximizing the margin between two linearly separable classes by discriminant function $\vec{w} . \vec{x} + b = 0$ in addition we have imposed constraints so that all instances are correctly classified. In our case:

$$\vec{w} . \vec{x} + b \leq -1 \, if \, y_i = -1 (lateral \ view) \tag{4}$$

$$\vec{w} . \vec{x} + b \geq +1 \, if \, y_i = +1 (frontal \ view) \tag{5}$$

Further, we have extended linear SVM by applying the Cubic Kernel function in our classification problem.

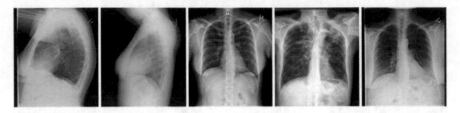

Fig. 3 Samples of chest radiographs from NLM-NIH CXR dataset

4 Experiments

4.1 Dataset and Evaluation Protocol:

To evaluate our proposed method, we have used the publically available Chest X ray data set from National Institute of Health, USA. Total 7470 image which comprises 3821 Frontal View and 3649 Lateral View are considered in our experiments. Some samples from the dataset are shown in Fig. 3.

To estimate the genrilized error of our method, we have applied 10 fold cross validation technique instead of traditional classification training-testing procedure. At the beginning, the complete data set is divided into 10 subparts. Each part given the opportunity to serve as training and testing, when one serves as test set other subparts serves as training set. This method is repeated 10 times, in such way that each sub part will serve for both training as well as for testing. The classification accuracy can be given by:

$$Accuracy = \frac{Total\ Correctly\ Classified\ Chest\ Radiographs}{Total\ Radiographs} \tag{6}$$

4.2 Results and Discussion

Results: Using tenfold cross validation with Cubic Support Vector Machine we have evaluated the efficiency of DWT LBP descriptors for the task of chest x-ray view classification. We have presented the results for chest x-ray view classification in Table 1. From the Table 1 it can be noted that, we have received 98% accuracy for both the classes i.e. Frontal and Lateral views. 2% error rate is noted, which can be further improved. For in depth analysis, in Fig. 4 we have presented. Receiver Operating Curve (ROC) for the chest- x-ray view classification. The AUC values justifies the better performance of our method. We have also compared our work with recently reported previous work in Table 2. Xue et al. [10] presented an approach for view classification of chest x-ray images based on combinations of several features, with SVM classifier, but it requires large dimension of features, we have achieved

Table 1 Accuracy in % for classification of chest X-ray view based on DWT LBP descriptors

Chest X-Ray view	Accuracy (%)
Frontal	98
Lateral	98
Average	98

Fig. 4 ROC Curve for Chest X-ray view classification using DWT-LBP descriptors

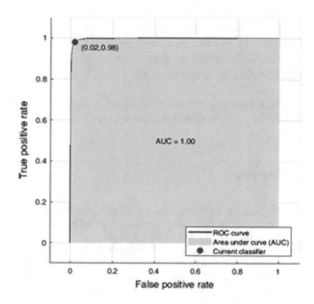

Table 2 Comparison with recent previous work

Method Proposed by	Feature Extraction Technique	Feature Dimension	Accuracy in %
Xue et al. [10]	Image profile + Contour-based shape feature + Pyramid of histograms of orientation gradients	1276	99.90
Santosh et al. [11]	Angular relational signature	32	99.86
Proposed method	DWT LBP descriptors	60	98.00

slightly low accuracy with only 5% of its feature dimension. Santosh et al. [11] presented angular relational signature based feature descriptor and evaluated with SVM, MLP and RF classifiers. Their method is effective and simple but requires the binarization, which is quite difficult when considering low resolutions and noisy images. Our method is based on multiresolution analysis therefore it is immune to noise and also does not requires much preprocessing.

5 Conclusion

Identifying the correct view of input chest x-ray is very essential step for fully automatic Computer Aided Diagnosis system. In this paper, we have presented hybrid feature descriptors based on Discrete Wavelet Transform and Local Binary Patterns. Support Vector machine with cubic kernel is employed for the classification task. The efficacy of our method is evaluated with large publically available dataset of 7470 images of chest x-ray from U.S. National Library of Medicine (NLM), National Institutes of Health (NIH) and achieved the 98% accuracy. In future, we planned to evaluate the different deep learning algorithms for the task of Chest-x ray view classification by considering the pediatric chest x-rays.

Acknowledgements We thank toNational Library of Medicine (NLM), National Institutes of Health (NIH) for providing the Chest Radiogrpah Dataset.

References

1. Stephen, Okeke, Mangal Sain, Uchenna Joseph Maduh, and Do-Un Jeong. 2019. An efficient deep learning approach to pneumonia classification in healthcare. *Journal of Healthcare Engineering*, Article ID 4180949, 7 pages.
2. Jaeger, Stefan, Alexandros Karargyris, Sema Candemir, Les Folio, Jenifer Siegelman, Fiona Callaghan, Zhiyun Xue, et al. 2013. Automatic tuberculosis screening using chest radiographs. *IEEE Transactions on Medical Imaging* 33 (2): 233–245.
3. Pieka, E., and H.K. Huang. 1992. Orientation correction for chest images. *Journal of Digital Imaging* 5 (3): 185–189.
4. Boone, John M., Sadananda Seshagiri, and Robert M. Steiner. 1992. Recognition of chest radiograph orientation for picture archiving and communications systems display using neural networks. *Journal of Digital Imaging* 5 (3): 190.
5. Arimura, Hidetaka, Shigehiko Katsuragawa, Qiang Li, Takayuki Ishida, and Kunio Doi. 2002. Development of a computerized method for identifying the posteroanterior and lateral views of chest radiographs by use of a template matching technique. *Medical Physics* 29 (7): 1556–1561.
6. Lehmann, Thomas M., O. Güld, Daniel Keysers, Henning Schubert, Michael Kohnen, and Berthold B. Wein. 2003. Determining the view of chest radiographs. *Journal of Digital Imaging* 16 (3): 280–291.
7. Lehmann, Thomas Martin, Mark Oliver Gueld, Daniel Keysers, Henning Schubert, Andrea Wenning, and Berthold B. Wein. 2003. Automatic detection of the view position of chest radiographs. In *Medical imaging 2003: image processing*, vol. 5032, 1275–1283. International Society for Optics and Photonics.
8. Kao, E-Fong, Chungnan Lee, Twei-Shiun Jaw, Jui-Sheng Hsu, and Gin-Chung Liu. 2006. Projection profile analysis for identifying different views of chest radiographs. *Academic Radiology* 13 (4): 518–525.
9. Kao, E-Fong, Wei-Chen Lin, Jui-Sheng Hsu, Ming-Chung Chou, Twei-Shiun Jaw, and Gin-Chung Liu. 2011. A computerized method for automated identification of erect posteroanterior and supine anteroposterior chest radiographs. *Physics in Medicine & Biology* 56 (24): 7737.
10. Xue, Z., D. You, S. Candemir et al. 2015. Chest x-ray image view classification. In *Proceedings of the computer-based medical systems IEEE 28th international symposium*, São Paulo, Brazil, June 2015.
11. Santosh, K.C., and Laurent Wendling. 2018. Angular relational signature-based chest radiograph image view classification. *Medical & Biological Engineering & Computing*, 1–12.

12. Ojala, T., M. Pietikäinen, and T. Mäenpää. 2002. Multiresolution gray-scale androtation invariant texture classification with local binarypatterns. *IEEE Transactions on Pattern Analysis & Machine Intelligence* 7: 971–987.
13. Vapnik, Vladimir. 2013. *The nature of statistical learning theory*. Springer Science & Business Media.
14. Mallat, Stephane G. 1989. A theory for multiresolution signal decomposition: the wavelet representation. *IEEE Transactions on Pattern Analysis & Machine Intelligence* 7: 674–693.

Comparative Study of Latent Fingerprint Image Segmentation Techniques Based on Literature Review

Neha Chaudhary, Harivans Pratap Singh and Priti Dimri

Abstract Latent fingerprints are the fingerprints that are left by the criminals unintentionally on items touched by the fingers. These types of fingerprints are not often directly visible by naked eyes. Segmentation is a very important part of the fingerprint identification system (AFIS). The fingerprint segmentation algorithms separate the foreground (friction ridge pattern) from background. In this paper different segmentation algorithms are presented that are DTV, ADTV, ATV, Ridge Template Correlation method, Segmentation based on statistical characteristics of gray and orientation field information theory, Adaptive Latent Fingerprint Segmentation using Feature Selection and Random Decision Forest Classification, Latent Fingerprint Image Segmentation using Fractal Dimension Features and WELME are discussed and compared their performance. This study evaluates the effectiveness, advantages, limitations and applications of various segmentation methods that are being used in latent fingerprinting segmentation techniques.

1 Introduction

Fingerprints are generally used for identification purposes due to their easier accessibility, consistency, and Uniqueness. The main three categories of the fingerprint image are a plain fingerprint, rolled fingerprint and the latent fingerprint. Plain fingerprints are the fingerprint that is captured by pressing the finger on a flat surface. Rolled fingerprints are the fingerprint that is capture by rolling a finger one side to another. Latent fingerprints are the fingerprint that is left by the criminals unintentionally on items touched by the fingers. These types of fingerprints not often directly visible by naked eyes but it can be identified by some special technique like dusting with fine powder and then collected that pattern by transparent tape (Fig. 1).

Segmentation is the main step of the automatic fingerprint recognition system. Segmentation [1] decomposes the fingerprint image into two parts foreground and background. The foreground is the component in which we are interested, or we can say it is the part of the fingerprint. The noisy data of the fingerprint images are called

N. Chaudhary (✉) · H. P. Singh · P. Dimri
Uttarakhand Technical University, Dehradun, India

© Springer Nature Singapore Pte Ltd. 2020
Y.-C. Hu et al. (eds.), *Ambient Communications and Computer Systems*, Advances in Intelligent Systems and Computing 1097,
https://doi.org/10.1007/978-981-15-1518-7_33

391

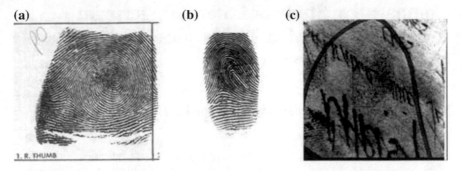

Fig. 1 Types of fingerprint images [3]. **a** Rolled fingerprint **b** plain fingerprint **c** latent fingerprint

background that will be discarded. Accurate segmentation is very important for feature extraction like singular point or minutiae. When feature extracted without using a segmentation algorithm then it detects more false feature and also it takes a long time to process the full image. The segmentation algorithm is used to separate the fingerprint image into the foreground and background, minimizing the false feature detection and increase the matching accuracy. In our work, we have discussed different segmentation algorithms. There are several fingerprint segmentation approaches are known from the literature. Which are divided into two parts (1) pixels-wise segmentation (2) block-wise segmentation. In the pixel-wise method analyzing the pixel feature for classifying the pixels. The most used features are frequency domain, orientation features, and greyscale and so on. In the block-wise fingerprint, the method divides the fingerprint image into the block in a not overlapping manner and then categories each block in the background and foreground according to block feature. Basically two approaches are used for fingerprint segmentation [2–5], frequency field estimation [6, 7] and orientation field estimation [5, 7–10], but these approaches not work effetely on latent fingerprint because structured noise is presented in latent fingerprint. This paper evaluates the effectiveness, advantages, limitations and applications of various segmentation methods that are being used in latent fingerprinting segmentation techniques.

2 Existing Segmentation Algorithm

In literature several approaches are available for fingerprint segmentation such as, Adaptive Total Variation (ATV) Model [11], Directional Total Variation(DTV) Model [12], Adaptive Directional Total-Variation(ADTV) [13] Model, Ridge Template Correlation method [14], Segmentation based on statistical characteristics of gray and orientation field information theory [15], Latent fingerprint image segmentation using fractal dimension features and weighted extreme learning machine ensemble [16], Adaptive latent fingerprint segmentation using feature selection and random decision forest classification [17].

2.1 Adaptive Total Variation Model

The total variant model [2], with the constant fidelity not relevant for the latent fingerprint so considers the TV model with varying fidelity. In the Adaptive total variation (ATV) model [11] the weight coefficient is adjusted adaptively with L1 fidelity that depends on the background noise level. This can be estimated by texture analysis. The total variation model with fidelity L1 represented by TV-L1.The total variation model is appropriate for image decomposition and selection of features of the multiscale image. This model decomposes the input image (y) into two parts texture (s) and cartoon (t). Cartoon(t)—It contains piece-wise smooth components in image y. Texture(s)—It contains texture components in image y.

The decomposition:

$$y = s + t; \tag{1}$$

is obtained by solving the following variation problem:

$$\min_t \int |\nabla t| + \int \lambda(x)|t - y|\mathrm{d}x \tag{2}$$

∇t is the gradient value of t. s, t and y image functions of gray-scale intensity values in R2.$|t - y|$ and $\int |\nabla t|$ are the fidelity item and total variation of t, respectively and $\lambda(x)$ is the spatially varying parameter. Basically in the TV-L1 model the value of λ used to select the features. If the values of λ is small then s consist most fine structure and t consist of the different noises in the background. If the size of λ increases then small noises and fingerprint appear in s and large part of image y appears in t.

Figure 2 shows that when the value of $\lambda = 0.10$ very small the most of the structure appears in s and the enlightenment appears in the t. When the value of $\lambda = 0.20$ increased then the arch appears in t and small structure appears in s. When $\lambda = 0.30$ most of the fine structure appears in t and the noise appears in s.

The adaptive model gives a good result of latent fingerprints but for ugly latent fingerprints, it does not provide a satisfactory result.

2.2 Directional Total Variation Model

The DTV model [12] provide batter result as compared to the ATV model by including special-dependent texture orientation in total variance computation. DTV model based on the TV-L2 model (L2 fidelity). TV-L2 was designed for image segmentation it separates the noise from the image. TV-L2 model is similar to the TV-L1 model but it includes special-dependent texture orientation in total variance computation and provides batter segmentation result as compared to the TV-L1 model. It breaks the input image(y) into two parts texture(s) and cartoon (t) as similar to TV-L1. The

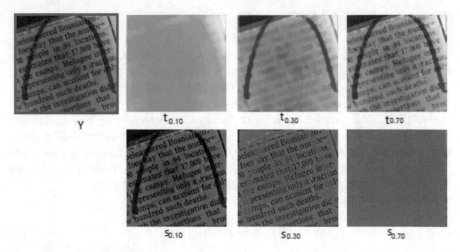

Fig. 2 Selection of image using TV-L1 model of latent fingerprint, y is input image it decomposes the image in two component s and t and value of λ shows in subscript [11]

total variant model with L2 fidelity can be defined with spatially variant DTV:

$$\min_t \int \left| \nabla t . \vec{a}(x) \right| dx + \frac{\lambda}{2} y - t^2 \qquad (3)$$

Spatially varying orientation vector is $\vec{a}(x)$ that adjusted to local texture orientation. According to the direction of $\vec{a}$ decides which signal captured by the texture outputs. Figure 3 shows four different directions of $\vec{a}$. At the particular direction of $\vec{a}$ we are concerned that in this direction minimizing the total variation of t and in other direction allowing the existence of total variation of task a result the texture that have orthogonal direction of $\vec{a}$ completely ignored from s and if the direction of $\vec{a}$ is matching with texture in that condition it will be fully captured by s and in some other direction it is weakened from s.

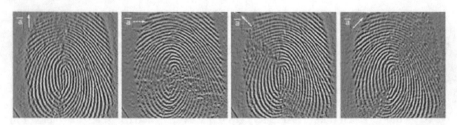

Fig. 3 Texture output s for $\vec{a}$ in four different from left to right [12]

2.3 Adaptive Directional Total Variation

The ADTV model [13] is the combination of ATV and the DTV model. This model is used for image segmentation. The ADTV model used classical TV-L1, orientation and scale two unique features of the fingerprint. ADTV decompose the input image into two parts foreground and background. The foreground contains the fingerprint and the background contains the structured noise. The ADTV model uses orientation and scales two varying parameters. These two parameters are chosen according to the textural orientation and background noise.

2.4 Segmentation Based on Statistical Characteristics of Gray and Orientation Field Information Theory

This segmentation method is the combination of statistical characteristics of gray and orientation field information theory [15]. This method uses the main features of the statistical characteristics of gray (variance and mean). For different types of foreground, it has a different relationship between variance and mean value. The statistical characteristics of gray have a small variance for the blank area in a foreground and background area, a larger variance for foreground and unwanted region in backgrounds. After it, the algorithm uses the orientation field information for segmentation. In the orientation field method calculates the point and block orientation. This method of segmentation is very useful but not work effetely where the images arc too dry and wet.

2.5 Ridge Template Correlation Method

Ridge Template correlation [14] is the novel approach for the segmentation of the latent fingerprint. It models an idle ridge selection with a sinusoid that is adjusted to frequency, direction, and local contrast. In this method, the image is divided into the number of blocks and then each block compares to the generated templates, and check the similarity between the block and ridge templates and find the goodness of fit score that score is further used to allocate the quality level to block. In this method, we can easily segment the fingerprint image if the block shows the negative correlation with ridge templates that means it is not the part of the fingerprint otherwise it is the part of the fingerprint.

Figure 4 shows the workflow of the ridge templates correlation method. In this method, normalization is used to find the coarse ROI containing ridge area and ridge templates are used for determining the goodness of fit scores. The threshold value is chosen according to the quality of latent fingerprint images so that it can differentiate foreground and background.

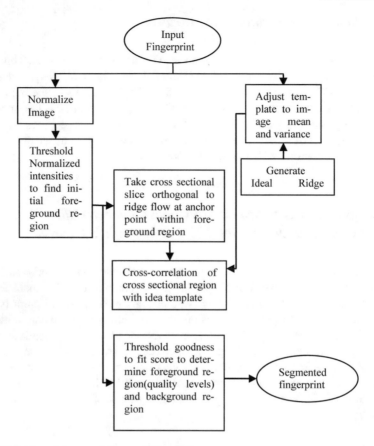

Fig. 4 Ridge templates correlation workflow [14]

2.6 Adaptive Latent Fingerprint Segmentation Using Feature Selection and Random Decision Forest Classification

The fingerprint is the sequence of the valley and ridges. This method [17] focuses on automatic latent fingerprint segmentation to differentiate non-ridge and ridge pattern. In this research author take advantage of three more researches: (1) For automatic latent fingerprint segmentation, a machine learning algorithm used that combines saliency, gradient, ridge, intensity and quality features. (2) A modified RELIF formulation to find the optimal features of a fingerprint. (3) Spectral Image Validation and Verification based metric for finding the effect of the segmentation algorithm. In this method, the latent fingerprint image is divided into the local block then extracts the features of each local block by using five different feature extraction methods. Modified RELIF formulation finds the optimal features and a random decision forest classifier algorithm classify each local block into the foreground and background region.

2.7 Latent Fingerprint Image Segmentation Using Fractal Dimension Features and Weighted Extreme Learning Machine Ensemble (WELME)

This method used fractal dimension features and WELME for latent fingerprint segmentation. In this paper, the authors extract the patches from the latent fingerprint image and construct a feature vector from the fractal dimension feature. Feature vector works as an input of WELME that classifies the patches into non-fingerprint and fingerprint classes. Patch Assembler used to assemble all the fingerprint patches together, considered as foreground and non-fingerprint patches together, considered as background.

3 Comparison of Different Algorithms

Different segmentation algorithms for latent fingerprint are discussed in Table 1.

4 Conclusion and Future Work

Human intervention is required to achieve high accuracy in fingerprints segmentation and matching using AFIS, as AFIS posses some problems. Comparative evaluation of selected techniques suggests that DTV Model [12], ATV Model [11] and ALFS using Feature Selection and Random Decision Forest Classification produces better results for poor quality latent fingerprints. Feature work involves additional Literature available on latent fingerprint segmentation techniques shall also be studied, analyzed & compared with the techniques which are studies here in order to identify the best suitability of the segmentation techniques which can be used based on the sample of latent fingerprint to generate best results.

Table 1 Comparison of the latent fingerprint segmentation algorithm

References	Measuring parameters	Complexity	Advantages	Limitation	Applications
[11]	Features (variance, mean and coherence) and $\lambda =$ Fidelity weight coefficient	For different fingerprint image the selection of λ value	Good for latent fingerprint segmentation	For worse latent fingerprint not gives an exact result	Decomposition and segmentation of fingerprint image
[12]	Spatially varying orientation vector $\vec{a}(x)$	Aligning the Spatially varying orientation vector $\vec{a}(x)$ and local fingerprint ridge orientation	Provide good results as compared to Adaptive total variance model	Not segment multiple overlapped fingerprint	Image decomposition with orientation texture
[13]	Spatially varying Parameter $\lambda(x)$ and spatially varying orientation vector $\vec{a}(x)$	Selection of λ value and aligning the $\vec{a}(x)$ with local fingerprint ridge orientation	Identify regions with coherent orientations along one single direction	Incapable of handling regions with overlapped fingerprints	Segmentation and enhancement of Fingerprint image
[14]	Ridge templates	Time consuming and lengthy algorithm	It reduces the area of detected fingerprint	When detection of false minutiae is high it gives wrong segmentation result	Segmentation of latent fingerprint
[15]	Variance and mean value	Selection of threshold value	Reduce processing time and improve segmentation accuracy	Not appropriate for too wet and too dry fingerprint image	Segmentation of latent fingerprint
[16]	Patches classification and Feature Extraction	Minimize FDR, selection of optimal weight vector	Improvement in FDR and accuracy of segmentation	Low classification accuracy of patches from ugly latent fingerprint	Latent fingerprint segmentation
[17]	Classification, feature extraction	Feature extraction, false block detection	High performance on NSIT SD-27,SD-4 Database	Low matching accuracy	Automatic latent fingerprint Segmentation

References

1. Mehtre, B.M., N.N. Murthy, S. Kapoor, and B. Chatterjee. 1987. Segmentation of fingerprint images using the directional image. *Pattern Recognition* 20 (4): 429–435.
2. Chen, T., W. Yin, X.S. Zhou, D. Comaniciu, and T.S. Huang. 2006. Total variation models for variable lighting face recognition. *IEEE Transactions on Pattern Analysis and Machine Intelligence* 28 (9): 1519–1524.
3. Cao, K., E. Liu, and A.K. Jain. 2014. Segmentation and enhancement of latent fingerprints: A coarse to fine ridge structure dictionary. *IEEE Transactions on Pattern Analysis and Machine Intelligence* 36 (9): 1847–1859.
4. Chen, X., J. Tian, J. Cheng, and X. Yang. 2004. Segmentation of fingerprint images using linear classifier. *EURASIP Journal on Advances in Signal Processing* 2004 (4): 978695.
5. Zhu, E., J. Yin, C. Hu, and G. Zhang. 2006. A systematic method for fingerprint ridge orientation estimation and image segmentation. *Pattern Recognition* 39 (8): 1452–1472.
6. Jiang, X. 2000. Fingerprint image ridge frequency estimation by higher order spectrum. In *Proceedings 2000 International Conference on Image Processing (Cat. No. 00CH37101)* vol. 1, 462–465. IEEE.
7. Chikkerur, S., A.N. Cartwright, and V. Govindaraju. 2007. Fingerprint enhancement using stft analysis. *Pattern Recognition* 40 (1): 198–211.
8. Hong, L., Y. Wan, and A. Jain. 1998. Fingerprint image enhancement: Algorithm and performance evaluation. *IEEE Transactions on Pattern Analysis and Machine Intelligence* 20 (8): 777–789.
9. Bazen, A.M., and S.H. Gerez. 2002. Systematic methods for the computation of the directional fields and singular points of fingerprints. *IEEE Transactions on Pattern Analysis and Machine Intelligence* 24 (7): 905–919.
10. Wang, Y., J. Hu, and D. Phillips. 2007. A fingerprint orientation model based on 2D Fourier expansion (FOMFE) and its application to singular-point detection and fingerprint indexing. *IEEE Transactions on Pattern Analysis and Machine Intelligence* 29 (4): 573–585.
11. Zhang, J., R. Lai, and C.C.J. Kuo. 2012. Latent fingerprint segmentation with adaptive total variation model. In *2012 5th IAPR International Conference on Biometrics (ICB)*, 189–195. IEEE.
12. Zhang, J., R. Lai, and C.C.J. Kuo, 2012. Latent fingerprint detection and segmentation with a directional total variation model. In *2012 19th IEEE International Conference on Image Processing*, 1145–1148. IEEE.
13. Zhang, J., R. Lai, and C.C.J. Kuo. 2013. Adaptive directional total-variation model for latent fingerprint segmentation. *IEEE Transactions on Information Forensics and Security* 8 (8): 1261–1273.
14. Short, N.J., M.S. Hsiao, A.L. Abbott, and E.A, Fox. 2011. *Latent Fingerprint Segmentation Using Ridge Template Correlation*.
15. Xue, J., and H. Li. 2012. Fingerprint image segmentation based on a combined method. In *Proceedings 2012 IEEE International Conference on Virtual Environments Human-Computer Interfaces and Measurement Systems (VECIMS)*, 207–208. IEEE.
16. Ezeobiejesi, J., and B. Bhanu. 2016. Latent fingerprint image segmentation using fractal dimension features and weighted extreme learning machine ensemble. In *Proceedings of the IEEE Conference on Computer Vision and Pattern Recognition Workshops*, 146–154.
17. Sankaran, A., A. Jain, T. Vashisth, M. Vatsa, and R. Singh. 2017. Adaptive latent fingerprint segmentation using feature selection and random decision forest classification. *Information Fusion* 34: 1–15.

Hybrid Domain Feature-Based Image Super-resolution Using Fusion of APVT and DWT

Prathibha Kiran and Fathima Jabeen

Abstract Digital image-processing technique is used in image super-resolution for a variety of applications. In this paper, we propose hybrid domain feature-based image super-resolution using a fusion of Average Pixel Values Technique (APVT) and Discrete Wavelet Transform (DWT). The low-resolution (LR) images are considered and converted into high-resolution (HR) images using a novel technique of APTV by inserting an average of rows between rows and average of columns between columns to get HR images. The DWT is applied on HR images to obtain four bands. The HR images are downsampled, and to enhance image quality, histogram equalization (HE) is utilized. The LL band and HE matrix are added to obtain new LL band. The inverse DWT is applied on four bands to derive SR image. It is observed that the performance of the proposed method is better than existing methods.

Keywords Image processing · High resolution · Histogram equalization · DWT

1 Introduction

Nowadays, there is tremendous growth in digital image processing discipline and has brought up the demand of high-resolution (HR) images for proper analysis in different domains. The image is processed and displayed on computers. The digital image processing is motivated by the two major facts that the improvisation of pictorial information for human perception and further processing for efficient storage, transmission, and autonomous application. Typical applications include robotics, industrial inspection, remote sensing, image transmission, medical imaging, and surveillance. The high-resolution image contains major information and aids in early detection of diseases or abnormalities in medical images.

Image resolution is asserted in different ways specifically such as (i) spatial resolution is typically described in pixels per inch (PPI), and it refers to how many pixels

P. Kiran (✉)
VTU-RRC, Bangalore, India

F. Jabeen
Islamiah Institute of Technology, Bangalore, India

© Springer Nature Singapore Pte Ltd. 2020
Y.-C. Hu et al. (eds.), *Ambient Communications and Computer Systems*, Advances in
Intelligent Systems and Computing 1097,
https://doi.org/10.1007/978-981-15-1518-7_34

are displayed per inch of image. The greater number of pixels in the image results in more pixel information referred as high spatial resolution and correspondingly better quality. (ii) Brightness resolution is defined as the degree to which how fully a pixel in a digital image represents the brightness of corresponding points in the original image. In example, 8 bit depth image can be represented by 256 grayscale levels. (iii) Temporal resolution describes the number of times an object is sampled or how often data are obtained from the same area.

The three basic components of SR image reconstruction methods are as follows:

(i)   Firstly, motion compensation method tries to acquire data in all motion states and uses additional information during image reconstruction to transform data to reference motion state and modeled by the vectors which represents the estimated motion.
(ii) Interpolation refers to estimate unknown values from known values and tries to achieve a best approximation of pixel intensity based on values at surrounding pixel.
(iii) To eliminate the sensor and optical blurring and also noise introduced by them.

Development of the image super-resolution algorithms undergone three important phases—the interpolation, the super-resolution reconstruction, and learning-based super-resolution methods. The primary objective of the proposed research work is to develop the intelligent algorithms for the SR of medical images suitable for FPGA implementation. The proposed research work has the following major objectives:

- Classification and assessment of SR algorithms.
- Discuss the merits and demerits of the analyzed methods.
- Develop efficient approaches to highlight the important visual details in medical images using latest SR schemes such as fuzzy logic, artificial intelligence, wavelet transforms, and nature-inspired intelligent techniques.
- Simulate the algorithm using MATLAB software.
- Exploit a diverse set of examples of medical images under various conditions.
- Develop the architectures for the algorithm suitable for FPGA implementation.
- Simulate and synthesis the methods by realizing them in hardware description languages.
- Compare the MATLAB and hardware approaches with the other existing methods and arrive at conclusions.

These objectives provide the researcher with necessary practical backgrounds for medical image SR techniques and allow them to choose right SR scheme.

Contributions: In this paper, hybrid domain feature-based image super-resolution using fusion of APVT and DWT is proposed. The LR images are converted into HR images by inserting average values of rows and columns between rows and columns. The HR images are further enhanced in quality by HE technique. The DWT is tested on HR images, and LL band is added to HE image to obtain new LL band. The IDWT is used on new LL along with LH, HL, and HH of HR image to obtain SR image.

Organization: The details of organization of the paper are as follows. Section 1 includes introduction and objectives of proposed research work. Section 2 discusses the literature survey for SR enhancement technique, motivation, and conclusive summary. The proposed flowchart model is demonstrated in Sect. 3. Section 4 discusses the tabulation of performance parameters of the proposed model and comparisons with existing approaches. Finally, the conclusions are discussed in Sect. 5.

2 Literature Survey

In this section, the reviews of SR enhancement techniques motivation for the proposed work and conclusive summary are discussed. Medical image SR has been a very active field of research due to its advantages in the health sector, specifically for the doctors in order to make correct decisions during image analysis.

Subhasis Chaudhary has provided the elaborated description and advanced image SR methods. Sung et al. presented the technical overview of SR reconstruction schemes such as estimation techniques, interpolation methods, and de-blurring process.

Trinh et al. [1] proposed an example-based method for de-noising and super-resolution of medical images. In this scheme, HR version of the image is evaluated by sparse positive linear representation, and the corresponding coefficients are assessed based on the similarity between input image and the standard images in database. The databases are collected for the same modality to ensure good quality and utilized for the reconstruction of HR image. Final output construction of HR image is depicted as non-negative sparse linear combination of reference HR images in the database. Further, neighbor-embedding SR method which is employed assumes same value patches of HR and LR images and forms same contour values for different feature spaces. To minimize the error, it computes the reconstruction weights of each low-resolution patches neighbor in a LR training image. By preserving the local features, the HR image is estimated from the training image.

Tomasz et al. [2] proposed an efficient FPGA implementation of a high-quality, super-resolution algorithm with real-time performance. The Algorithm produces satisfactory quality output based on the standard software using FPGA. The interpolation algorithm developed in this scheme is implemented in an FPGA device which provides superior output quality which is non-iterative in nature. In the proposed approach, utilization of hardware description languages such as VHDL/Verilog with optimization schemes further improves the performance.

Gungor et al. [3] developed image super-resolution based on Bayesian non-parametric approach. The method uses Gibbs sampling and Online Variational Bayes algorithms. In comparison with various models from the literature research, the results were tested on both benchmarks and natural images. Gibbs sampling approach was employed to achieve high visual quality, but it was impervious for large-scale data. Hence an advanced approach known as an online Variation Bayes (VB) algorithm was developed which discovers high-quality dictionaries in shorter duration. Thus to improve super-resolution, advanced techniques like Gibbs sampling, Bayesian nonparametric factor analysis model, and online variation inference alternate algorithms were developed.

Dai-Viet et al. [4] presented an approach for enhancing spatial resolution of medical images embedded with noise referred as an example-based super-resolution. In this method with the aid of sparsity of patches and already existing databases constructed from example images, HR image is recovered from a low-resolution image. Thus, the problem of non-negative sparse representation is solved. The basic concept of Earth Movers Distance (EMD) was instigated to estimate the resemblance between the image patches. It was used in optimization problem which only selects the most homogeneity candidates. It is observed from the experimental analysis that the proposed approach achieves high-efficient LR image with the presence of noise. To select the image patch 'L', EMD-based distance was developed which can be employed only for normal distributions and highly time consuming. Thus to estimate the closeness between two patches, a fast EMD method was developed as a standard method. Jingxu et al. [5] initiated a deep learning method and a gradient transformation approach for the super-resolution method. CNN method was employed, which uses gradient transformation networks to instruct and transform the gradients. Final results demonstrate that the proposed approach obtains high restoration quality of reconstructed HR with few artifacts. An effective method to train the network known as mean squared error (MSE) is employed, and back propagation algorithm is implemented to minimize the loss function. High-resolution image is estimated by optimizing the energy function. It is noticed that the modified gradients possess higher PSNR and Structure Similarity Index (SSIM) scores in contrast with unchanged gradients. Thus, the proposed mechanism generates transparent HR image with sharp edges and attains the exorbitant scores in evaluation measures.

Haijun et al. [6] recommended a sophisticated mechanism known as Gaussian Process Regression (GPR) which aids in learning the distribution of variables from an observed space to hidden space where similar data points are closer together. When GPR is applied to the example, learning-based SR, the time complexity of GP is high for large datasets, and it was not suitable for SR reconstruction. Therefore, all adjacent samples were merged into dense training subset to achieve reduction in computational complexity. Student-t likelihood approach proves statistically more compatible in contrast to Gaussian likelihood. A most effective method known as Student-t likelihood is employed, which uses dictionary atoms to choose a subset from on the original set of GPR. Junjun et al. [7] suggested a new technique known as Regularized Anchored Neighborhood Regression. In this scheme, it generates spellbound HR image from a given LR input. In this sophisticated method, local constraint values are used to select dictionary atoms. Based on the correlation of input LR image, the local constraint values assign distinct freedom to each atom. Due to this, the proposed model negotiated the problem of mapping functions between LR and HR images and leads to enrichment of texture in the output images. Further study in the behavior of projection of dictionary atoms LANR-NM model is employed. To minimize the noise corrupted in images, the NLM filter enhancement approach is employed which also enhances the evaluation measures of the original model. Oliver et al. [8] introduced a mechanism which involves hardware device utilization such as FPGA for real-time super-resolution. It involves merging of overlapping LR images to obtain a single HR image. It employs weighted mean super-resolution algorithm to enhance the quality of the image along with the actual most secure and vigorous multi-frame super-resolution algorithm. FPGA-based solution achieves reasonable output frames with the incorporated frequency range. However, the authors are unsuccessful to outline the comparison between MATLAB and hardware implementation results.

The motivation behind this research work is to highlight the salient features in the medical image that are not clearly visible owing to different illumination, deficiency in acquisition, transmission etc. SR schemes improve the interpretability of images to human viewers as well as provides better input for other automated image-processing techniques. SR techniques play a vital role in situations such as imaging under dark or poor lighting conditions, images captured with poor dynamic range in the image sensor, and images corrupted by noise during transmission. The major limitations of an image or video SR schemes are tuning of parameters, lack of integrated algorithms, and necessity of quantitative standards. Further, most of the SR algorithms highly depend on the input images instead of adapting to local features. Currently, SR in medical image research requires better reconstruction of higher-quality images than possible with the currently available research outputs. Another important limitation of SR techniques is computationally expensive algorithms realized in software implementation.

3 Implementation of the Proposed Model

In the process of recording or acquisition of a digital image, there exist natural losses in spatial resolution. This loss is caused by the optical distortions (out of focus, diffraction limit, etc.), motion blur due to limited shutter speed, noise that occurs within the sensor or during transmission, and insufficient sensor density, etc. Thus, the recorded image usually suffers from blur, noise, and aliasing effects. Although the main concern of an SR algorithm is to reconstruct HR images from under sampled LR images, it covers image restoration techniques that produce high-quality images from noisy, blurred images. Therefore, the goal of SR techniques is to restore an HR image from several degraded and aliased LR images. Development of algorithm includes analysis of state-of-the-art super-resolution schemes proposed by other researchers.

A problem related to SR is the image restoration, which is a well-established area of image-processing applications. The goal of image restoration is to recover a degraded (e.g., blurred, noisy) image, but it does not change the size of image. In fact, restoration and SR reconstruction are closely related theoretically, and SR reconstruction can be considered as a second-generation problem of image restoration.

Another problem related to SR reconstruction is image interpolation that has been used to increase the size of a single image. Although this field has been extensively studied, the quality of an image magnified from an aliased LR image is inherently limited even though the ideal sinc basis function is employed. That is, single image interpolation cannot recover the high-frequency components lost or degraded during the LR sampling process. For this reason, image interpolation methods are not considered as SR techniques. To achieve further improvements in this field, the next step requires the utilization of multiple datasets in which additional data constraints from several observations of the same scene can be used. The fusion of information from various observations of the same scene allows the reconstruction of the SR of the scene.

The demonstration of proposed SR model based on spatial domain fused with an LL band of DWT is discussed in the present section. The proposed flow diagram is as shown in Fig. 1. The LR images are converted into HR images using APVT. The HR images are converted into SR images using HE technique, DWT and IDWT.

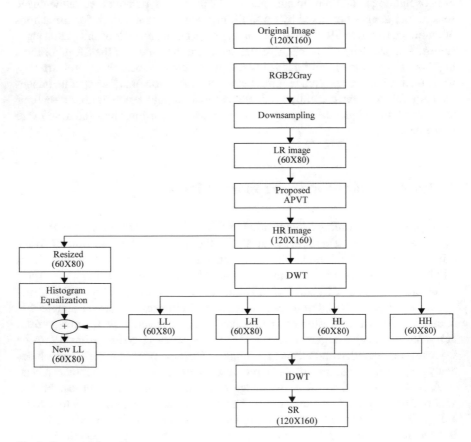

Fig. 1 Proposed flowchart

(a)

(b)

Fig. 2 Original images: **a** images of building and **b** fossil images

3.1 Original Image

The images, viz. buildings and fossils, are employed to examine the proposed method, and the corresponding images are shown in Fig. 2. The original images considered are of size 120 × 160 with image format JPEG.

3.2 Pre-processing

The color images are converted into grayscale images to increase speed of computation and also to reduce hardware complexity in real-time systems. The process of down sampling is carried to obtain low-resolution images of size 60 × 80.

3.3 Average Pixel Value Technique (APVT)

The LR images are converted into HR images using proposed Average Pixel Value Technique (APVT). The demonstration of converting the LR image to HR image is given in Fig. 3. The matrix size of 5 × 5 for LR is considered for demonstration. The average between each column of Fig. 3a is computed and inserted between two columns as shown in Fig. 3b. The average between two rows of Fig. 3b is computed and inserted between two rows as shown in Fig. 3c. The last row and last column of Fig. 3c are deleted to obtain 8 × 8 matrix. Repeat boundary coefficients in Fig. 3d obtain 10 × 10 matrixes, i.e., HR matrix.

251.75	162	72.25	73.75	82
217	67.25	69	67.25	68
91.5	70.5	72.75	65.75	70.25
79.5	73	73.75	67.5	64.75
67	66	65	69	56.25

(a) Original 5x5 matrix

251.75	206.875	162	117.125	72.25	73	73.75	77.875	82
217	142.125	67.25	68.125	69	68.125	67.25	67.625	68
91.5	81	70.5	71.625	72.75	69.25	65.75	68	70.25
79.5	76.25	73	73.375	73.75	70.625	67.5	66.125	64.75
67	66.5	66	65.5	65	67	69	62.625	56.25

(b) Average between each column

251.75	206.875	162	117.125	72.25	73	73.75	77.875	82
234.375	174.5	114.625	92.625	70.625	70.5625	70.5	72.75	75
217	142.125	67.25	68.125	69	68.125	67.25	67.625	68
154.25	111.562	68.875	69.875	70.875	68.6875	66.5	67.8125	69.125
91.5	81	70.5	71.625	72.75	69.25	65.75	68	70.25
85.5	78.625	71.75	72.5	73.25	69.9375	66.625	67.0625	67.5
79.5	76.25	73	73.375	73.75	70.625	67.5	66.125	64.75
73.25	71.375	69.5	69.4375	69.375	68.8125	68.25	64.375	60.5

(c) Average between each row

251.75	206.875	162	117.125	72.25	73	73.75	77.875
234.375	174.5	114.625	92.625	70.625	70.5625	70.5	72.75
217	142.125	67.25	68.125	69	68.125	67.25	67.625
154.25	111.562	68.875	69.875	70.875	68.6875	66.5	67.8125
91.5	81	70.5	71.625	72.75	69.25	65.75	68
85.5	78.625	71.75	72.5	73.25	69.9375	66.625	67.0625
79.5	76.25	73	73.375	73.75	70.625	67.5	66.125
73.25	71.375	69.5	69.4375	69.375	68.8125	68.25	64.375

(d) Delete last row and last column to obtain 8x8 matrix

251.75	251.75	206.875	162	117.125	72.25	73	73.75	77.875	77.875
251.75	251.75	206.875	162	117.125	72.25	73	73.75	77.875	77.875
234.375	234.375	174.5	114.625	92.625	70.625	70.5625	70.5	72.75	72.75
217	217	142.125	67.25	68.125	69	68.125	67.25	67.625	67.625
154.25	154.25	111.562	68.875	69.875	70.875	68.6875	66.5	67.8125	67.8125
91.5	91.5	81	70.5	71.625	72.75	69.25	65.75	68	68
85.5	85.5	78.625	71.75	72.5	73.25	69.9375	66.625	67.0625	67.0625
79.5	79.5	76.25	73	73.375	73.75	70.625	67.5	66.125	66.125
73.25	73.25	71.375	69.5	69.4375	69.375	68.8125	68.25	64.375	64.375
73.25	73.25	71.375	69.5	69.4375	69.375	68.8125	68.25	64.375	64.375

(e) Repeat boundary coefficient to obtain HR 10x10 matrixes

Fig. 3 Conversion of LR to HR demonstration: **a** original 5 × 5 matrix, **b** average between each column, **c** average between each row, **d** delete last row and last column to obtain 8 × 8 matrix, and **e** repeat boundary coefficient to obtain HR 10 × 10 matrixes

3.4 Two-Dimensional Discrete Wavelet Transform (DWT)

The implementation of the transformation involves the utilization of filters such as the low-pass and high-pass filters contemporaneously on rows and columns of an image. It generates the following four bands, viz. approximation band, horizontal band, vertical band, and diagonal band. The approximation band is generated based on low-pass filter on both rows and columns, and it has significant information of an image. The horizontal band is generated by using high- and low-pass filters on rows and columns of an image and has horizontal details. The vertical band is produced using low- and high-pass filters on rows and columns and has vertical edge details. The diagonal band is generated using low- and high-pass filters and has diagonal details. The original image is obtained using only approximation band based on the reconstruction process as the approximation band consists of significant information of the original image.

The spatial domain of HR images of size 120 × 60 is resized to 60 × 80. The DWT is tested on HR images to derive low- and high-frequency bands such as LL, LH, HL, and HH of each size 60 × 80. The low-frequency LL band is considered for further processing as it has significant information about HR images. The spatial domain HR image of size 60 × 80 and a low-frequency LL band of size 60 × 80 is fused using addition technique to generate new LL band for DWT. The IDWT applies to the new LL band along with LH, HL, and HH bands to generate effective SR image of size 120 × 160.

3.5 Histogram Equalization

The HR image which is resized into 60 × 80 is fed to histogram equalizer. The enrichment of contrast of an image is assessed by using histogram equalization technique. The original contrast of an image is as shown in Fig. 4a. Histogram equalization technique is tested on original image which improves the contrast of an

(a)   **(b)**

Fig. 4 Super-resolution image: **a** original image and **b** enhanced image

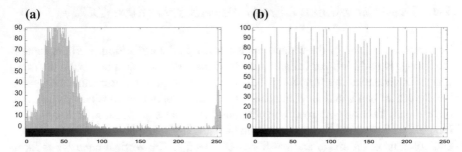

Fig. 5 Image enhancement by histogram equalization: **a** representation of original image histogram and **b** representation of enhanced image histogram

image and is as shown in Fig. 4b. The corresponding histogram of an original image is as shown in Fig. 5a and enhanced contrast images in Fig. 5b, respectively. The intensity values in the histogram are present densely between the values zero and hundred as shown in Fig. 5a. It is noticed that the pixel intensity values are widely distributed for all intensity levels, i.e., between zero and 255, as shown in Fig. 5b. Thus, it can be inferred that image enhancement can be achieved using histogram equalization technique.

The HR image of size 60×80 is added using an LL image of size 60×80 to derive new LL band of size 60×80. The IDWT is applied on new LL, LH, HL, and HH to obtain SR images.

4 Results

In the present section, the definitions of performance parameters such as peak signal-to-noise ratio (PSNR) and Structure Similarity Index (SSIM) and also the performance study of proposed model for diverse images are discussed.

4.1 Definitions

(i) PSNR: It is a quality measure metric in image which computes the peak signal-to-noise ratio, and the expression is given in Eq. (1)

$$\text{PSNR} = 20 \log_{10}\left(\frac{\text{MAX}_f}{\sqrt{\text{MSE}}}\right) \tag{1}$$

(ii) SSIM: It is a quality metric that measures the similarity between the two images.

4.2 Performance Analysis of Proposed Model

The Performance parameter values are dependent on types of images. The demonstrations of the proposed model are tested for buildings and fossil images and are presented with respective PSNR and SSIM for different kinds of building images in Tables 1 and 2 for LR and HR images.

It is evident that the PSNR and SSIM are high with respect to building images in contrast to fossil images, and the building images are clearly visible.

Table 1 Performance parameters for building images

	Reference Image	LR Image	HR Image	PSNR	SSIM
1				25.4298	0.7508
2				25.0926	0.7686
3				25.0017	0.8257
4				26.9860	0.7640
5				27.2065	0.7570

Table 2 Performance parameters for fossil images

	Reference Image	LR Image	HR Image	PSNR(db)	SSIM
1				22.6163	0.6617
2				22.5122	0.6499
3				20.8675	0.6664
4				30.8150	0.8970
5				22.6998	0.7380

4.3 Comparison of Proposed Model to Existing Model

In this section, the resulted SR images are presented and compared with different approaches. Table 3 represents a comparison of image performance measures, namely

Table 3 Comparison of proposed model to existing model

S. No.	Image	Author	PSNR	
			Existing	Proposed
1	Mandrill	Ahmad and Shafique Qureshi	21.26	26.3373
2	Stadium	Alvarez-Ramos et al.	30.26	30.7345
3	Nebuta	Yasutaka Matsuo et al.	22.6	23.4219

PSNR of diverse methods. Further, our propounded model is compared with enduring model.

The test images are considered for comparison such as Mandrill, Stadium and Nebuta images. The LR images are converted into HR images by using our proposed method. To better understand the significance and difference of our result we have used objective metrics PSNR. It is noticed that PSNR values are high for stadium images in contrast to another Mandrill and Nebuta images. It is also observed from the Table 3 that even in case of Mandrill and Nebuta the PSNR value are high when compared to the existing methods. The proposed model accomplishes high PSNR values with better quality and shows significant improvements over other existing methods, and reconstructed images are visually appealing.

5 Conclusions

Hybrid domain based image super-resolution method is presented using fusion of APVT and DWT. The LR image of size 60×80 is considered and converted into HR images of size 120×160 using novel technique of APVT. To procure HR images in spatial domain, the average pixel intensity values of two rows and columns of LR images are calculated and inserted between rows and columns. Further, the HR images are resized to 60×80, and HE procedure is utilized to enrich the contrast in images. The DWT is tested on HR images, and LL band of size 60×80 is added to HE image of size 60×80 to obtain new LL band. The final SR images are obtained using IDWT on new LL, LH, HL, and HH bands. The preferred method substantiated better results in terms of PSNR and SSIM values and achieves higher quality which is noticeable from empirical analysis. In future, this method can be improved by combining DWT with other transfer domain techniques. As is evident from the literature of review presented that there exist numerous methods for super-resolution of medical images suitable for FPGA implementation. However, the following substantiation shows that the proposed development of intelligent methods for super-resolution of medical images suited for FPGA implementation approach is still a challenging such as the super-resolution reconstruction constraints provide less and less useful information as the magnification factor increases. The major cause of this phenomenon is the spatial averaging over the photosensitive area. Also, the information content is fundamentally limited by the dynamic range of the images. The strong class-based algorithms can provide additional information than the simple smoothness priors that are used in existing super-resolution algorithms. The interpolation and reconstruction super-resolution algorithms largely depend on the input image data. The SR adopted in medical image de-noising algorithms degrades the reconstructed image for high-level noise density. This is due to the fact that majority of the de-noising algorithms are nonlinear in nature. The SR algorithms are iterative and computationally complex. Therefore, dedicated hardware design is highly essential to address this issue.

References

1. Trinh, Dinh-Hoan, Marie Luong, Françoise Dibos, Jean-Marie Rocchisani, Canh-Duong Pham, and Truong Q. Nguyen. 2014. Novel example-based method for super-resolution and de-noising of medical images. *IEEE Transactions on Image Processing* 23 (4): 1882–1895.
2. Szydzik, Tomasz, Gustavo M. Callico, and Antonio Nunez. 2011. Efficient FPGA implementation of a high-quality super-resolution algorithm with real-time performance. *IEEE Transactions on Consumer Electronics* 57 (2): 664–672.
3. Polatkan, Gungor, Mingyuan Zhou, Lawrence Carin, David Blei, and Ingrid Daubechies. 2015. A Bayesian nonparametric approach toimage super-resolution. *IEEE Transaction on Pattern Analysis and Machine Intelligence* 37 (2): 346–358.
4. Tran, Dai-Viet, Sebastein Li-Thaiao-Te, Marie Luong, Thuong Le-Tien, Francoise Dibos, Jean-Marie Rocchisani. 2016. Example-based super-resolution for enhancing spatial resolution of medical images. In *IEEE 38th Annual International Conference of the IEEE Engineering in Medicine and Biology Society*, 457–460.
5. Chen, Jingxu, Xiaohai He, Hanggang Chen, Qizhi Teng, and Linbo Qing. 2016. Single image super-resolution based on deep learning and gradient transformation. In *2016 IEEE 13th International Conference on Signal Processing (ICSP)*, 663–667.
6. Wang, Haijun, Xinbo Gao, Kaibing Zhang, and Jie Li. 2017. Single image super-resolution using gaussian process regression with dictionary-based sampling and student-t likelihood. *IEEE Transactions on Image Processing* 26 (7): 3556–3568.
7. Jiang, Junjun, Xiang Ma, Chen Chen, Tao Lu, and Zhongyuan Wang. 2017. Single image super-resolution via locally regularized anchored neighborhood and nonlocal means. *IEEE Transactions on Multimedia* 19 (1): 15–26.
8. Bowen, Oliver, and Christos-Savvas Bouganis. 2008. Real-time image super-resolution using an FPGA. In *2008 International Conference on Field Programmable Logic and Applications (FPL-2008)*, 89–94.

A Novel BPSO-Based Optimal Features for Handwritten Character Recognition Using SVM Classifier

Sonali Bali, Shekhar Sharma and Preeti Trivedi

Abstract The handwritten character recognition is a line of research within the image processing for which there have been developed many techniques and methodologies. Its main objective is to identify a character from a digitized image that is represented as a set of pixels. This paper presents a handwritten character recognizing framework with the help of rows and column-wise segmentation, feature extraction using discrete wavelet transform (DWT) and histogram of oriented gradients (HOG) features and support vector machine (SVM)-based classifier. Further research is extended for optimal feature selection using BPSO for the training of SVM to reduce the machine overhead maintain near to same accuracy. Achieved accuracy of the proposed method is 99.23%.

Keywords DWT · HOG · OCR · BPSO · SVM

1 Introduction

Handwritten structures are introduced by Mesopotamians to develop a system using symbols representing quantities, phonemes, ideas to write on different media, a memory of information. Writing allows both to communicate over great distances, but also over time. Without writing, it would not have access to the knowledge of men like Plato, Socrates, Galileo or Newton, and our world would be very different [1].

In the digital age, handwritten documents (and more generally paper documents) seem to play a less and less important role. This focuses on digital media at the expense of paper to register and share our knowledge. Fifteen years ago still used people to write down our friends' phone numbers, recipes, meeting notes or agenda. Today, all these activities can be carried out on digital media that facilitate the sharing of this content with other people. However, handwriting, although gradually replaced by digital applications, remains extremely present in our world. Continue to write our checks, our mailing addresses, our postcards, fill out our forms and put our ideas on rough sheets [2].

S. Bali (✉) · S. Sharma · P. Trivedi
Shri Govindram Seksaria Institute of Technology and Science, Indore, India

© Springer Nature Singapore Pte Ltd. 2020
Y.-C. Hu et al. (eds.), *Ambient Communications and Computer Systems*, Advances in Intelligent Systems and Computing 1097,
https://doi.org/10.1007/978-981-15-1518-7_35

Software capable of digitizing documents is called handwritten character recognition systems. The scanning of documents also allows their exploitation without degrading the original, as in the case of old documents.

Handwritten character recognition is a process for recovering symbols from scanned text images. The images consist of pixel matrices. The task of a handwritten character recognition is to segment images, words, characters and then perform character recognition. It results in a textual transcription of the image by which automatic treatments are possible: search for words, names or summary. In general, handwritten character recognition concerns the processing of a digitized document.

Handwritten character recognition technology, widely used in commercial applications since the 1970s, is now used to automate tasks such as passport processing, secure document processing (checks, financial documents, invoices), postal tracking, editing, packaging of consumer goods (batch codes, package codes, expiry dates) and clinical applications. Handwritten character recognition readers and software can be used, as well as intelligent cameras and machine vision systems with additional capabilities such as barcode reading and product inspection [3].

Although the developed OCRs are more and more efficient, none of them surpass or equal a human operator in terms of the recognition rate. Researchers have developed error correction strategies, such as the use of lexicons or grammatical rules, to improve the results. These improvements result in industrial applications with a low error rate: Millions of letters are routed automatically using OCR, and the majority of checks are also processed entirely by OCR [4].

These OCRs are nevertheless limited to very specific documents with a very limited dictionary. To date, there is no OCR capable of handling heterogeneous documents, while offering usable performances in an industrial context. Handwritten character recognition is a scientific field in which there are many systems that have already proven themselves. In general, these systems transform, using image processing algorithms, an image into a sequence of data vectors. This sequence of data vectors can then be processed by a statistical sequence classification method, also called a dynamic classifier. In our study support vector machine classifier is used for handwritten character for prediction.

2 Proposed Method

The detailed methodology is explained as (Fig. 1).

2.1 *Image Acquisition*

In the acquisition phase, the character from the database is put in the form of an image in the format jpg, gif or jif (types supported by the system).

Fig. 1 Flow diagram of proposed work

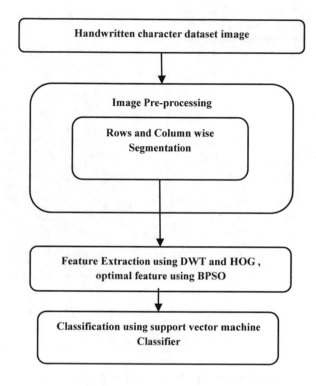

2.2 Pre-processing

Pre-processing performed on word or character image are intended to improve recognition by normalizing images or reducing the amount of noise in the signal. Binarization reduces the amount of information contained in the image which facilitates recognition. The purpose of normalization is to standardize the appearance of the images, thus reducing inter-word and inter-character variations. Pre-processing is not required, but the use of pre-processing has been shown to improve recognition. The presence of some pre-processing may also be related to the extraction of characteristics [5].

Segmentation is another phase of pre-processing. Its purpose is to locate and extract as precisely as possible the information to recognize. In this research work, the segmentation is done using rows and column-wise separation operations [6, 7].

2.2.1 Rows and Column-wise Segmentation

The data is presented as a symmetric matrix as $Y = (Y_{ij})_{1 \le i, j \le n}$ of size $n \times n$ where each box Y_{ij} represents the similarity between the images i and j. Let assume that there are K^* breaks in rows (and therefore in columns), denoted $1 \le t_1^* < \cdots < t_k^* \le n-1$ representing the boundaries of the desired $(K^* + 1)^2$ blocks (see Fig. 2). For any pair

Fig. 2 Schematic
representation of the matrix
[8]

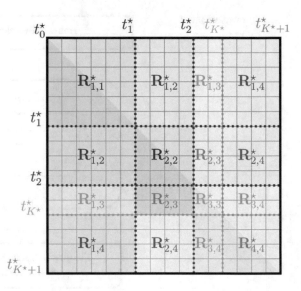

$(k, l) \in \{1, \ldots, K^* + 1\}^2$, it is denoted by R_{kl}^* block defined by Siddharth et al. [8]:

$$R_{kl}^* = \{(i, j) \in \{1, \ldots, n\}^2 | t_{k-1}^* + 1 \leq i \leq t_k^* \text{ and } t_{l-1}^* + 1 \leq j \leq t_l^*\} \quad (1)$$

With the assumption of $t_0^* = 0$ and $t_{K^*+1}^* = n$.

2.3 Feature Extraction Using DWT and HOG

2.3.1 Discrete Wavelet Transform

The wavelet analysis was introduced by Morlet to improve the windowing solution that Gabor had proposed to overcome the limitations of Fourier transformation.

In fact, the Fourier transformation has some disadvantages that can be summarized in three points:

- It does not allow precise characterization of a signal simultaneously in the frequency domain and the spatial domain;
- The result obtained by transformation eliminates all information concerning the spatial domain: The beginning or the end of a component of the signal is no longer localizable in the parametric model;
- The spatial frequency of a signal being inversely proportional to its period, if information obtains on a low-frequency signal, the observation interval must be large (and vice versa).

The advantage of using wavelets is their greater similarity with real signals, which are generally used, which vary in a waveform and which concentrate a large part of their information load in certain places and not in a uniform way. In this sense, they become much more efficient at the moment of wanting to extract certain characteristics of an image thanks to their translational and dilatational capacity that allows them to be similarly shaped to the forms that want to determine, discriminated against the others. If it is wanted to locate a certain figure, it would be enough to choose a mother wavelet that resembles it as much as possible and apply with this wavelet the two-dimensional transform to the image.

$$\psi_{\tau,s}(t) = \frac{1}{\sqrt{|s|}} \psi\left(\frac{t - \tau}{s}\right) \tag{2}$$

2.3.2 Histogram of Oriented Gradients (HOG)

The oriented gradient [9] is a descriptor using the contour information. For this situation, every window can be portrayed by the local distribution of the introductions and amplitudes. This distribution is defined by means of a histogram. This is modeled by partitioning the location window into an arrangement of cells on which the sufficiency of the gradient for every orientation interval is incorporated over the pixels of the cell.

The gradients of the image can be calculated at each point (x, y) by a Sobel operator,

$$G(x, y) = \sqrt{G_x(x, y)^2 + G_y(x, y)^2} \tag{3}$$

where G_x and G_y represent, respectively, the gradient in the x and y directions, the orientation of the contour at this point is therefore,

$$\theta(x, y) = \arctan\frac{G_y(x, y)}{G_x(x, y)} \tag{4}$$

The contours are then classified into K intervals (bins). The value of the kth interval is given by,

$$\psi_k(x, y) = \begin{cases} G(x, y) \sin \theta(x, y) \in \text{bin}_k \\ 0 \qquad\qquad\qquad\quad \text{otherwise} \end{cases} \tag{5}$$

The calculation of $\psi_k(x, y)$ on a region R of the image can be done efficiently using the concept of the integral image.

$$\text{SAT}_k(R) = \sum_{(x,y)\in R} \psi_k(x, y) \tag{6}$$

For generating the descriptor associated with a character image, divide the $K = 9$ intervals corresponding to the intervals equated to $[0 \ldots \pi]$. The concatenating global histogram and the normalized histograms of the 12 blocks define the signature of the character area to be recognized.

Since the histograms of the oriented gradient are basic descriptors of windows, the choice of the detection window is crucial. This requires a good reset, whose tolerance depends on the position of the window but also on the size of the cells.

2.3.3 Feature Selection Using PSO

Since PSO was introduced, it has been successfully applied to so-called continuous optimization problems (optimizing nonlinear continuous functions) [11]. Nevertheless, its application to discrete problems requires a phase of adaptation of model equations of the PSO. In this context, Eberhart and Kennedy (1997) also developed the discrete binary version of the PSO. The latter, called binary PSO (BPSO), differs essentially from the classical PSO (continuous) on two characteristics [12]:

- The positions of the particle i are represented binary vectors

$$X_i = (x_{i1}, x_{i2}, \ldots, x_{iD}), x_{ij} \in \{0, 1\} \tag{7}$$

- The particle velocities are rather defined in terms of probabilities, a bit will change to 1. According to this definition, velocity should be limited in the interval $[0, 1]$ by applying the following sigmoid function:

$$\text{sig}(v_{ij}^t) = \frac{1}{1 + e^{-v_{ij}^t}} \tag{8}$$

So the new position is obtained using the equation below:

$$x_{ij}^t = \begin{cases} 1 \text{ if } r_{ij} < \text{sig}(v_{ij}^t) \\ 0 \text{ Otherwise} \end{cases} \tag{9}$$

where r_{ij} is a uniform random number in the range $[0, 1]$.

To prevent $\text{sig}(v_{ij}^t)$ from approaching zero or one, the constant $V_{\max}$ is used and often set at 4, i.e., $v_{ij}^t \in [-4, 4]$.

The BPSO Algorithm for the Attribute Selection

The attribute selection problem consists of finding a Boolean assignment to propositional variables that maximize the number of clauses that can be satisfied simultaneously. The problem can be expressed as follows.

Let C be a set of m clauses over n propositional variables. Seek to determine an interpretation among the 2^n possible, which can satisfy the largest number of clauses.

The application of the BPSO for the attribute selection begins with random initialization of the swarm in the search space. For each particle (of dimension n), a solution of the problem randomly defined its position (a Boolean assignment of the propositional variables) and its speed (direction for a future displacement in the search space). In addition, the global social neighborhood topology (star topology) selected. The particle is thus attracted to the best particle (the global optimum). At each iteration of the algorithm, each particle moves next (9). Once the particles have been moved, the new positions are evaluated using the objective function, which calculates the number of clauses satisfied. Best personal positions (P_i) as well as the best position in the swarm (G_{BEST}) are updated. The algorithm stops when the maximum number of iterations is reached or all the clauses are satisfied ($G_{BEST} = m$).

Pseudo-code of the BPSO algorithm is given as follows:

1. Initialization

   a. Define a size for the swarm (N_p) and randomize the position (X_i), the best position (P_i) and the velocity (V_i) of each particle
   b. Evaluate these particles, then calculate the best position in the swarm (G_{BEST})
   c. Initialize the coefficients of inertia and attraction (c_0, c_1, c_2).

2. As long as [the maximum number of iterations is not reached or (G_{BEST}) different from the number of clauses (m)]

   For each particle $i = 1$ to N_p do

a. Change its speed
b. Calculate its sigmoid function (8)
c. Move to a new position using (8)
d. Evaluate the performance using its current position X_i^t and compare it with:

   - Its best personal performance: If $F(X_i^t) > F(P_i)$, then $P_i = X_i^t$
   - The best performance of the swarm: If $F(X_i^t) > F(G_{BEST})$ then $G_{BEST} = X_i^t$.

2.4 Classification Using Support Vector Machine (SVM)

Support vector machine, or SVMs, is the set of supervised machine learning algorithm designed to solve out regression and discrimination problems. SVMs are the generalization of linear classifiers.

The objective is to find a decision boundary (hyperplane separator) that can separate the cloud of points into two regions by making a minimum of errors, i.e., find the optimal hyperplane.

SVMs can be used to solve binary discrimination problems, that is, decide which class a sample belongs to. The solution to this problem is the construction of a function f which, at an input vector $x \in X$ matches an output $f(x)$: It is then decided that x is of class $+1$ if $f(x) > 0$ and of class -1 if $f(x) < 0$. It is a linear classifier. The decision boundary $f(x) = 0$ is a separating hyperplane.

Let H be a hyperplane, w its normal vector, and b its offset from the origin (see Fig. 3). The hyperplane H is then given by Gaur and Yadav [3]:

$$f(x) = w^T x + b = 0 \tag{10}$$

The goal of the SVM learning algorithm is to find out the parameters w and b of the best hyperplane through a learning set:

$$X \times Y = \{(x_1, y_1), \dots, (x_i, y_i)\} \in \mathbb{R}^N \times \{+1, -1\} \tag{11}$$

where y_i are the respective labels of x_i and N is a size of learning set.

Fig. 3 Margin illustration and support vectors [10]

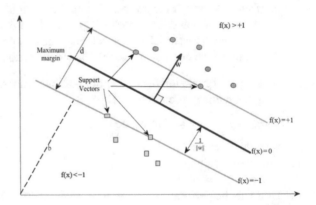

Fig. 4 Input image

Fig. 5 Gray image

Fig. 6 Binary image

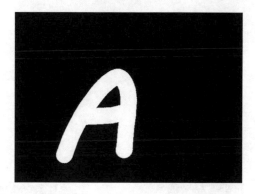

Fig. 7 Cropped gray image

3   Simulation and Results

Figures 10, 11 and 12 represent the accuracy, precision and sensitivity, for the BPSO-DWT, HOG-based approach. The proposed recognition system performs well on the benchmark dataset Center of Excellence for Document Analysis and Recognition

Fig. 8 Resized gray image

Fig. 9 HOG features

Fig. 10 Accuracy graph for proposed approach

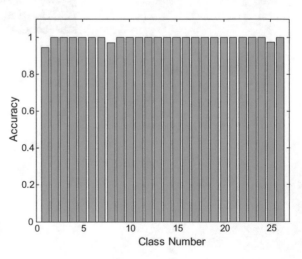

(CEDAR). Figures 4, 5, 6, 7, 8 and 9 represent the input image, gray image, binary image, cropped gray image, resize gray image and HOD features.

The system is evaluated under database with 30 images in each class, and there are total of 26 class has been taken. Twenty images from each class have been taken for reduction in training overhead. Experimental results claim the accuracy

Fig. 11 Precision graph for proposed approach

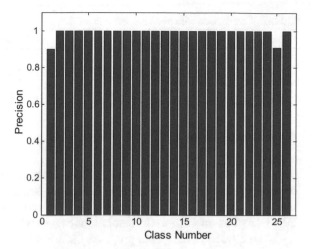

Fig. 12 Sensitivity graph for proposed approach

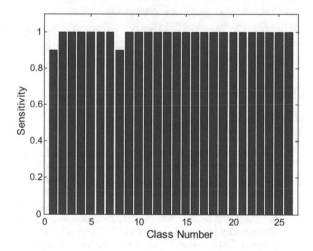

and sensitivity according to class. Average accuracy is claimed with this method is 99.23% (Tables 1 and 2).

Table 1 Result for proposed work

Accuracy	99.23%
Average F-score	90%
Precision	99.27%
Sensitivity	99.23%

Table 2 Comparison with previous research work

Method	No. of features	Accuracy (%)
Yadav et al. [11] (neural network)	–	86.74
Proposed approach (SVM + DWT + HOG)	10,404	99.23
Proposed approach (BPSO-based feature selection)	7456	97.45

4 Conclusion

This paper designed a system for recognizing the handwritten character from a certain database. Segmentation is performed using row and column pixel values. Once ROI has been extracted, feature extraction is done using hybrid structure of DWT and HOG. DWT ensures the multi-resolution features which are further used for training SVM. Additionally, this paper proposes the novel framework for feature selection to reduce the machine overhead while training the SVM. The proposed research work provides the recognition accuracy of 99.23%. With the optimal features using BPSO achieved accuracy is 97.45%. Even though accuracy is slightly lesser than proposed one, the feature size of BPSO method is marginally low to claim lower machine overhead.

References

1. Lecolinet, E., and O. Baret. 1994. Cursive word recognition: methods and strategies. In *Fundamentals in handwriting recognition*, 235–263. Berlin, Heidelberg: Springer.
2. Kolman, E., and M. Margaliot. 2008. A new approach to knowledge-based design of recurrent neural networks. *IEEE Transactions on Neural Networks* 19 (8): 1389–1401.
3. Gaur, A., and S. Yadav. 2015. Handwritten Hindi character recognition using k-means clustering and SVM. In 2015 *4th International symposium on emerging trends and technologies in libraries and information services (ETTLIS)*, 65–70. IEEE.
4. Barve, S. 2012. Artificial neural network based on optical character recognition. *International Journal of Engineering Research & Technology (IJERT)*, *1* (4). ISSN 2278-0181.
5. Barve, S. Optical character recognition using artificial neural network. *International Journal of Advanced Technology & Engineering Research (IJATER)*, 2 (2). ISSN NO 2250-3536.
6. Mehfuz, S., and Gauri katiyar. 2012. Intelligent Systems for off-line handwritten character recognition: a review. *International Journal of Emerging Technology and Advanced Engineering*, 2 (4). ISSN 2250-2459.
7. Patel, Dileep Kumar, Tanmoy Som, Sushil Kumar Yadav, and Manoj Kumar Singh. 2012. Handwritten character recognition using multiresolution technique and euclidean distance metric. *Journal of Signal and Information Processing*, 3(2): 208–214.
8. Siddharth, Kartar Singh, Mahesh Jangid, Renu Dhir, and Rajneesh Rani. 2011. Handwritten Gurmukhi character recognition using statistical and background directional distribution. *International Journal of Computer Science Engineering (IJCSE)* 3: 2332–2345.
9. Newell, Andrew J., and Lewis D. Griffin. 2011. Multiscale histogram of oriented gradient descriptors for robust character recognition. In *2011 International conference on document analysis and recognition*, 1085–1089. IEEE.
10. Duda, Richard O., Peter E. Hart, and David G. Stork. 2001. *Pattern classification*, 2nd ed. New York: Wiley.

11. Yadav, Suman Avdhesh, Smita Sharma, and Shipra Ravi Kumar. A robust approach for offline English character recognition. In 2015 *International conference on futuristic trends on computational analysis and knowledge management (ABLAZE)*, 121–126. IEEE.
12. Khanesar, Mojtaba Ahmadieh, Mohammad Teshnehlab, and Mahdi Aliyari Shoorehdeli. 2007. A novel binary particle swarm optimization. In *2007 Mediterranean conference on control & automation*, 1–6. IEEE.

Printed in the United States
By Bookmasters